D1604779

THE STABILITY OF MATTER IN QUANTUM MECHANICS

Research into the stability of matter has been one of the most successful chapters in mathematical physics, and is a prime example of how modern mathematics can be applied to problems in physics.

A unique account of the subject, this book provides a complete, self-contained description of research on the stability of matter problem. It introduces the necessary quantum mechanics to mathematicians, and aspects of functional analysis to physicists. The topics covered include electrodynamics of classical and quantized fields, Lieb–Thirring and other inequalities in spectral theory, inequalities in electrostatics, stability of large Coulomb systems, gravitational stability of stars, basics of equilibrium statistical mechanics, and the existence of the thermodynamic limit.

The book is an up-to-date account for researchers, and its pedagogical style makes it suitable for advanced undergraduate and graduate courses in mathematical physics.

Elliott H. Lieb is a Professor of Mathematics and Higgins Professor of Physics at Princeton University. He has been a leader of research in mathematical physics for 45 years, and his achievements have earned him numerous prizes and awards, including the Heineman Prize in Mathematical Physics of the American Physical Society, the Max-Planck medal of the German Physical Society, the Boltzmann medal in statistical mechanics of the International Union of Pure and Applied Physics, the Schock prize in mathematics by the Swedish Academy of Sciences, the Birkhoff prize in applied mathematics of the American Mathematical Society, the Austrian Medal of Honor for Science and Art, and the Poincaré prize of the International Association of Mathematical Physics.

Robert Seiringer is an Assistant Professor of Physics at Princeton University. His research is centered largely on the quantum-mechanical many-body problem, and has been recognized by a Fellowship of the Sloan Foundation, by a U.S. National Science Foundation Early Career award, and by the 2009 Poincaré prize of the International Association of Mathematical Physics.

THE STABILITY OF MATTER IN QUANTUM MECHANICS

ELLIOTT H. LIEB AND ROBERT SEIRINGER
Princeton University

CAMBRIDGE UNIVERSITY PRESS
Cambridge, New York, Melbourne, Madrid, Cape Town, Singapore, São Paulo, Delhi

Cambridge University Press
The Edinburgh Building, Cambridge CB2 8RU, UK

Published in the United States of America by Cambridge University Press, New York

www.cambridge.org
Information on this title: www.cambridge.org/9780521191180

First published 2010

Printed in the United Kingdom at the University Press, Cambridge

A catalog record for this publication is available from the British Library

Library of Congress Cataloging in Publication data
Lieb, Elliott H.
The stability of matter in quantum mechanics / Elliott H Lieb, Robert Seiringer.
p. cm.
Includes bibliographical references and index.
ISBN 978-0-521-19118-0 (hardback)
1. Thomas-Fermi theory. 2. Quantum theory. 3. Matter – Properties. 4. Structural stability.
I. Seiringer, Robert. II. Title.
QC173.4.T48L543 2010
530.12 – dc22 2009031810

ISBN 978-0-521-19118-0 Hardback

To

Christiane, Letizzia and Laura

Contents

Preface

The fundamental theory that underlies the physicist's description of the material world is quantum mechanics – specifically Erwin Schrödinger's 1926 formulation of the theory. This theory also brought with it an emphasis on certain fields of mathematical analysis, e.g., Hilbert space theory, spectral analysis, differential equations, etc., which, in turn, encouraged the development of parts of pure mathematics.

Despite the great success of quantum mechanics in explaining details of the structure of atoms, molecules (including the complicated molecules beloved of organic chemists and the pharmaceutical industry, and so essential to life) and macroscopic objects like transistors, it took 41 years before the most fundamental question of all was resolved: Why doesn't the collection of negatively charged electrons and positively charged nuclei, which are the basic constituents of the theory, implode into a minuscule mass of amorphous matter thousands of times denser than the material normally seen in our world? Even today hardly any physics textbook discusses, or even raises this question, even though the basic conclusion of stability is subtle and not easily derived using the elementary means available to the usual physics student. There is a tendency among many physicists to regard this type of question as uninteresting because it is not easily reducible to a quantitative one. Matter is either stable or it is not; since nature tells us that it is so, there is no question to be answered. Nevertheless, physicists firmly believe that quantum mechanics is a 'theory of everything' at the level of atoms and molecules, so the question whether quantum mechanics predicts stability cannot be ignored. The depth of the question is further revealed when it is realized that a world made of bosonic particles would be unstable. It is also revealed by the fact that the seemingly innocuous interaction of matter and electromagnetic radiation at ordinary, every-day energies – quantum electrodynamics – should be a settled, closed subject, but it is not and it can be understood only in the context of perturbation theory. Given these observations, it is clearly important to know that at least the quantum-mechanical part of the story is well understood.

It is this stability question that will occupy us in this book. After four decades of development of this subject, during which most of the basic questions have gradually been answered, it seems appropriate to present a thorough review of the material at this time.

Schrödinger's equation is not simple, so it is not surprising that some interesting mathematics had to be developed to understand the various aspects of the stability of matter. In particular, aspects of the spectral theory of Schrödinger operators and some new twists on classical potential theory resulted from this quest. Some of these theorems, which play an important role here, have proved useful in other areas of mathematics.

The book is directed towards researchers on various aspects of quantum mechanics, as well as towards students of mathematics and students of physics. We have tried to be pedagogical, recognizing that students with diverse backgrounds may not have all the basic facts at their finger tips. Physics students will come equipped with a basic course in quantum mechanics but perhaps will lack familiarity with modern mathematical techniques. These techniques will be introduced and explained as needed, and there are many mathematics texts which can be consulted for further information; among them is [118], which we will refer to often. Students of mathematics will have had a course in real analysis and probably even some basic functional analysis, although they might still benefit from glancing at [118]. They will find the necessary quantum-mechanical background self-contained here in chapters two and three, but if they need more help they can refer to a huge number of elementary quantum mechanics texts, some of which, like [77, 22], present the subject in a way that is congenial to mathematicians.

While we aim for a relaxed, leisurely style, the proofs of theorems are either completely rigorous or can easily be made so by the interested reader. It is our hope that this book, which illustrates the interplay between mathematical and physical ideas, will not only be useful to researchers but can also be a basis for a course in mathematical physics.

To keep things within bounds, we have purposely limited ourselves to the subject of stability of matter in its various aspects (non-relativistic and relativistic mechanics, inclusion of magnetic fields, Chandrasekhar's theory of stellar collapse and other topics). Related subjects, such as a study of Thomas–Fermi and Hartree–Fock theories, are left for another day.

Our thanks go, first of all, to Michael Loss for his invaluable help with some of this material, notably with the first draft of several chapters. We also thank László Erdős, Rupert Frank, Heinz Siedentop, Jan Philip Solovej and Jakob Yngvason for a critical reading of parts of this book.

Elliott Lieb and Robert Seiringer
Princeton, 2009

The reader is invited to consult the web page http://www.math.princeton.edu/books/ where a link to errata and other information about this book is available.

CHAPTER 1

Prologue

1.1 Introduction

The basic constituents of ordinary matter are electrons and atomic nuclei. These interact with each other with several kinds of forces – electric, magnetic and gravitational – the most important of which is the electric force. This force is attractive between oppositely charged particles and repulsive between like-charged particles. (The electrons have a negative electric charge $-e$ while the nuclei have a positive charge $+Ze$, with $Z = 1, 2, \dots, 92$ in nature.) Thus, the strength of the attractive electrostatic interaction between electrons and nuclei is proportional to Ze^2, which equals $Z\alpha$ in appropriate units, where α is the dimensionless **fine-structure constant**, defined by

$$\alpha = \frac{e^2}{\hbar c} = 7.297\,352\,538 \times 10^{-3} = \frac{1}{137.035\,999\,68}, \tag{1.1.1}$$

and where c is the speed of light, $\hbar = h/2\pi$ and h is Planck's constant.

The basic question that has to be resolved in order to understand the existence of atoms and the stability of our world is:

Why don't the point-like electrons fall into the (nearly) point-like nuclei?

This problem of classical mechanics was nicely summarized by Jeans in 1915 [97]:

> "There would be a very real difficulty in supposing that the (force) law $1/r^2$ held down to zero values of r. For the force between two charges at zero distance would be infinite; we should have charges of opposite sign continually rushing together and, when once together, no force would be adequate to separate them... Thus the matter in the universe would tend to shrink into nothing or to diminish indefinitely in size."

A sensitive reader might object to Jeans' conclusion on the grounds that the non-zero radius of nuclei would ameliorate the collapse. Such reasoning is beside the point, however, because the equilibrium separation of charges observed in nature is not the nuclear diameter (10^{-13} cm) but rather the atomic size (10^{-8} cm) predicted by Schrödinger's equation. Therefore, as concerns the problem of understanding stability, in which equilibrium lengths are of the order of 10^{-8} cm, there is no loss in supposing that all our particles are point particles.

To put it differently, why is the energy of an atom with a point-like nucleus not $-\infty$? The fact that it is not is known as **stability of the first kind**; a more precise definition will be given later. The question was successfully answered by quantum mechanics, whose exciting development in the beginning of the twentieth century we will not try to relate – except to note that the basic theory culminated in Schrödinger's famous equation of 1926 [156]. This equation explained the new, non-classical, fact that as an electron moves close to a nucleus its kinetic energy necessarily increases in such a way that the minimum total energy (kinetic plus potential) occurs at some positive separation rather than at zero separation.

This was one of the most important triumphs of quantum mechanics!

Thomson discovered the electron in 1897 [180, 148], and Rutherford [155] discovered the (essentially) point-like nature of the nucleus in 1911, so it took 15 years from the discovery of the problem to its full solution. But it took almost three times as long, 41 years from 1926 to 1967, before the second part of the stability story was solved by Dyson and Lenard [44].

The second part of the story, known as **stability of the second kind**, is, even now, rarely told in basic quantum mechanics textbooks and university courses, but it is just as important. Given the stability of atoms, is it obvious that bulk matter with a large number N of atoms (say, $N = 10^{23}$) is also stable in the sense that the energy and the volume occupied by $2N$ atoms are twice that of N atoms? Our everyday physical experience tells us that this additivity property, or linear law, holds but is it also necessarily a consequence of quantum mechanics? Without this property, the world of ordinary matter, as we know it, would not exist.

Although physicists largely take this property for granted, there were a few that thought otherwise. Onsager [145] was perhaps the first to consider this kind of question, and did so effectively for classical particles with Coulomb interactions but with the addition of hard cores that prevent particles from getting too close together. The full question (without hard cores) was addressed by Fisher and Ruelle in 1966 [66] and they generalized Onsager's results to smeared out charges. In 1967 Dyson and Lenard [44] finally succeeded in showing that stability of the second kind for truly point-like quantum particles with Coulomb forces holds but, surprisingly, that it need not do so. That is, the *Pauli exclusion principle*, which will be discussed in Chapter 3, and which has no classical counterpart, was essential. Although matter would not collapse without it, the linear law would *not* be satisfied, as Dyson showed in 1967 [43]. Consequently, stability of the second kind does *not* follow from stability of the first kind! If the electrons and nuclei were all bosons (which are particles that do not satisfy the exclusion principle), the energy would not satisfy a linear law but rather decrease like $-N^{7/5}$; we will return to this astonishing discovery later.

The Dyson–Lenard proof of stability of the second kind [44] was one of the most difficult, up to that time, in the mathematical physics literature. A challenge was to find an essential simplification, and this was done by Lieb and Thirring in 1975 [134]. They introduced new mathematical inequalities, now called Lieb–Thirring (LT) inequalities (discussed in Chapter 4), which showed that a suitably modified version of the 1927 approximate theory of Thomas and Fermi [179, 62] yielded, in fact, a lower bound to the exact quantum-mechanical answer. Since it had already been shown, by Lieb and Simon in 1973 [129, 130], that this Thomas–Fermi theory possessed a linear lower bound to the energy, the many-body stability of the second kind immediately followed.

The Dyson–Lenard stability result was one important ingredient in the solution to another, but related problem that had been raised many years earlier. Is it true that the 'thermodynamic limit' of the free energy per particle exists for an infinite system at fixed temperature and density? In other words, given that the energy per particle of some system is bounded above and below, independent of the size of the system, how do we know that it does not oscillate as the system's size increases? The existence of a limit was resolved affirmatively by Lebowitz and Lieb in 1969 [103, 116], and we shall give that proof in Chapter 14.

There were further surprises in store, however! The Dyson–Lenard result was not the end of the story, for it was later realized that there were other sources of instability that physicists had not seriously thought about. Two, in fact. The eventual solution of these two problems leads to the conclusion that, ultimately, stability requires more than the Pauli principle. It also requires an upper bound on both the physical constants α and $Z\alpha$.[1]

One of the two new questions considered was this. What effect does Einstein's relativistic kinematics have? In this theory the Newtonian kinetic energy of an electron with mass m and momentum $\boldsymbol{p}$, $\boldsymbol{p}^2/2m$, is replaced by the much weaker $\sqrt{\boldsymbol{p}^2c^2 + m^2c^4} - mc^2$. So much weaker, in fact, that the simple atom is stable only if the relevant coupling parameter $Z\alpha$ is not too large! This fact was known in one form or another for many years – from the introduction of Dirac's 1928 relativistic quantum mechanics [39], in fact. It was far from obvious, therefore, that many-body stability would continue to hold even if $Z\alpha$ is kept small (but fixed, independent of N). Not only was the linear N-dependence in doubt but also stability of the first kind was unclear. This was resolved by Conlon in 1984 [32], who showed that stability of the second kind holds if $\alpha < 10^{-200}$ and $Z = 1$.

Clearly, Conlon's result needed improvement and this led to the invention of interesting new inequalities to simplify and improve his result. We now know that stability of the second kind holds if and only if *both* α and $Z\alpha$ are not too large. The bound on α itself was the new reality, previously unknown in the physics literature.

Again new inequalities were needed when it was realized that magnetic fields could also cause instabilities, even for just one atom, if $Z\alpha^2$ is too large. The understanding of this strange, and totally unforeseen, fact requires the knowledge that the appropriate Schrödinger equation has '*zero-modes*', as discovered by Loss and Yau in 1986 [139] (that is, square integrable, time-independent solutions with zero kinetic energy). But stability of the second kind was still open until Fefferman showed in 1995 [57, 58] that stability of the second kind holds if $Z = 1$ and α is very small. This result was subsequently improved to robust values of $Z\alpha^2$ and α by Lieb, Loss and Solovej in 1995 [123].

The surprises, in summary, were that stability of the second kind requires bounds on the fine-structure constant and the nuclear charges. In the relativistic case, smallness of α and of $Z\alpha$ is necessary, whereas in the non-relativistic case with magnetic fields, smallness of α and of $Z\alpha^2$ is required.

---

[1] If $Z \geq 1$, which it always is in nature, a bound on $Z\alpha$ implies a bound on α, of course. The point here is that the necessary bound on α is independent of Z, even if Z is arbitrarily small. In this book we shall not restrict our attention to integer Z.

Given these facts, one can ask if the *simultaneous* introduction of relativistic mechanics, magnetic fields, and the quantization of those fields in the manner proposed by M. Planck in 1900 [149], leads to new surprises about the requirements for stability. The answer, proved by Lieb, Loss, Siedentop and Solovej [127, 119], is that in at least one version of the problem no new conditions are needed, except for expected adjustments of the allowed bounds for $Z\alpha$ and α.

While we will visit all these topics in this book, we will not necessarily follow the historical route. In particular, we will solve the non-relativistic problem by using the improved inequalities invented to handle the relativistic problem, without the introduction of Thomas–Fermi theory. The Thomas–Fermi story is interesting, but no longer essential for our understanding of the stability of matter. Hence we will mention it, and sketch its application in the stability of matter problem, but we will not treat it thoroughly, and will not make further use of it. Some earlier pedagogical reviews are in [108, 115].

1.2 Brief Outline of the Book

An elementary introduction to quantum mechanics is given in **Chapter 2**. It is a thumbnail sketch of the relevant parts of the subject for readers who might want to refresh their memory, and it also serves to fix notation. Readers familiar with the subject can safely skip the chapter.

Chapter 3 discusses the many-body aspects of quantum mechanics and, in particular, introduces the concept of stability of matter in Section 3.2. The chapter also contains several results that will be used repeatedly in the chapters to follow, like the monotonicity of the ground state energy in the nuclear charges, and the fact the bosons have the lowest possible ground state energy among all symmetry classes.

A detailed discussion of Lieb–Thirring inequalities is the subject of **Chapter 4**. These inequalities play a crucial role in our understanding of stability of matter. They concern bounds on the moments of the negative eigenvalues of Schrödinger type operators, which lead to lower bounds on the kinetic energy of many-particle systems in terms of the corresponding semiclassical expressions. This chapter, like Chapters 5 and 6, is purely mathematical and contains analytic inequalities that will be applied in the following chapters.

Electrostatics is an old subject whose mathematical underpinning goes back to Newton's discussion in the *Principia* [144] of the gravitational force, which behaves in a similar way except for a change of sign from repulsive to attractive. Nevertheless, new inequalities are essential for understanding many-body systems, and these are given in **Chapters 5 and 6**. The latter chapter contains a proof of the Lieb–Oxford inequality [125], which gives a bound on the indirect part of the Coulomb electrostatic energy of a quantum system.

Chapter 7 contains a proof of stability of matter of non-relativistic fermionic particles. This is the same model for which stability was first shown by Dyson and Lenard [44] in 1967. The three proofs given here are different and very short given the inequalities derived in Chapters 4–6. As a consequence, matter is not only stable but also extensive, in the sense that the volume occupied is proportional to the number of particles. The instability of the same model for bosons will also be discussed.

The analogous model with relativistic kinematics is discussed in **Chapter 8**, and stability for fermions is proved for a certain range of the parameters α and $Z\alpha$. Unlike in the non-relativistic case, where the range of values of these parameters was unconstrained, bounds on these parameters are essential, as will be shown. The proof of stability in the relativistic case will be an important ingredient concerning stability of the models discussed in Chapters 9, 10 and 11.

The influence of spin and magnetic fields will be studied in **Chapter 9**. If the kinetic energy of the particles is described by the Pauli operator, it becomes necessary to include the magnetic field energy for stability. Again, bounds on various parameters become necessary, this time α and $Z\alpha^2$. It turns out that zero modes of the Pauli operator are a key ingredient in understanding the boundary between stability and instability.

If the kinetic energy of relativistic particles is described by the Dirac operator, the question of stability becomes even more subtle. This is the content of **Chapter 10**. For the Brown–Ravenhall model, where the physically allowed states are the positive energy states of the free Dirac operator, there is always instability in the presence of magnetic fields. Stability can be restored by appropriately modifying the model and choosing as the physically allowed states the ones that have a positive energy for the Dirac operator *with* the magnetic field.

The effects of the quantum nature of the electromagnetic field will be investigated in **Chapter 11**. The models considered are the same as in Chapters 9 and 10, but now the electromagnetic field will be quantized. These models are caricatures of quantum electrodynamics. The chapter includes a self-contained mini-course on the electromagnetic field and its quantization. The stability and instability results are essentially the same as for the non-quantized field, except for different bounds on the parameter regime for stability.

How many electrons can an atom or molecule bind? This question will be addressed in **Chapter 12**. The reason for including it in a book on stability of matter is to show that for a lower bound on the ground state energy only the minimum of the number of nuclei and the number of electrons is relevant. A large excess charge can not lower the energy.

Once a system becomes large enough so that the gravitational interaction can not be ignored, stability fails. This can be seen in nature in terms of the gravitational collapse of stars and the resulting supernovae, or as the upper mass limit of cold stars. Simple models of this gravitational collapse, as appropriate for white dwarfs and neutron stars, will be studied in **Chapter 13**. In particular, it will be shown how the critical number of particles for collapse depends on the gravitational constant G, namely $G^{-3/2}$ for fermions and G^{-1} for bosons, respectively.

The first 13 chapters deal essentially with the problem of showing that the lowest energy of matter is bounded below by a constant times the number of particles. The final **Chapter 14** deals with the question of showing that the energy is really proportional to the number of particles, i.e., that the energy per particle has a limit as the particle number goes to infinity. Such a limit exists not only for the ground state energy, but also for excited states in the sense that at positive temperature the thermodynamic limit of the free energy per particle exists.

CHAPTER 2

Introduction to Elementary Quantum Mechanics and Stability of the First Kind

In this second chapter we will review the basic mathematical and physical facts about quantum mechanics and establish physical units and notation. Those readers already familiar with the subject can safely jump to the next chapter.

An attempt has been made to make the presentation in this chapter as elementary as possible, and yet present the basic facts that will be needed later. There are many beautiful and important topics which will not be touched upon such as self-adjointness of Schrödinger operators, the general mathematical structure of quantum mechanics and the like. These topics are well described in other works, e.g., [150].

Much of the following can be done in a Euclidean space of arbitrary dimension, but in this chapter the dimension of the Euclidean space is taken to be three – which is the physical case – unless otherwise stated. We do this to avoid confusion and, occasionally, complications that arise in the computation of mathematical constants. The interested reader can easily generalize what is done here to the $\mathbb{R}^d$, $d > 3$ case. Likewise, in the next chapters we mostly consider N particles, with spatial coordinates in $\mathbb{R}^3$, so that the total spatial dimension is $3N$.

2.1 A Brief Review of the Connection Between Classical and Quantum Mechanics

Considering the range of validity of quantum mechanics, it is not surprising that its formulation is more complicated and abstract than classical mechanics. Nevertheless, classical mechanics is a basic ingredient for quantum mechanics. One still talks about position, momentum and energy which are notions from Newtonian mechanics.

The connection between these two theories becomes apparent in the semiclassical limit, akin to passing from wave optics to geometrical optics. In its Hamiltonian formulation, classical mechanics can be viewed as a problem of geometrical optics. This led Schrödinger to guess the corresponding wave equation. We refrain from fully explaining the semiclassical limit of quantum mechanics. For one aspect of this problem, however, the reader is referred to Chapter 4, Section 4.1.1.

We turn now to classical dynamics itself, in which a point particle is fully described by giving its **position** $\boldsymbol{x} = (x^1, x^2, x^3)$ in $\mathbb{R}^3$ and its **velocity** $\boldsymbol{v} = \mathrm{d}\boldsymbol{x}/\mathrm{d}t = \dot{\boldsymbol{x}}$ in $\mathbb{R}^3$ at any time t, where the dot denotes the derivative with respect to time.[1] Newton's law of motion says that along any mechanical trajectory its acceleration $\dot{\boldsymbol{v}} = \ddot{\boldsymbol{x}}$ satisfies

$$m\ddot{\boldsymbol{x}} = \boldsymbol{F}(\boldsymbol{x}, \dot{\boldsymbol{x}}, t), \tag{2.1.1}$$

where $\boldsymbol{F}$ is the **force** acting on the particle and m is the **mass**. With $\boldsymbol{F}(\boldsymbol{x}, \dot{\boldsymbol{x}}, t)$ given, the expression (2.1.1) is a system of second order differential equations which together with the initial conditions $\boldsymbol{x}(t_0)$ and $\boldsymbol{v}(t_0) = \dot{\boldsymbol{x}}(t_0)$ determine $\boldsymbol{x}(t)$ and thus $\boldsymbol{v}(t)$ for all times. If there are N particles interacting with each other, then (2.1.1) takes the form

$$m_i\ddot{\boldsymbol{x}}_i = \boldsymbol{F}_i, \quad i = 1, \dots, N, \tag{2.1.2}$$

where $\boldsymbol{F}_i$ denotes the sum of all forces acting on the ith particle and $\boldsymbol{x}_i$ denotes the position of the ith particle. As an example, consider the force between two charged particles, whose respective charges are denoted by Q_1 and Q_2, namely the **Coulomb force** given (in appropriate units, see Section 2.1.7) by

$$\boldsymbol{F}_1 = Q_1 Q_2 \frac{\boldsymbol{x}_1 - \boldsymbol{x}_2}{|\boldsymbol{x}_1 - \boldsymbol{x}_2|^3} = -\boldsymbol{F}_2. \tag{2.1.3}$$

If $Q_1 Q_2$ is positive the force is repulsive and if $Q_1 Q_2$ is negative the force is attractive. Formula (2.1.3) can be written in terms of the **potential energy function**

$$V(\boldsymbol{x}_1, \boldsymbol{x}_2) = \frac{Q_1 Q_2}{|\boldsymbol{x}_1 - \boldsymbol{x}_2|}, \tag{2.1.4}$$

noting that

$$\boldsymbol{F}_1 = -\nabla_{\boldsymbol{x}_1} V \quad \text{and} \quad F_2 = -\nabla_{\boldsymbol{x}_2} V. \tag{2.1.5}$$

As usual, we denote the gradient by $\nabla = (\partial/\partial x^1, \partial/\partial x^2, \partial/\partial x^3)$.

[1] We follow the physicists' convention in which vectors are denoted by boldface letters.

2.1.1 Hamiltonian Formulation

Hamilton's formulation of classical mechanics is the entry to quantum physics. **Hamilton's equations** are

$$\dot{\boldsymbol{x}} = \frac{\partial H}{\partial \boldsymbol{p}}, \quad \dot{\boldsymbol{p}} = -\frac{\partial H}{\partial \boldsymbol{x}} \tag{2.1.6}$$

where $H(\boldsymbol{x}, \boldsymbol{p})$ is the **Hamilton function** and $\boldsymbol{p}$ the **canonical momentum** of the particle. Assuming that

$$F(\boldsymbol{x}) = -\nabla V(\boldsymbol{x}) \tag{2.1.7}$$

for some potential V then, in the case that the canonical momentum is given by

$$\boldsymbol{p} = m\boldsymbol{v}, \tag{2.1.8}$$

Eq. (2.1.6) with

$$H = \frac{\boldsymbol{p}^2}{2m} + V(\boldsymbol{x}) \tag{2.1.9}$$

yields (2.1.1). The function

$$T(\boldsymbol{p}) = \frac{\boldsymbol{p}^2}{2m} \tag{2.1.10}$$

is called the **kinetic energy** function. A simple computation using Eq. (2.1.6) shows that along each mechanical trajectory the function $H(\boldsymbol{x}(t), \boldsymbol{p}(t))$ is a constant which we call the **energy**, E.

2.1.2 Magnetic Fields

Not in all cases is the canonical momentum given by (2.1.8). An example is the motion of a charged particle of mass m and charge $-e$ in a magnetic field $\boldsymbol{B}(\boldsymbol{x})$ in addition to a potential, $V(\boldsymbol{x})$. The **Lorentz force** on such a particle located at $\boldsymbol{x}$ and having velocity $\boldsymbol{v}$ is[2]

$$F_{\text{Lorentz}} = -\frac{e}{c}\boldsymbol{v} \wedge \boldsymbol{B}(\boldsymbol{x}). \tag{2.1.11}$$

[2] We use the symbol $\wedge$ for the vector product on $\mathbb{R}^3$, instead of $\times$, since the latter may be confused with $\boldsymbol{x}$.

The Hamilton function is then given by[3]

$$H(\boldsymbol{x}, \boldsymbol{p}) = \frac{1}{2m}\left(\boldsymbol{p} + \frac{e}{c}\boldsymbol{A}(\boldsymbol{x})\right)^2 + V(\boldsymbol{x}), \tag{2.1.12}$$

where $\boldsymbol{A}(\boldsymbol{x})$, called the **vector potential**, determines the **magnetic field** (more properly called the **magnetic induction**) $\boldsymbol{B}(\boldsymbol{x})$ by the equation

$$\boxed{\nabla \wedge \boldsymbol{A}(\boldsymbol{x}) = \text{curl}\, \boldsymbol{A}(\boldsymbol{x}) = \boldsymbol{B}(\boldsymbol{x}).} \tag{2.1.13}$$

The fact that an arbitrary magnetic field can be written this way as a curl is a consequence of the fact that Maxwell's equations dictate that all physical magnetic fields satisfy

$$\nabla \cdot \boldsymbol{B}(\boldsymbol{x}) = \text{div}\, \boldsymbol{B}(\boldsymbol{x}) = 0. \tag{2.1.14}$$

The parameter c in (2.1.12) is the **speed of light**, which equals 299 792 458 meters/sec.

The canonical momentum $\boldsymbol{p}$ is now *not* equal to mass times velocity but rather

$$m\boldsymbol{v} = \boldsymbol{p} + \frac{e}{c}\boldsymbol{A}(\boldsymbol{x}). \tag{2.1.15}$$

It is a simple calculation to derive the Lorentz law for the motion of an electron in an external magnetic field using (2.1.6) with (2.1.12) as the Hamilton function.

The energy associated with this $\boldsymbol{B}$ field (i.e., the amount of work needed to construct this field or, equivalently, the amount of money we have to pay to the electric power company) is[4]

$$\boxed{\mathcal{E}_{\text{mag}}(\boldsymbol{B}) := \frac{1}{8\pi}\int_{\mathbb{R}^3} |\boldsymbol{B}(\boldsymbol{x})|^2 \mathrm{d}\boldsymbol{x}.} \tag{2.1.16}$$

The units we use are the conventional *absolute electrostatic units.* For further discussion of units see Section 2.1.7.

---

[3] We note that we use the convention that the electron charge equals $-e$, with $e > 0$, and hence the proper form of the kinetic energy is given by (2.1.12). In the formula $(\boldsymbol{p} - e\boldsymbol{A}(\boldsymbol{x})/c)^2/(2m)$, which is usually found in textbooks, e denotes a generic charge, which can be positive or negative.

[4] The equation $a := b$ (or $b =: a$) means that a is defined by b.

Since the only requirement on $\boldsymbol{A}$ is that it satisfy (2.1.13) we have a certain amount of freedom in choosing $\boldsymbol{A}$. It would appear that the $\boldsymbol{A}$ has three degrees of freedom (the three components of the vector $\boldsymbol{A}$) but in reality there are only two since $\boldsymbol{B}$ has only two degrees of freedom (because $\operatorname{div} \boldsymbol{B} = 0$). If we assume that $\mathcal{E}_{\text{mag}}(\boldsymbol{B}) < \infty$ (which should be good enough for physical applications) then we can choose the field $\boldsymbol{A}$ such that (2.1.13) holds and

$$\operatorname{div} \boldsymbol{A} = 0 \quad \text{and} \quad \int_{\mathbb{R}^3} |\boldsymbol{A}(\boldsymbol{x})|^6 \, \mathrm{d}\boldsymbol{x} < \infty. \tag{2.1.17}$$

A proof of this fact is given in Lemma 10.1 in Chapter 10. The condition $\operatorname{div} \boldsymbol{A} = 0$ is of no importance to us until we get to Chapter 10. All results prior to Chapter 10 hold irrespective of this condition. Its relevance is explained in Section 10.1.1 on gauge invariance.

2.1.3 Relativistic Mechanics

It is straightforward to describe relativistic mechanics in the Hamiltonian formalism. The Hamilton function for a free relativistic particle is

$$T_{\text{rel}}(\boldsymbol{p}) := c\sqrt{\boldsymbol{p}^2 + m^2c^2} - mc^2, \tag{2.1.18}$$

from which the relation between $\boldsymbol{p}$ and the velocity, $\boldsymbol{v}$, is found to be

$$\boldsymbol{v} = \frac{\partial T_{\text{rel}}(\boldsymbol{p})}{\partial \boldsymbol{p}} = \frac{c\,\boldsymbol{p}}{\sqrt{\boldsymbol{p}^2 + m^2c^2}}. \tag{2.1.19}$$

Note that $|\boldsymbol{v}| < c$. We can, of course, include a magnetic field in this relativistic formalism simply by replacing $\boldsymbol{p}$ by $\boldsymbol{p} + \frac{e}{c}\boldsymbol{A}(\boldsymbol{x})$.

A potential can be added to this $T_{\text{rel}}(\boldsymbol{p})$ so that the Hamilton function becomes

$$H_{\text{rel}}(\boldsymbol{p}, \boldsymbol{x}) = T_{\text{rel}}(\boldsymbol{p}) + V(\boldsymbol{x}). \tag{2.1.20}$$

Hamilton's equations (2.1.6) then yield a mathematically acceptable theory, but it has to be admitted that it is not truly a relativistic theory from the physical point of view. The reason is that the theory obtained this way is not invariant under Lorentz transformations, i.e., the equations of motion (and not merely the solutions of the equations) are different in different inertial systems. We shall not attempt to explain this further, because we shall not be concerned with true relativistic invariance in this book. In any case, 'energy' itself is not a relativistically invariant quantity (it is only a component of a 4-vector). We shall, however, be concerned with the kind of mechanics defined by the Hamilton function H_{rel} in (2.1.20) because this dynamics *is* an interesting *approximation* to a truly relativistic mechanics.

2.1.4 Many-Body Systems

There is no difficulty in describing many-body systems in the Hamiltonian formalism – with either relativistic or non-relativistic kinematics. As an example, consider the problem of N electrons and M static nuclei interacting with each other via the Coulomb force. The electrons have charge $-e$, and are located at positions $\underline{X} = (\boldsymbol{x}_1, \dots, \boldsymbol{x}_N)$, $\boldsymbol{x}_i \in \mathbb{R}^3$ for $i = 1, \dots, N$. The M nuclei have charges $e\underline{Z} = e(Z_1, \dots, Z_M)$ and are located at $\underline{\boldsymbol{R}} = (\boldsymbol{R}_1, \dots, \boldsymbol{R}_M)$ with $\boldsymbol{R}_i \in \mathbb{R}^3$ for $i = 1, \dots, M$. Then the **potential energy function** of this system is $e^2 V_C(\underline{X}, \underline{\boldsymbol{R}})$, with

$$\boxed{V_C(\underline{X}, \underline{\boldsymbol{R}}) = W(\underline{X}, \underline{\boldsymbol{R}}) + I(\underline{X}) + U(\underline{\boldsymbol{R}}),} \tag{2.1.21}$$

where

$$W(\underline{X}, \underline{\boldsymbol{R}}) = -\sum_{i=1}^{N} \sum_{j=1}^{M} \frac{Z_j}{|\boldsymbol{x}_i - \boldsymbol{R}_j|} \tag{2.1.22}$$

$$I(\underline{X}) = \sum_{1 \le i < j \le N} \frac{1}{|\boldsymbol{x}_i - \boldsymbol{x}_j|} \tag{2.1.23}$$

$$U(\underline{\boldsymbol{R}}) = \sum_{1 \le i < j \le M} \frac{Z_i Z_j}{|\boldsymbol{R}_i - \boldsymbol{R}_j|}. \tag{2.1.24}$$

The three terms have the following meaning: $W(\underline{X}, \underline{\boldsymbol{R}})$ is the **electron–nucleus attractive Coulomb interaction**, $I(\underline{X})$ is the **electron–electron repulsive interaction** and $U(\underline{\boldsymbol{R}})$ is the **nucleus–nucleus repulsive interaction**. The total force acting on the ith electron is thus given by

$$\boldsymbol{F}_i = -e^2 \nabla_{\boldsymbol{x}_i} V_C(\underline{X}, \underline{\boldsymbol{R}}). \tag{2.1.25}$$

The Hamilton function is the sum of kinetic energy and potential energy

$$H(\underline{X}, \underline{P}) = T(\underline{P}) + e^2 V_C(\underline{X}, \underline{\boldsymbol{R}}), \tag{2.1.26}$$

where

$$T(\underline{\boldsymbol{P}}) = \sum_{j=1}^{N} \frac{\boldsymbol{p}_j^2}{2m}, \quad \underline{\boldsymbol{P}} = (\boldsymbol{p}_1, \dots, \boldsymbol{p}_N) \tag{2.1.27}$$

in the non-relativistic case (or with the obvious change in the relativistic case). If the nuclei are dynamic, one also has to add the nuclear kinetic energy, of course.

In the case of static nuclei, $\underline{\boldsymbol{R}}$ are simply fixed parameters. We point out that when we study stability of the quantum analogue of this system, it will be essential to look for bounds that are *independent* of $\underline{\boldsymbol{R}}$.

2.1.5 Introduction to Quantum Mechanics

On atomic length scales, position and momentum can no longer describe the state of a particle. They both play an important role as observables but to describe the **state of a quantum mechanical particle** one requires a complex valued function $\psi : \mathbb{R}^3 \to \mathbb{C}$, called the **wave function.** In the remainder of the present chapter we limit the discussion to a single particle. The discussion of N-particle wave functions, $\psi : \mathbb{R}^{3N} \to \mathbb{C}$ is deferred to Chapter 3.

In order to fix the state of an electron one has to specify infinitely many numbers (i.e., a whole function) – not just the six numbers $\boldsymbol{p}$ and $\boldsymbol{x}$ of classical mechanics. The function $\boldsymbol{x} \mapsto |\psi(\boldsymbol{x})|^2$ is interpreted as a probability density and hence we require the **normalization condition**

$$\boxed{\int_{\mathbb{R}^3} |\psi(\boldsymbol{x})|^2 \mathrm{d}\boldsymbol{x} = 1.} \tag{2.1.28}$$

The classical energy is replaced by an energy functional, $\mathcal{E}(\psi)$, of the wave function of the system:

$$\mathcal{E}(\psi) = T_\psi + V_\psi, \tag{2.1.29}$$

where

$$\boxed{T_\psi = \frac{\hbar^2}{2m} \int_{\mathbb{R}^3} |\nabla \psi(\boldsymbol{x})|^2 \mathrm{d}\boldsymbol{x},} \tag{2.1.30}$$

and

$$V_\psi = \int_{\mathbb{R}^3} V(\boldsymbol{x})|\psi(\boldsymbol{x})|^2 \mathrm{d}\boldsymbol{x}. \tag{2.1.31}$$

The functional T_ψ is called the **expectation value of the kinetic energy** or, in short, the **kinetic energy** of ψ. Similarly, V_ψ is called the **expectation value of the potential energy** or, in short, the **potential energy** of ψ. The constant $\hbar$ (pronounced h-bar) is

$$\hbar = \frac{h}{2\pi} = 1.055 \times 10^{-27} \text{ grams cm}^2 \text{ sec}^{-1},$$

where $h = 6.626 \times 10^{-27}$ grams cm^2 sec^{-1} is a constant of nature called **Planck's constant**.

A comparison of (2.1.30) and (2.1.10) shows that the transition from classical to quantum mechanics is accomplished by replacing the classical momentum $\boldsymbol{p}$ by the operator $-i\hbar\nabla$. We shall frequently denote $-i\hbar\nabla$ by $\boldsymbol{p}$. In this notation we have

$$T_\psi = \frac{1}{2m} \int_{\mathbb{R}^3} |(\boldsymbol{p}\psi)(\boldsymbol{x})|^2 \mathrm{d}\boldsymbol{x}.$$

Note that for a complex vector $\boldsymbol{v}$ such as $\boldsymbol{p}\psi(\boldsymbol{x})$, the quantity $|\boldsymbol{v}|^2$ denotes the sum of the squares of the absolute values of the components.

Associated with the kinetic energy (2.1.30) is the operator, called the **free Hamilton operator**, or simply the free Hamiltonian for brevity,

$$H_0 = -\frac{\hbar^2}{2m}\Delta = -\frac{\hbar^2}{2m}\sum_{i=1}^{3} \frac{\partial^2}{\partial (x^i)^2}, \tag{2.1.32}$$

and associated with $\mathcal{E}$ is the **Hamiltonian**

$$H = H_0 + V$$

which acts on functions ψ by

$$(H\psi)(\boldsymbol{x}) = -\frac{\hbar^2}{2m}(\Delta\psi)(\boldsymbol{x}) + V(\boldsymbol{x})\psi(\boldsymbol{x}). \tag{2.1.33}$$

The derivatives in (2.1.33) can be taken to be in the distributional sense.[5] For nice functions ψ

$$\mathcal{E}(\psi) = (\psi, H\psi), \tag{2.1.34}$$

where the **inner product** of f and g is defined by

$$(f, g) = \int_{\mathbb{R}^3} \overline{f}(\boldsymbol{x})g(\boldsymbol{x})\mathrm{d}\boldsymbol{x}. \tag{2.1.35}$$

(The notation $\overline{f}$ denotes the complex conjugate, sometimes also written as f^*.) Again, for nice functions ψ and ϕ,

$$\begin{aligned}(\psi, H\phi) &= \int_{\mathbb{R}^3} \overline{\psi}(\boldsymbol{x})(H\phi)(\boldsymbol{x})\mathrm{d}\boldsymbol{x} \\ &= \int_{\mathbb{R}^3} \left(\frac{\hbar^2}{2m}\overline{(\nabla\psi)}(\boldsymbol{x}) \cdot (\nabla\phi)(\boldsymbol{x}) + V(\boldsymbol{x})\overline{\psi}(\boldsymbol{x})\phi(\boldsymbol{x})\right)\mathrm{d}\boldsymbol{x}. \end{aligned} \tag{2.1.36}$$

One of the technical problems in quantum mechanics is that the left side of (2.1.36) does not always make sense for arbitrary ϕ's. For the purpose of this book *we shall always interpret* $(\psi, H\phi)$ *as the right side of* (2.1.36), which is well defined if ψ and ϕ are in $H^1(\mathbb{R}^3)$.[6] Note that T_ψ is always positive and is always well defined for $\psi \in H^1(\mathbb{R}^3)$. See [164].

Returning to the Coulomb law in (2.1.3), we see that the energy function for a **hydrogenic atom** is given by

$$\mathcal{E}(\psi) = \int_{\mathbb{R}^3} \left(\frac{\hbar^2}{2m}|\nabla\psi(\boldsymbol{x})|^2 - \frac{Ze^2}{|\boldsymbol{x}|}|\psi(\boldsymbol{x})|^2\right)\mathrm{d}\boldsymbol{x} \tag{2.1.37}$$

---

[5] The functions that appear in quantum mechanics are not always differentiable in the classical sense. To define the following concepts rigorously it is important to know what a distributional derivative is. These matters are discussed fully elsewhere, see, e.g., [118]. Nevertheless, the reader who is willing to accept things on faith can just assume that all derivatives are classical and will still be able to follow the presentation.

[6] The most important spaces relevant for this book are $L^2(\mathbb{R}^d)$ consisting of functions $f(\boldsymbol{x})$ such that $\|f\|_2^2 := \int_{\mathbb{R}^d} |f(\boldsymbol{x})|^2\mathrm{d}\boldsymbol{x} < \infty$, and $H^1(\mathbb{R}^d)$, which consists of functions that are square integrable and whose distributional derivatives are also square integrable *functions*. Again, we refer to [118] for further details.

and the corresponding Hamiltonian is

$$H = -\frac{\hbar^2}{2m}\Delta - \frac{Ze^2}{|\boldsymbol{x}|}. \tag{2.1.38}$$

We refer to this as a hydrogenic atom because it is the Hamiltonian for a one-electron atom; hydrogen corresponds to the case $Z = 1$. It will be discussed further in Section 2.2.2.

A convenient way of rewriting the kinetic energy of a function $\psi \in L^2(\mathbb{R}^d)$ (for any $d \geq 1$) is via Fourier transforms. Recall that the **Fourier transform** of a function $\psi(\boldsymbol{x})$ is, formally, defined as[7]

$$\widehat{\psi}(\boldsymbol{k}) = \int_{\mathbb{R}^d} \psi(\boldsymbol{x}) e^{-2\pi i \boldsymbol{x}\cdot\boldsymbol{k}} \mathrm{d}\boldsymbol{x}. \tag{2.1.39}$$

(We say 'formally' since the integral is absolutely convergent only if $\psi \in L^1(\mathbb{R}^d)$. Nevertheless, $\widehat{\psi}$ is well defined for all $\psi \in L^2(\mathbb{R}^d)$. See [118] for details.) Then ψ is given in terms of the inverse Fourier transform as

$$\psi(\boldsymbol{x}) = \int_{\mathbb{R}^d} \widehat{\psi}(\boldsymbol{k}) e^{2\pi i \boldsymbol{x}\cdot\boldsymbol{k}} \mathrm{d}\boldsymbol{k}.$$

Recall also Plancherel's identity,

$$\int_{\mathbb{R}^d} |\psi(\boldsymbol{x})|^2 \mathrm{d}\boldsymbol{x} = \int_{\mathbb{R}^d} |\widehat{\psi}(\boldsymbol{k})|^2 \mathrm{d}\boldsymbol{k}. \tag{2.1.40}$$

The (non-relativistic) kinetic energy of a function ψ can then be expressed as

$$\int_{\mathbb{R}^d} |\nabla\psi(\boldsymbol{x})|^2 \mathrm{d}\boldsymbol{x} = \int_{\mathbb{R}^d} (2\pi\boldsymbol{k})^2 |\widehat{\psi}(\boldsymbol{k})|^2 \mathrm{d}\boldsymbol{k}. \tag{2.1.41}$$

So any $\psi \in L^2(\mathbb{R}^d)$ is also in $H^1(\mathbb{R}^d)$ if and only if the right side of (2.1.41) is finite. Putting it differently, the operator $\boldsymbol{p}$ acts as multiplication by $2\pi\boldsymbol{k}$ in Fourier space, and $\psi \in L^2(\mathbb{R}^d)$ is in $H^1(\mathbb{R}^d)$ if and only if $|\boldsymbol{k}|\widehat{\psi}(\boldsymbol{k})$ is in $L^2(\mathbb{R}^d)$.

[7] A different convention for the Fourier transform that is often used is $\widehat{\psi}(\boldsymbol{k}) = (2\pi)^{-d/2} \int_{\mathbb{R}^d} \psi(\boldsymbol{x}) e^{-i\boldsymbol{x}\cdot\boldsymbol{k}} \mathrm{d}\boldsymbol{x}$.

Using Fourier space, it is straightforward to define the relativistic kinetic energy (2.1.18) in the quantum case. Namely,

$$(\psi, c\sqrt{\boldsymbol{p}^2 + m^2c^2}\,\psi) = \int_{\mathbb{R}^3} c\sqrt{(2\pi \boldsymbol{k})^2 + m^2c^2}|\widehat{\psi}(\boldsymbol{k})|^2 \mathrm{d}\boldsymbol{k}. \qquad (2.1.42)$$

The space of $L^2(\mathbb{R}^d)$ functions for which the right side of (2.1.42) is finite is called $H^{1/2}(\mathbb{R}^d)$, for obvious reasons. Again, see [118] for further details.

Magnetic fields can be introduced in quantum mechanics in the same way as in classical mechanics. The Hamilton function (2.1.12) of one particle in a magnetic field becomes

$$H = \frac{1}{2m}\left(\boldsymbol{p} + \frac{e}{c}\boldsymbol{A}(\boldsymbol{x})\right)^2 + V(\boldsymbol{x}) \qquad (2.1.43)$$

in the quantum case, with $\boldsymbol{p} = -i\hbar\nabla$, as before. One can also consider magnetic fields together with relativity, in which case the kinetic energy becomes

$$c\sqrt{\left(\boldsymbol{p} + \frac{e}{c}\boldsymbol{A}(\boldsymbol{x})\right)^2 + m^2c^2}. \qquad (2.1.44)$$

The square root of a positive operator can be defined via the spectral theorem. (See, e.g., [150].) Alternatively, it can also be defined in terms of the Green's function, or resolvent, $[(\boldsymbol{p} + (e/c)\boldsymbol{A}(\boldsymbol{x}))^2 + t]^{-1}$ for $t > 0$. In fact, the formula

$$\sqrt{x} = \frac{1}{\pi}\int_0^\infty \frac{x}{x+t}\frac{\mathrm{d}t}{\sqrt{t}}$$

valid for $x > 0$, can be used to define the square root of any non-negative operator, such as $\left(\boldsymbol{p} + \frac{e}{c}\boldsymbol{A}(\boldsymbol{x})\right)^2 + m^2c^2$.

2.1.6 Spin

Elementary particles have an internal degree of freedom called **spin** which is characterized by a specific number that can take one of the values $S = 0, 1/2, 1, 3/2, \ldots$. ($S = 0$ is usually called 'spinless'.) A particle with spin S carries with it an internal Hilbert space of dimension $2S + 1$, that is, its wave function is an element of $L^2(\mathbb{R}^3) \otimes \mathbb{C}^{2S+1} =: L^2(\mathbb{R}^3; \mathbb{C}^{2S+1})$. For example, electrons are spin $1/2$ particles and what this means is that the wave function is really a pair of ordinary complex-valued functions

$$\psi(\boldsymbol{x}, \sigma), \quad \sigma = 1, 2. \tag{2.1.45}$$

Another way of writing this is a two-component vector called a **spinor**

$$\psi(\boldsymbol{x}) = \begin{pmatrix} \psi_1(\boldsymbol{x}) \\ \psi_2(\boldsymbol{x}) \end{pmatrix}. \tag{2.1.46}$$

The normalization condition is now given by

$$\int_{\mathbb{R}^3} \langle \psi(\boldsymbol{x}), \psi(\boldsymbol{x}) \rangle \mathrm{d}\boldsymbol{x} = 1, \tag{2.1.47}$$

where

$$\langle \psi, \phi \rangle = \overline{\psi_1}\phi_1 + \overline{\psi_2}\phi_2 \tag{2.1.48}$$

is the inner product on $\mathbb{C}^2$.

In the absence of magnetic fields, the kinetic energy acts separately on each of the two components of ψ in a manner similar to the normalization condition, i.e.,

$$T_\psi = T_{\psi_1} + T_{\psi_2}. \tag{2.1.49}$$

The second formulation (2.1.46) is convenient for discussing spin $1/2$ in terms of the three **Pauli matrices** $\boldsymbol{\sigma} = (\sigma^1, \sigma^2, \sigma^3)$, with[8]

$$\sigma^1 = \begin{pmatrix} 0 & 1 \\ 1 & 0 \end{pmatrix}, \quad \sigma^2 = \begin{pmatrix} 0 & -i \\ i & 0 \end{pmatrix}, \quad \sigma^3 = \begin{pmatrix} 1 & 0 \\ 0 & -1 \end{pmatrix}. \tag{2.1.50}$$

The following relations are easily verified:

$$\sigma^j \sigma^k = -\sigma^k \sigma^j, \quad \text{for } j \neq k \tag{2.1.51}$$

and

$$\sigma^j \sigma^k = i\sigma^l, \tag{2.1.52}$$

where the indices (j, k, l) are any cyclic permutation of the numbers $(1, 2, 3)$. Given a spinor we can form the three dimensional vector

$$\langle \psi, \boldsymbol{\sigma}\psi \rangle = (\langle \psi, \sigma^1\psi \rangle, \langle \psi, \sigma^2\psi \rangle, \langle \psi, \sigma^3\psi \rangle). \tag{2.1.53}$$

[8] The reader should not confuse the three Pauli matrices $\boldsymbol{\sigma}$ with the integers σ labeling the spin components.

In terms of the Pauli matrices, the angular momentum operators associated with the **electron spin** are $S = (\hbar/2)\sigma$. This is discussed in every standard quantum mechanics textbook and is not important for us here.

In the presence of a magnetic field, the kinetic energy of an electron has to be modified as[9]

$$\frac{1}{2m}\int_{\mathbb{R}^3} \left|\left(p + \frac{e}{c}A(x)\right)\psi\right|^2 \mathrm{d}x + \mu_0 \int_{\mathbb{R}^3} \langle\psi, \sigma\cdot B(x)\psi\rangle \mathrm{d}x. \qquad (2.1.54)$$

The constant μ_0 is called the **Bohr magneton** and is given by

$$\mu_0 = \frac{\hbar e}{2mc} = 9.274 \times 10^{-21}\mathrm{erg\ gauss}^{-1}. \qquad (2.1.55)$$

The second term in (2.1.54) is called the **Zeeman energy**. Using the electron spin operators, it can alternatively by written as the expectation value of $(e/mc)S\cdot B(x)$. It is very important in our daily lives since it is responsible for the magnetization of a piece of iron.

The reader might wonder for which ψ and A the expression (2.1.54) is well defined. If we assume that each component of the vector potential A is in $L^2_{\mathrm{loc}}(\mathbb{R}^3)$ then the first term in (2.1.54) is defined for all those functions $\psi \in L^2(\mathbb{R}^3)$ with $(\partial_{x^j} + i(e/\hbar c)A^j)\psi \in L^2(\mathbb{R}^3)$ for $j = 1, 2, 3$. They form a function space which is denoted by $H^1_A(\mathbb{R}^3)$. For further properties of this space see [118].

There is no difficulty in extending the above definition to the spin case. For any spinor ψ with $(\partial_{x^j} + i(e/\hbar c)A^j)\psi \in L^2(\mathbb{R}^3;\mathbb{C}^2)$ for $j = 1, 2, 3$, the following expression makes sense and serves as a definition of the kinetic energy in (2.1.54):

$$\frac{1}{2m}\int_{\mathbb{R}^3} \left|\sigma\cdot\left(p + \frac{e}{c}A(x)\right)\psi(x)\right|^2 \mathrm{d}x. \qquad (2.1.56)$$

If the vector potential A is sufficiently smooth, an integration by parts and the use of the commutation relations (2.1.51) and (2.1.52) shows that the above expression can be rewritten as (2.1.54). Thus the Zeeman-term is hidden in (2.1.56). The advantage of (2.1.56) is that no smoothness assumption on A has to be made. Moreover, the positivity of the kinetic energy is apparent from the formulation (2.1.56).

[9] For convenience, we use the same absolute value symbol for the length of a vector in $\mathbb{C}^2$ or $\mathbb{C}^2 \otimes \mathbb{C}^3$ as we used earlier for the length of a vector in $\mathbb{C}^3$.

The Hamiltonian associated with (2.1.54) and (2.1.56) is given by

$$\frac{1}{2m}\left(\boldsymbol{p}+\frac{e}{c}\boldsymbol{A}(\boldsymbol{x})\right)^2+\mu_0\boldsymbol{\sigma}\cdot\boldsymbol{B}(\boldsymbol{x})=\frac{1}{2m}\left[\boldsymbol{\sigma}\cdot\left(\boldsymbol{p}+\frac{e}{c}\boldsymbol{A}(\boldsymbol{x})\right)\right]^2. \tag{2.1.57}$$

This operator is called the **Pauli operator**[10] and will be discussed in detail in Chapter 9.

An important concept is **gauge invariance**. Physical quantities, like the ground state energy, depend on $\boldsymbol{A}$ only through $\boldsymbol{B}$. This can be seen as follows. If there are two vector potentials $\boldsymbol{A}_1$ and $\boldsymbol{A}_2$, with curl $\boldsymbol{A}_1=$ curl $\boldsymbol{A}_2=\boldsymbol{B}$, then $\boldsymbol{A}_1=\boldsymbol{A}_2+\nabla\chi$ for some function χ. This follows from the fact that a curl-free vector field is necessarily a gradient field.[11] We emphasize that this is true only on the *whole* of $\mathbb{R}^3$ and not on punctured domains, such as the exterior of an infinitely extended cylinder. Gauge invariance of the Hamiltonian $H_{\boldsymbol{A}}$ means that

$$H_{\boldsymbol{A}+\nabla\chi}=U(\chi)H_{\boldsymbol{A}}U(\chi)^\dagger,$$

where $U(\chi)$ is the unitary multiplication operator $e^{i(e/\hbar c)\chi(\boldsymbol{x})}$, and $U^\dagger$ denotes the **adjoint** of U.

2.1.7 Units

Anyone who has studied electromagnetism knows that the problem of choosing suitable units can be a nightmare. This is even more so when we want to include $\hbar$ and simultaneously make the system convenient for quantum mechanics. One solution, favored by many physics texts, is to include all physical units $(m, e, \hbar, c)$ in all equations. While this is certainly clear it is cumbersome and somewhat obscures the main features of the equations.

Except for the gravitational constant G, which will be discussed in Chapter 13, there are four (dimensional) physical constants that play a role in this book. These are

- $m=$ mass of the electron $=9.11\times10^{-28}$ grams
- $e=(-1)\times$ charge of the electron $=4.803\times10^{-10}$ grams$^{1/2}$ cm$^{3/2}$ sec^{-1}

[10] A closely related operator that is often used is the simpler **Pauli–Fierz operator** [64]. It is the same as (2.1.57) except that the term $(e^2/2mc^2)\boldsymbol{A}(\boldsymbol{x})^2$ is omitted. This operator is not gauge invariant (see Chapter 10) although it is useful when the $\boldsymbol{A}$ field is small. However, the absence of the $\boldsymbol{A}^2$ term can, if taken literally, lead to instabilities.

[11] A formula for χ is given by the line integral $\chi(\boldsymbol{x})=\int_0^{\boldsymbol{x}}(\boldsymbol{A}_1-\boldsymbol{A}_2)\cdot d\boldsymbol{s}$.

- $\hbar$ = Planck's constant divided by $2\pi = 1.055 \times 10^{-27}$ grams cm^2 sec^{-1}
- c = speed of light $= 3.00 \times 10^{10}$ cm sec^{-1}.

(Note that $\hbar$ has the dimension of energy times time.) These are the conventional cgs (centimeter, gram, second) electrostatic units. In particular, the energy needed to push two electrons from infinite separation to a distance of r centimeters is e^2/r.

In our choice of units for this book we were guided by the idea, which really originates in relativistic quantum mechanics, that the electron's charge is the quantity that governs the coupling between electromagnetism and dynamics (classical or quantum, relativistic or non-relativistic). When $e = 0$ all the particles in the universe are free and independent, so one wants to highlight the dependence of all physical quantities on e. To emphasize the role of e, we introduce the *only* dimensionless number that can be made from our four constants – the **fine-structure constant:**

$$\alpha := e^2/\hbar c = 1/137.04 = 7.297 \times 10^{-3}.$$

We should think of $\hbar$ and c as fixed and α as measuring the strength of the interaction, namely the electron charge squared. Our intention is to expose the role of e clearly and therefore we avoid using units of length, etc. that involve e in their definition.

Next, we need units for length, energy and time, and the only ones we can form that do *not* involve e are, respectively:

$$\lambda_C = \frac{\hbar}{mc} = \frac{1}{2\pi} \times \textbf{Compton wavelength of the electron} = 3.86 \times 10^{-11}\ \text{cm} \tag{2.1.58}$$

as the unit of length,

$$mc^2 = \textbf{rest mass energy of the electron} = 8.2 \times 10^{-7}\ \text{ergs} \tag{2.1.59}$$

as the unit of energy, and

$$\frac{\lambda_C}{c} = \frac{\hbar}{mc^2} = 1.29 \times 10^{-19}\ \text{sec} \tag{2.1.60}$$

as the unit of time. In other words, we shall set $\hbar = m = c = 1$, and thus $e = \sqrt{\alpha}$. *These will be the units used throughout the book.*

Thus, our quantum mechanical wave function $\psi(\boldsymbol{x})$ equals $\lambda_C^{-3/2}\tilde{\psi}(\boldsymbol{y})$ with $\boldsymbol{x}$ being given in terms of the dimensionless $\boldsymbol{y}$ by $\boldsymbol{x} = \lambda_C \boldsymbol{y}$. The subsequent expressions will be written in terms of $\tilde{\psi}$ (which will henceforth be called ψ, and the dimensionless argument $\boldsymbol{y}$ will henceforth be called $\boldsymbol{x}$).

In non-relativistic quantum mechanics (without magnetic fields) the speed of light does not appear and then the 'natural' units of length and energy are the **Bohr radius** and the **Rydberg**:

$$a_{\text{Bohr}} = \hbar^2/me^2 = \lambda_C/\alpha = 5.29 \times 10^{-9}\,\text{cm} \tag{2.1.61}$$

$$\text{Rydberg} = \frac{e^4 m}{2\hbar^2} = \frac{1}{2}mc^2\alpha^2 = 2.66 \times 10^{-5}\,mc^2. \tag{2.1.62}$$

These units involve e, however. While they are 'natural' for chemistry and atomic physics, they are not really the most convenient for dealing with relativistic dynamics and with the interaction of particles with the electromagnetic field. Thus, our unit of length, the electron Compton wavelength, is about 0.007 Bohr radii and our unit of energy is about 19 000 Rydbergs. These units have the advantage that e appears only in α and nowhere else.

Turning to units of the magnetic field, the equation for the Lorentz force (2.1.11) fixes the unit of $\boldsymbol{B}$. In these units, the energy required to create a magnetic field $\boldsymbol{B}$ is given by (2.1.16). If we still want to use $(8\pi)^{-1}\int |\boldsymbol{B}(\boldsymbol{x})|^2 \mathrm{d}\boldsymbol{x}$ as the energy of the electromagnetic field in our new units, $\boldsymbol{B}/\sqrt{\hbar c}$ has to have dimension of length^{-2}. From this it follows that the vector potential $\boldsymbol{A}/\sqrt{\hbar c}$ has the dimension of length^{-1}. Thus, we choose

$$\lambda_C^{-1}\sqrt{\hbar c} \quad \text{as the unit for } \boldsymbol{A}$$

$$\lambda_C^{-2}\sqrt{\hbar c} \quad \text{as the unit for } \boldsymbol{B}.$$

The argument of both vector fields is the dimensionless quantity $\lambda_C^{-1}\boldsymbol{x}$.

Our unit for $\boldsymbol{B}$ is unnaturally large, for it is 3.77×10^8 Tesla or 3.77×10^{12} Gauss, where 1 Gauss equals 1 gram$^{1/2}$ cm$^{-1/2}$ sec^{-1}, which is of the order of the magnetic field on the surface of many neutron stars. By comparison, it is an achievement to produce 10 Tesla = 10^5 Gauss in the laboratory.

Thus, for an atom with a single electron (i.e., a hydrogenic atom), *our non-relativistic Hamiltonian* (in units of mc^2, with $\boldsymbol{x}$ being the length in units of the Compton wavelength and with $\boldsymbol{p} = -i\nabla$) becomes

$$\boxed{H = \frac{1}{2}\boldsymbol{p}^2 - \frac{Z\alpha}{|\boldsymbol{x}|}.} \tag{2.1.63}$$

In these units (2.1.37) becomes

$$\mathcal{E}(\psi) = \int_{\mathbb{R}^3} \left(\frac{1}{2} |(\nabla\psi)(\boldsymbol{x})|^2 - \frac{Z\alpha}{|\boldsymbol{x}|} |\psi(\boldsymbol{x})|^2 \right) \mathrm{d}\boldsymbol{x}. \tag{2.1.64}$$

Similarly, the 'relativistic' Hamiltonian becomes

$$\boxed{H = \sqrt{\boldsymbol{p}^2 + 1} - 1 - \frac{Z\alpha}{|\boldsymbol{x}|}.} \tag{2.1.65}$$

The kinetic energy (2.1.54), including the interaction with a magnetic field and the energy of this field, becomes

$$\frac{1}{2}\int_{\mathbb{R}^3} \left|(\nabla + i\sqrt{\alpha}\boldsymbol{A}(\boldsymbol{x}))\psi(\boldsymbol{x})\right|^2 \mathrm{d}\boldsymbol{x} + \frac{1}{2}\sqrt{\alpha}\int_{\mathbb{R}^3} \langle \psi(\boldsymbol{x}), \boldsymbol{\sigma}\cdot\boldsymbol{B}(\boldsymbol{x})\psi(\boldsymbol{x})\rangle\, \mathrm{d}\boldsymbol{x}$$
$$+ \frac{1}{8\pi}\int_{\mathbb{R}^3} |\boldsymbol{B}(\boldsymbol{x})|^2 \mathrm{d}\boldsymbol{x}. \tag{2.1.66}$$

The corresponding Hamiltonian is

$$\boxed{H = \frac{1}{2}\left(\boldsymbol{p} + \sqrt{\alpha}\boldsymbol{A}(\boldsymbol{x})\right)^2 + \frac{\sqrt{\alpha}}{2}\boldsymbol{\sigma}\cdot\boldsymbol{B}(\boldsymbol{x}) + \frac{1}{8\pi}\int_{\mathbb{R}^3} |\boldsymbol{B}(\boldsymbol{x})|^2 \mathrm{d}\boldsymbol{x}.} \tag{2.1.67}$$

2.2 The Idea of Stability

The expected value of the energy in a state ψ is $\mathcal{E}(\psi)$ and we can ask about its range of values. Obviously $\mathcal{E}(\psi)$ can be made arbitrarily large and positive, but can it be arbitrarily negative? (Recall that we are interested in normalized ψ, i.e., $\int |\psi|^2 = 1$.) The answer in the classical case is *yes!* That is, the function $H(\boldsymbol{x}, \boldsymbol{p})$ is not bounded from below since $-1/|\boldsymbol{x}|$ can be arbitrarily negative and $\boldsymbol{p}$ can be zero. *Any classical Coulomb system with point charges (in which the charges are not all positive or all negative) is an unlimited source of energy.* In contrast, the lowest energy of an electron in a quantum-mechanical atom is finite – as we shall see. It is one Rydberg for hydrogen ($Z = 1$) and is of the order of 10^4 Rydbergs in large atoms where $Z \approx 90$.

As mentioned before, one could object that the nucleus is not really a point particle and hence the Coulomb potential is not really an infinite energy source.

However, the radius of a nucleus is about 10^{-13} centimeters, which must be compared with the 10^{-8} centimeter radius of the hydrogen atom, i.e., the Bohr radius. The distinction between a point nucleus and a 10^{-13} centimeter nucleus would be significant only if energies of the order of 10^5 times the energy of hydrogen played a significant role, which they do not. Such energies are effectively 'infinite' on the scale we are concerned with. The somewhat astonishing fact that an electron is forced to stay away from the nucleus (i.e., it is mostly to be found at the Bohr radius instead of the nuclear radius) is one of the most important features of quantum mechanics. Planck's constant introduces a new length scale, the Bohr radius and, concomitantly, introduces a scale of energy. This scale of energy was not present in pre-quantum physics and it is one Rydberg.

Thus the question is raised whether quantum mechanics sets lower bounds on the energy of electrons in atoms, i.e., is

$$E_0 = \inf \left\{ \mathcal{E}(\psi) : \int_{\mathbb{R}^3} |\psi(\boldsymbol{x})|^2 \mathrm{d}\boldsymbol{x} = 1 \right\} \tag{2.2.1}$$

finite and, if so, what is its value? Note that if the infimum is a minimum (i.e., $E_0 = \mathcal{E}(\psi_0)$ for some ψ_0) then E_0 is the lowest energy value the system can attain and is therefore called the **ground state energy**, and ψ_0 is called a **ground state**. *We will call E_0 the ground state energy even if the infimum in* (2.2.1) *is not attained by any function ψ*. Indeed, it is always the case that a minimum is not attained if there are too many electrons in an atom (see Chapter 12).

The finiteness of E_0 is **stability of the first kind**. That quantum mechanics achieves this sort of stability is of great importance, for it resolves one of the crucial problems of classical physics, and it will be discussed extensively in the chapters to follow.

Assuming for the moment that E_0 in (2.2.1) is attained for some ψ_0, a simple variational calculation leads (see Chapter 11 in [118]) to the (stationary, or time-independent) **Schrödinger equation** for ψ_0:

$$H\psi_0 = E_0\psi_0. \tag{2.2.2}$$

In general, E_0 is not the only value for which Eq. (2.2.2) has a solution. There are usually infinitely many of them and they are called the **eigenvalues** of H. (See [118, Sects. 11.5, 11.6].) They label all the stationary states of the atom and the difference between two eigenvalues determines the frequency of light which is emitted when an electron falls from a higher stationary state to a lower one.

In this book, we shall not be concerned with the question of whether E_0 is an actual eigenvalue of the Hamiltonian H, except for Chapter 12 on the ionization problem. Even less will we be concerned with the existence of other stationary states. Rather we investigate whether E_0 is bounded (for stability of the first kind) or bounded by the number of particles (for stability of the second kind, which will be defined in Chapter 3, Section 3.2).

2.2.1 Uncertainty Principles: Domination of the Potential Energy by the Kinetic Energy

Any inequality in which the kinetic energy T_ψ dominates some kind of integral of ψ (but not involving $\nabla\psi$) is called an **uncertainty principle**. The historical reason for this strange appellation is that such an inequality implies that one cannot make the potential energy very negative without also making the kinetic energy large, i.e., one cannot localize a particle simultaneously in both configuration and momentum space. The most famous uncertainty principle, historically, is Heisenberg's: For $\psi \in H^1(\mathbb{R}^d)$ and $\|\psi\|_2^2 = (\psi, \psi) = 1$,

$$(\psi, \boldsymbol{p}^2\psi) \geq \frac{d^2}{4}(\psi, \boldsymbol{x}^2\psi)^{-1}. \tag{2.2.3}$$

The proof of this inequality (which uses the fact that $\nabla \cdot \boldsymbol{x} - \boldsymbol{x} \cdot \nabla = d\,\mathbb{I}$ in $\mathbb{R}^d$) can be found in many textbooks and we shall not give it here because (2.2.3) is not actually very useful. The quantity $(\psi, \boldsymbol{x}^2\psi)^{-1}$ is a poor indicator of the magnitude of $T_\psi = (1/2)(\psi, \boldsymbol{p}^2\psi)$. If the particle is concentrated near the origin, then T_ψ is large, but ψ can easily be modified in an arbitrarily small way (in $H^1(\mathbb{R}^d)$ norm, i.e., in such a way that both the difference between the function and its modification and the gradient of this difference is small in $L^2(\mathbb{R}^d)$ norm) so that $(\psi, \boldsymbol{x}^2\psi)$ becomes huge without $(\psi, \boldsymbol{p}^2\psi)$ becoming small. In other words, by a tiny modification of ψ, $(\psi, \boldsymbol{x}^2\psi)^{-1}$ can be made small even if T_ψ is large. To see this, take any fixed function ψ and then replace it by $\psi_{\boldsymbol{y}}(\boldsymbol{x}) = \sqrt{1-\varepsilon^2}\psi(\boldsymbol{x}) + \varepsilon\psi(\boldsymbol{x} - \boldsymbol{y})$ with $\varepsilon \ll 1$ and $|\boldsymbol{y}| \gg 1$. To a very good approximation, $\psi_{\boldsymbol{y}} = \psi$ but, as $|\boldsymbol{y}| \to \infty$, $\|\psi_{\boldsymbol{y}}\|_2 \to 1$ and $(\psi_{\boldsymbol{y}}, \boldsymbol{x}^2\psi_{\boldsymbol{y}}) \to \infty$. Thus, the right side of (2.2.3) goes to zero as $|\boldsymbol{y}| \to \infty$ while $T_{\psi_{\boldsymbol{y}}} \approx T_\psi$ does not go to zero.

A much more useful inequality is **Sobolev's inequality**. Recall the definition of the $L^p(\mathbb{R}^d)$ norms

$$\|\psi\|_p = \left[\int_{\mathbb{R}^d} |\psi(\boldsymbol{x})|^p \mathrm{d}\boldsymbol{x}\right]^{1/p}$$

for $1 \leq p < \infty$. For $p = \infty$, one uses $\|\psi\|_\infty = \sup_{x \in \mathbb{R}^d} |\psi(x)|$. Any function ψ whose gradient is in $L^2(\mathbb{R}^d)$ and that vanishes at infinity (meaning that the measure of the set where $|\psi(x)| \geq \mu$ is finite for any $\mu > 0$), with $d \geq 3$ (these conditions are important), is automatically in $L^{2d/(d-2)}(\mathbb{R}^d)$, although not necessarily in any other $L^p(\mathbb{R}^d)$ space. There is the inequality

$$2\,T_\psi = \|\nabla\psi\|_2^2 = \int_{\mathbb{R}^d} |\nabla\psi(x)|^2 \mathrm{d}x \geq S_d \|\psi\|_{2d/(d-2)}^2 \tag{2.2.4}$$

for some positive constants S_d. For $d = 3$, the optimal constant is $S_3 = \frac{3}{4}(4\pi^2)^{2/3}$. See [118, Chapter 8] for further information.

For $d = 1$ and $d = 2$, on the other hand, we have

$$2\,T_\psi \geq S_{2,p} \|\psi\|_2^{-4/(p-2)} \|\psi\|_p^{2p/(p-2)} \quad \text{for all } 2 < p < \infty, \qquad d = 2 \tag{2.2.5}$$

$$2\,T_\psi \geq \|\psi\|_2^{-2} \|\psi\|_\infty^4, \qquad d = 1. \tag{2.2.6}$$

Moreover, when $d = 1$ and $\psi \in H^1(\mathbb{R}^1)$, ψ is not only bounded but it is also continuous.

An application of **Hölder's inequality**,

$$\int_{\mathbb{R}^d} f(x)g(x)\mathrm{d}x \leq \|f\|_p \|g\|_q \qquad \text{for } 1 \leq p \leq \infty,\ p^{-1} + q^{-1} = 1,$$

to (2.2.4) yields, for any potential $V \in L^{d/2}(\mathbb{R}^d)$, $d \geq 3$,

$$T_\psi \geq S_d \|\psi\|_{2d/(d-2)}^2 \geq S_d (\psi, |V|\psi) \|V\|_{d/2}^{-1}. \tag{2.2.7}$$

An immediate application of (2.2.7) is that

$$T_\psi + V_\psi \geq 0 \tag{2.2.8}$$

whenever $\|V\|_{d/2} \leq S_d$, and thus stability of the first kind holds for such potentials V.

A simple extension of (2.2.7) leads to a lower bound on the ground state energy for $V \in L^{d/2}(\mathbb{R}^d) + L^\infty(\mathbb{R}^d)$, $d \geq 3$, i.e. for potentials V that can be written as

$$V(x) = v(x) + w(x) \tag{2.2.9}$$

for some $v \in L^{d/2}(\mathbb{R}^d)$ and $w \in L^\infty(\mathbb{R}^d)$. There is then some constant λ such that $h(x) := -(v(x) - \lambda)_- = \min(v(x) - \lambda, 0) \leq 0$ satisfies $\|h\|_{d/2} \leq \frac{1}{2} S_d$ (we leave this as an exercise for the reader). In particular by (2.2.7)

$h_\psi = (\psi, h\psi) \geq -\frac{1}{2}T_\psi$. Then we have

$$\begin{aligned}\mathcal{E}(\psi) = T_\psi + V_\psi = T_\psi + (v-\lambda)_\psi + \lambda + w_\psi \geq T_\psi + h_\psi + \lambda + w_\psi \\ \geq \frac{1}{2}T_\psi + \lambda - \|w\|_\infty \end{aligned} \tag{2.2.10}$$

and we see that $\lambda - \|w\|_\infty$ is a lower bound to E_0. Furthermore (2.2.10) implies that *the total energy effectively bounds the kinetic energy*, i.e., we have

$$T_\psi \leq 2(\mathcal{E}(\psi) - \lambda + \|w\|_\infty). \tag{2.2.11}$$

So far we have considered the non-relativistic kinetic energy, $T_\psi = (1/2)(\psi, \boldsymbol{p}^2\psi)$. Similar inequalities hold for the relativistic case. Because $|\boldsymbol{p}| \geq \sqrt{\boldsymbol{p}^2 + m^2} - m \geq |\boldsymbol{p}| - m$, it suffices to consider the **ultra-relativistic energy** $T_\psi = (\psi, |\boldsymbol{p}|\psi)$. The relativistic analogues of (2.2.4)–(2.2.6) are (2.2.12) and (2.2.13) below. There are constants S_d' for $d \geq 2$ and $S_{1,p}'$ for $2 \leq p < \infty$ such that

$$T_\psi \geq S_n' \|\psi\|_{2d/(d-1)}^2, \qquad d \geq 2 \tag{2.2.12}$$

and

$$T_\psi \geq S_{1,p}' \|\psi\|_2^{-2/(p-2)} \|\psi\|_p^{2p/(p-2)} \quad \text{for all } 2 < p < \infty, \quad d = 1. \tag{2.2.13}$$

For $d = 3$, $S_3' = (2\pi^2)^{1/3}$ [118, Thm. 8.4].

The results of this section can be summarized in the following statement. *In all dimensions $d \geq 1$, the hypothesis that V is in the space*

$$\textit{non-relativistic} \begin{cases} L^{d/2}(\mathbb{R}^d) + L^\infty(\mathbb{R}^d) & \text{if } d \geq 3 \\ L^{1+\varepsilon}(\mathbb{R}^2) + L^\infty(\mathbb{R}^2) & \text{if } d = 2 \\ L^1(\mathbb{R}^1) + L^\infty(\mathbb{R}^1) & \text{if } d = 1 \end{cases} \tag{2.2.14}$$

$$\textit{relativistic} \begin{cases} L^d(\mathbb{R}^d) + L^\infty(\mathbb{R}^d) & \text{if } d \geq 2 \\ L^{1+\varepsilon}(\mathbb{R}^1) + L^\infty(\mathbb{R}^1) & \text{if } d = 1 \end{cases} \tag{2.2.15}$$

(for some $\varepsilon > 0$) leads to the following two conclusions:

$$E_0 \text{ is finite}$$

$$T_\psi \leq C\mathcal{E}(\psi) + D\|\psi\|_2^2$$

when $\psi \in H^1(\mathbb{R}^d)$ (non-relativistic) or $\psi \in H^{1/2}(\mathbb{R}^d)$ (relativistic) and for suitable constants C and D which depend only on V and not on ψ.

2.2.2 The Hydrogenic Atom

The hydrogenic atom is the simplest conceptually and computationally. It has one electron and one nucleus with charge $Z > 0$ (the nucleus of real hydrogen in nature is the proton, which has $Z = 1$). If we assume that the nucleus is fixed at the origin $\boldsymbol{x} = 0$, then the energy is given by (2.1.64) as

$$\mathcal{E}(\psi) = \int_{\mathbb{R}^3} \left(\frac{1}{2} |(\nabla\psi)(\boldsymbol{x})|^2 - \frac{Z\alpha}{|\boldsymbol{x}|} |\psi(\boldsymbol{x})|^2 \right) \mathrm{d}\boldsymbol{x}. \tag{2.2.16}$$

The Coulomb potential $V(\boldsymbol{x}) = -Z\alpha/|\boldsymbol{x}|$ is in $L^{3-\varepsilon}(\mathbb{R}^3) + L^\infty(\mathbb{R}^3)$ for any $\varepsilon > 0$, but not in $L^3(\mathbb{R}^3) + L^\infty(\mathbb{R}^3)$. Hence the hydrogenic atom is stable non-relativistically as a consequence of Sobolev's inequality, as discussed in the previous section. In fact, it is well known that

$$\int_{\mathbb{R}^3} \left(\frac{1}{2} |(\nabla\psi)(\boldsymbol{x})|^2 - \frac{Z\alpha}{|\boldsymbol{x}|} |\psi(\boldsymbol{x})|^2 \right) \mathrm{d}\boldsymbol{x} \geq -\frac{(Z\alpha)^2}{2} \int_{\mathbb{R}^3} |\psi(\boldsymbol{x})|^2 \mathrm{d}\boldsymbol{x}, \tag{2.2.17}$$

with equality if and only if $\psi(\boldsymbol{x}) = C\exp(-Z\alpha|\boldsymbol{x}|)$ for some constant $C \in \mathbb{C}$. In fact, this ψ can easily be shown to satisfy the Schrödinger equation $(-\Delta/2 - Z\alpha/|\boldsymbol{x}|)\psi(\boldsymbol{x}) = -(Z\alpha)^2\psi(\boldsymbol{x})/2$, and since it is a positive function, it must be the eigenfunction corresponding to the lowest eigenvalue. (See [118, Section 11.10] for details; see also Corollary 3.1 in Section 3.2.4.)

An equivalent formulation of the inequality in (2.2.17) is

$$\int_{\mathbb{R}^3} |\nabla\psi(\boldsymbol{x})|^2 \mathrm{d}\boldsymbol{x} \int_{\mathbb{R}^3} |\psi(\boldsymbol{x})|^2 \mathrm{d}\boldsymbol{x} \geq \left[\int_{\mathbb{R}^3} \frac{|\psi(\boldsymbol{x})|^2}{|\boldsymbol{x}|} \mathrm{d}\boldsymbol{x} \right]^2 \tag{2.2.18}$$

for all functions $\psi \in H^1(\mathbb{R}^3)$, which is obtained by choosing $Z\alpha = (\int |\psi|^2)^{-1} \int |\psi|^2/|\boldsymbol{x}|$ in (2.2.17). This inequality will be useful later in Chapter 9.

The analysis in the previous section shows that the relativistic case is borderline. That is, $1/|\boldsymbol{x}|$ fails to be in $L^3(\mathbb{R}^3) + L^\infty(\mathbb{R}^3)$, although it is in $L^{3-\varepsilon}(\mathbb{R}^3) + L^\infty(\mathbb{R}^3)$ for any $\varepsilon > 0$. We shall in fact see in Chapter 8 that there is a critical value of $Z\alpha$ up to which there is stability, and beyond which stability fails.

CHAPTER 3

Many-Particle Systems and Stability of the Second Kind

The main purpose of this book is to study the ground state energies of quantum-mechanical systems of particles interacting via electric and magnetic forces. The present chapter defines the mathematical framework of the problem. We shall define the kinds of wave functions that have to be considered and their permutation symmetries. The ground state energy will be defined as a variational problem, and the concept of stability of the second kind will be discussed.

3.1 Many-Body Wave Functions

3.1.1 The Space of Wave Functions

A **wave function** for N spinless particles is any function $\psi : \mathbb{R}^{3N} \to \mathbb{C}$ which is in $L^2(\mathbb{R}^{3N}) \cong \bigotimes^N L^2(\mathbb{R}^3)$ with **unit norm**, i.e.,

$$\|\psi\|_2^2 = \int_{\mathbb{R}^{3N}} |\psi(\boldsymbol{x}_1, \ldots, \boldsymbol{x}_N)|^2 \mathrm{d}\boldsymbol{x}_1 \cdots \mathrm{d}\boldsymbol{x}_N = 1. \tag{3.1.1}$$

Here $\boldsymbol{x}_i \in \mathbb{R}^3$ is the (spatial) coordinate of the ith particle. Our notation is that subscripts label the particles, while particle coordinates are denoted by superscripts, e.g., $\boldsymbol{x}_1 = (x_1^1, x_1^2, x_1^3)$. It is convenient to denote a point in $\mathbb{R}^{3N}$ by

$$\underline{X} := (\boldsymbol{x}_1, \ldots, \boldsymbol{x}_N) \tag{3.1.2}$$

and

$$\mathrm{d}\underline{X} := \mathrm{d}\boldsymbol{x}_1 \cdots \mathrm{d}\boldsymbol{x}_N. \tag{3.1.3}$$

An important historical point is to be noted here. It might have been thought that the correct generalization for N particles is to use N functions of one variable instead of one function of N variables. Such a 'wrong turn' did not happen historically, which is, after all, remarkable. Nevertheless there are simple, but interesting and useful functions of N variables that can be built out of N functions of one variable called *determinantal functions* and these will be discussed later. They are important in an approximation to quantum mechanics called *Hartree* and *Hartree–Fock* theories, which we will not discuss.

The interpretation of ψ is that $|\psi(\underline{X})|^2 = |\psi(\boldsymbol{x}_1, \ldots, \boldsymbol{x}_N)|^2$ is the **probability density** for finding particle 1 at $\boldsymbol{x}_1$, particle 2 at $\boldsymbol{x}_2$, etc. This interpretation makes sense in view of condition (3.1.1). In the case of one particle, one might be tempted to interpret $|\psi(\boldsymbol{x})|^2$ as being proportional to a mass or charge density. In the N-particle case this interpretation is impossible since ψ is not a function of one variable. There is, however, a natural function of one variable associated with ψ, namely, for $\boldsymbol{x} \in \mathbb{R}^3$,[1]

$$\varrho_\psi(\boldsymbol{x}) = \sum_{i=1}^{N} \varrho_\psi^i(\boldsymbol{x}), \tag{3.1.4}$$

where

$$\varrho_\psi^i(\boldsymbol{x}) = \int\limits_{\mathbb{R}^{3(N-1)}} |\psi(\boldsymbol{x}_1, \ldots, \boldsymbol{x}_{i-1}, \boldsymbol{x}, \boldsymbol{x}_{i+1}, \ldots, \boldsymbol{x}_N)|^2 \mathrm{d}\boldsymbol{x}_1 \cdots \widehat{\mathrm{d}\boldsymbol{x}_i} \cdots \mathrm{d}\boldsymbol{x}_N \tag{3.1.5}$$

which, on account of (3.1.1), is the probability density of finding particle i at $\boldsymbol{x}$. The notation $\widehat{\mathrm{d}\boldsymbol{x}_i}$ means that the integration over the ith coordinate is omitted. Obviously $\int \varrho_\psi^i = 1$. Thus $\varrho_\psi(\boldsymbol{x})$ is interpreted as the total electron density at the point $\boldsymbol{x}$ and is customarily called the **single** or **one-particle density**. It satisfies

$$\int\limits_{\mathbb{R}^3} \varrho_\psi(\boldsymbol{x}) \mathrm{d}\boldsymbol{x} = N. \tag{3.1.6}$$

If the ith particle has **electric charge** e_i the analogous **charge density** can be defined as

$$\mathcal{Q}_\psi(\boldsymbol{x}) = \sum_{i=1}^{N} \mathcal{Q}_\psi^i(\boldsymbol{x}), \tag{3.1.7}$$

where

$$\mathcal{Q}_\psi^i(\boldsymbol{x}) = e_i \varrho_\psi^i(\boldsymbol{x}). \tag{3.1.8}$$

[1] Note that ϱ is the *particle* density. It should not be confused with the mass density, which equals the particle mass times ϱ.

For non-relativistic quantum mechanics it is also required that ψ has finite kinetic energy, namely ψ is in $H^1(\mathbb{R}^{3N})$, i.e., both ψ and each of the $3N$ components of $\nabla\psi = (\nabla_{x_1}\psi, \dots, \nabla_{x_N}\psi)$ are functions in $L^2(\mathbb{R}^{3N})$. The gradient is defined in the sense of distributions (see Chapter 6 in [118]). The **kinetic energy** ψ is then defined to be

$$T_\psi = \sum_{i=1}^{N} T_\psi^i, \tag{3.1.9}$$

where

$$T_\psi^i = \frac{1}{2m_i} \int_{\mathbb{R}^{3N}} |(\nabla_{x_i}\psi)(\underline{X})|^2 \mathrm{d}\underline{X}, \tag{3.1.10}$$

and where m_i is the **mass** of the ith particle. Recall that for our choice of units, discussed in Section 2.1.7, the mass of an electron is 1. In the case of relativistic quantum mechanics ψ need only be in $H^{1/2}(\mathbb{R}^{3N})$. In this case the T_ψ^i above are replaced by

$$T_\psi^i = \int_{\mathbb{R}^{3N}} \left(\sqrt{(2\pi \boldsymbol{k}_i)^2 + m_i^2} - m_i\right) |\widehat{\psi}(\underline{\boldsymbol{K}})|^2 \mathrm{d}\underline{\boldsymbol{K}} \tag{3.1.11}$$

(see Chapter 8 for more details), where $\underline{\boldsymbol{K}} = (\boldsymbol{k}_1, \dots, \boldsymbol{k}_N)$, and

$$\widehat{\psi}(\underline{\boldsymbol{K}}) = \int_{\mathbb{R}^{3N}} \psi(\underline{X}) \exp\Big(-2\pi i \sum_{j=1}^{N} \boldsymbol{x}_j \cdot \boldsymbol{k}_j\Big) \mathrm{d}\underline{X}$$

denotes the Fourier transform of ψ.

Accordingly, the **potential energy** of ψ is defined to be

$$V_\psi = \int_{\mathbb{R}^{3N}} V(\underline{X}) |\psi(\underline{X})|^2 \mathrm{d}\underline{X}, \tag{3.1.12}$$

where the function $V(\underline{X})$ is the total potential energy function. In the case of a molecule, $V(\underline{X})$ is given by (2.1.21)–(2.1.24).

3.1.2 Spin

In addition to the spatial coordinates, $\boldsymbol{x}$, a particle may also have internal degrees of freedom. The most important of these is **spin**, but there could be others such as isospin, flavor, etc. For the problems addressed in this book the precise nature of the internal states is not important and we shall, therefore, adopt the following *convention*: If a particle has q internal states available to it we shall say that the particle has **q spin states** and label them by the integer

$$\sigma \in \{1, 2, \ldots, q\}.$$

Electrons have two spin states and these are conventionally designated by $\sigma = +1/2$ or $-1/2$. Here, $\sigma = 1$ or 2 will be used, as stated above. From the mathematical point of view it is interesting to retain an arbitrary q in our formulation. This is due to the fact that the Pauli principle (explained below) introduces interesting dependencies of various quantities on q. More than that, by incorporating the q dependence explicitly we will be able pass from fermions to bosons, as will be explained in the next subsection. Bosons will be seen to correspond to $q = N$, the number of particles. For this reason alone it is worth retaining the dependence on q.

Suppose there are N particles and the ith particle has q_i spin states. A wave function for these N particles can then be written as

$$\psi(\boldsymbol{x}_1, \sigma_1, \ldots, \boldsymbol{x}_N, \sigma_N) \tag{3.1.13}$$

where, for each choice of the sigmas (with $1 \leq \sigma_i \leq q_i$), the function in (3.1.13) is an element of $H^1(\mathbb{R}^{3N})$. Alternatively, ψ can be viewed as an $H^1(\mathbb{R}^{3N})$-function with values in $\mathbb{C}^Q$ with $Q := \prod_{i=1}^{N} q_i$. (This means simply that ψ can be thought of as Q functions in $H^1(\mathbb{R}^{3N})$, with $\sigma_1, \ldots, \sigma_N$ furnishing the labeling of these functions.) The set of these $\mathbb{C}^Q$-valued functions will be denoted by $H^1(\mathbb{R}^{3N}; \mathbb{C}^Q)$, and it is isomorphic to $\bigotimes_{i=1}^{N} H^1(\mathbb{R}^3; \mathbb{C}^{q_i})$. In analogy with (3.1.2) the spins are collectively denoted by

$$\underline{\sigma} := (\sigma_1, \ldots, \sigma_N)$$

and

$$\sum_{\underline{\sigma}} := \sum_{\sigma_1=1}^{q_1} \cdots \sum_{\sigma_N=1}^{q_N}.$$

With this notation (3.1.13) can be conveniently written as $\psi(\underline{X}, \underline{\sigma})$. Another convenient notation is

$$z_i = (\boldsymbol{x}_i, \sigma_i) \quad \text{and} \quad \underline{z} = (z_1, \ldots, z_N) \tag{3.1.14}$$

and, for any function $f(\boldsymbol{x}_i, \sigma_i)$,

$$\int f(z_i)\mathrm{d}z_i := \sum_{\sigma_i=1}^{q_i} \int_{\mathbb{R}^3} f(\boldsymbol{x}_i, \sigma_i)\mathrm{d}\boldsymbol{x}_i. \tag{3.1.15}$$

Using the above notation, (3.1.13) will sometimes be written as

$$\psi(z_1, \dots, z_N) \quad \text{or} \quad \psi(\underline{z}).$$

The replacement of the normalization condition (3.1.1) is

$$\|\psi\|_2^2 := \sum_{\underline{\sigma}} \int_{\mathbb{R}^{3N}} |\psi(\underline{X}, \underline{\sigma})|^2 \mathrm{d}\underline{X} = \int |\psi(\underline{z})|^2 \mathrm{d}\underline{z} = 1.$$

3.1.3 Bosons and Fermions (The Pauli Exclusion Principle)

As explained in the previous section, the coordinate of a particle is a point in $\mathbb{R}^3 \times \{1, \dots, q\}$, i.e., a point $\boldsymbol{x} \in \mathbb{R}^3$ and a point $\sigma \in \{1, \dots, q\}$. The value of q is a fixed property of the kind of particle, i.e., whether the particle is an electron, proton or whatever. Recall the notation $z_j = (\boldsymbol{x}_j, \sigma_j)$ for the jth particle.

It is a basic postulate of quantum mechanics that the allowed wave functions for a given particle species must belong to some definite permutation symmetry type which is characteristic of that species. There are two kinds of particles in nature, called **bosons** and **fermions**. Bosons are characterized by the fact that a wave function describing a system containing several identical bosons, e.g., positively charged pions, must be **totally symmetric** with respect to exchange of any pair of boson coordinates. That is, if z_i and z_j are the space-spin variables of two identical bosons,

$$\psi(\dots, z_i, \dots, z_j, \dots) = \psi(\dots, z_j, \dots, z_i, \dots).$$

Fermions, on the other hand, demand **total antisymmetry** of the wave functions. That is,

$$\psi(\dots, z_i, \dots, z_j, \dots) = -\psi(\dots, z_j, \dots, z_i, \dots),$$

if z_i and z_j are the coordinates of identical fermions. The antisymmetry property is usually rephrased by saying that fermions obey thc **Pauli exclusion principle**.

Terminology: It is common in the physics literature to refer to the choice of symmetry type as the **statistics** of the particles or of their wave function. More exactly, the requirement of symmetry under coordinate exchange is referred to as **Bose–Einstein statistics**, named after S. N. Bose who analyzed the states of photons (quantized particles of light) and A. Einstein who applied Bose's method to ordinary massive particles. The antisymmetry requirement is referred to as **Fermi–Dirac statistics**, after E. Fermi and P. A. M. Dirac.

In general, one considers several species of particles simultaneously. There is just one wave function describing such a system, and it has to be symmetric with respect to exchange of particle coordinates corresponding to the same species of bosons, and antisymmetric for particles corresponding to the same species of fermions. For example, a wave function describing both positively and negatively charged pions (which are bosons) has to be symmetric separately in the coordinates corresponding to the positively and negatively charged particles, but there is no symmetry requirement for exchange of coordinates corresponding to different species.

In the following chapters, we will mostly be concerned with wave functions describing electrons and different kinds of nuclei. While the electrons are fermions (with $q = 2$), nuclei can be either fermions or bosons. Nuclei of different charges are, in any case, independent, i.e., they belong to different species.

A simple example of a symmetric function of N variables is $\psi(\underline{z}) = \prod_{i=1}^{N} u(z_i)$, where u is any function in $L^2(\mathbb{R}^3; \mathbb{C}^q)$ of one variable, z.

The simplest example of an antisymmetric function is obtained by taking any N functions $u_i(z) \in L^2(\mathbb{R}^3; \mathbb{C}^q)$, $i = 1, \dots, N$ that are **orthonormal**, i.e., $\int \overline{u_i(z)} u_j(z) \mathrm{d}z = \delta_{i,j}$. The N-particle wave function

$$\psi(\underline{z}) = (N!)^{-1/2} \det\{u_i(z_j)\}_{i,j=1}^{N} \tag{3.1.16}$$

is antisymmetric and normalized. It is called a **determinantal function**, or **Slater determinant**. The subspace of $L^2(\mathbb{R}^{3N}; \mathbb{C}^{q^N})$ consisting of all antisymmetric functions is denoted by $\bigwedge^N L^2(\mathbb{R}^3; \mathbb{C}^q)$, where $\bigwedge$ stands for the **antisymmetric tensor product**. If $\{u_i(z)\}_{i=1}^{\infty}$ forms a basis for $L^2(\mathbb{R}^3; \mathbb{C}^q)$ then all the determinantal functions given by (3.1.16), using all possible choices of N of the functions u_i, form an orthonormal basis for $\bigwedge^N L^2(\mathbb{R}^3; \mathbb{C}^q)$.

Another fact of nature, which can only be explained using relativistic quantum field theory, is that bosons have an odd number q of spin states (one says the spin, defined to be $\frac{q-1}{2}$, is an integer) and the fermions have q even, i.e., the spin $\frac{q-1}{2}$ is $1/2, 3/2$, etc. This will not be of any importance in our discussion, however, because this restriction is not mathematically natural in our context. We shall consider all integers q in this book. In the literature one sometimes comes across academic $q = 1$ fermions, which are called **spinless fermions**. One can think of this as a situation in which the particles are $q = 2$ fermions, but the wave functions are restricted to be those in which $\psi = 0$ unless all σ_j are 1. In treating bosons here we shall only consider $q = 1$, but everything can be easily generalized to arbitrary q.

3.1.3.1 The case $q \geq N$

Suppose we have a Hamiltonian H that does not depend on spin. That is to say, it acts on $L^2(\mathbb{R}^{3N})$, for example the Coulomb Hamiltonian, which will be discussed in Section 3.2. This Hamiltonian can be extended in a trivial way to $H \otimes \mathbb{I}$, where $\mathbb{I}$ is the identity on the spin space $\mathbb{C}^{q^N}$. It acts on $L^2(\mathbb{R}^{3N}; \mathbb{C}^{q^N}) \cong L^2(\mathbb{R}^{3N}) \otimes \mathbb{C}^{q^N}$.[2]

Consider now the case where $q \geq N$, N being the number of particles. For fermions with q spin states, the physical wave functions are totally antisymmetric functions in $L^2(\mathbb{R}^{3N}; \mathbb{C}^{q^N})$. We shall now show that for $q \geq N$, this fermionic system is equivalent to considering $q = 1$, i.e., the original Hamiltonian H on $L^2(\mathbb{R}^{3N})$, *without any symmetry restrictions on the wave functions*, and without any spin degrees of freedom. This can be seen as follows.

Take any function $\phi(\boldsymbol{x}_1, \ldots, \boldsymbol{x}_N)$ of N space variables, and let

$$\psi(z_1, \ldots, z_N) = \frac{1}{\sqrt{N!}} \mathcal{A}\left[\phi(\boldsymbol{x}_1, \ldots, \boldsymbol{x}_N) f_1(\sigma_1) \cdots f_N(\sigma_N)\right],$$

where $f_i(\sigma) = 1$ for $\sigma = i$ and 0 otherwise, and where $\mathcal{A}$ denotes total antisymmetrization. That is,

$$\mathcal{A}[\chi(z_1, \ldots, z_N)] = \sum_{\pi \in S_N} \varepsilon_\pi \chi(z_{\pi(1)}, \ldots, z_{\pi(N)}), \qquad (3.1.17)$$

where the sum runs over all permutations of the numbers $1, \ldots, N$, and $\varepsilon_\pi = 1$ for even permutations, and $\varepsilon_\pi = -1$ for odd permutations. The function ψ is clearly an antisymmetric function in $L^2(\mathbb{R}^{3N}; \mathbb{C}^{q^N})$, and it is not identically zero if ϕ is not. In fact, the norm of ψ is the same as the norm of ϕ. This follows from the fact that the functions f_i form a set of orthonormal functions in $\mathbb{C}^q$.

[2] Note that $\bigotimes^N \mathbb{C}^q \cong \mathbb{C}^{q^N}$, not $\mathbb{C}^{qN}$, i.e., the dimension is q^N, not qN.

Moreover, for the same reason

$$(\psi, H \otimes \mathbb{I}\, \psi) = (\phi, H\phi)$$

if H is invariant under permutations. In particular, since ϕ was arbitrary, the infimum of $(\psi, H \otimes \mathbb{I}\, \psi)$ over all antisymmetric ψ equals the infimum of $(\phi, H\phi)$ over *all* ϕ. We shall use this fact later when discussing stability of bosonic systems.

We note that if H is permutation invariant, the ϕ that yields the lowest value of $(\phi, H\phi)$ among all possible (normalized) functions can be taken to have a definite symmetry type without loss of generality, just as an energy minimizing state for a rotation invariant H can be taken to have a definite angular momentum. 'Symmetry type' refers to representations of the permutation group, i.e., to the behavior of a function under permutations, and symmetric (i.e., bosonic) and antisymmetric (i.e., fermionic) are but two examples of symmetry types. There are other symmetry types than symmetric or antisymmetric but these will not be used in this book. (For more information on representations of the permutation group, we refer to any textbook on finite groups, e.g., [84].) The energy minimizer need not be bosonic or fermionic, however. Indeed, for systems with magnetic fields or systems under rotation it is known that the energy minimizer is, in general, neither bosonic nor fermionic [162]. In the absence of magnetic fields, the minimizer can be shown to be always bosonic; we will show this in Corollary 3.1 on page 59.

3.1.4 Density Matrices

Given a normalized function $\psi \in L^2(\mathbb{R}^{3N}; \mathbb{C}^{q^N}) \cong \bigotimes^{N} L^2(\mathbb{R}^3; \mathbb{C}^q)$ there is naturally associated with it an orthogonal projection Γ_ψ on $L^2(\mathbb{R}^{3N}; \mathbb{C}^{q^N})$ whose action on a function $\phi \in L^2(\mathbb{R}^{3N}; \mathbb{C}^{q^N})$ is given by

$$(\Gamma_\psi \phi)(\underline{z}) = (\psi, \phi)\psi(\underline{z}) \tag{3.1.18}$$

with $(\psi, \phi) = \int \overline{\psi}(\underline{z})\phi(\underline{z})\mathrm{d}\underline{z}$.[3] Clearly $\Gamma_\psi \Gamma_\psi = \Gamma_\psi$ and Γ_ψ is self-adjoint, i.e., $(\phi', \Gamma_\psi \phi) = (\Gamma_\psi \phi', \phi)$ for all ϕ and ϕ'. This Γ_ψ has two other important

[3] A notation that is often used is Dirac's bra and ket notation $\Gamma_\psi = |\psi\rangle\langle\psi|$. In this notation the inner product is given by $\langle\psi|\phi\rangle$ instead of (ψ, ϕ), and the matrix elements of an operator H are denoted by $\langle\psi|H|\phi\rangle$ instead of $(\psi, H\phi)$.

properties:

$$(\phi, \Gamma_\psi \phi) \geq 0 \quad \textbf{positive semidefinite},$$
$$\operatorname{Tr} \Gamma_\psi = 1 \quad \textbf{unit trace}. \tag{3.1.19}$$

The symbol Tr denotes the **trace**, which can be computed in any orthonormal basis $\{\phi_\alpha\}$ as $\sum_\alpha (\phi_\alpha, \Gamma_\psi \phi_\alpha)$. By choosing $\phi_1 = \psi$ and the other ϕ_α to be orthogonal to ψ, Eq. (3.1.19) is easily deduced.

The projection Γ_ψ has an additional property. It has one eigenvector with eigenvalue 1, namely ψ itself, $\Gamma_\psi \psi = \psi$, and all other eigenvalues are zero. It is easy to prove that any linear, self-adjoint operator Γ that is positive semidefinite and has unit trace, and has one eigenvalue equal to 1 has the form (3.1.18). The ψ appearing in (3.1.18) is the unit eigenvector of Γ, i.e., the ψ such that $\Gamma \psi = \psi$.

Any operator of the form (3.1.18) is called a **pure state density matrix**. A general **density matrix** is any linear, self-adjoint positive semidefinite operator with unit trace – the condition that 1 is an eigenvalue is dropped. In some sense such an operator can be viewed as a generalization of the concept of a wave function; this generalization is not only useful but it is actually forced upon us, as will be seen shortly.

The two conditions (3.1.19) on a density matrix Γ will be written as

$$0 \leq \Gamma \leq \mathbb{I}, \qquad \operatorname{Tr} \Gamma = 1, \tag{3.1.20}$$

where $\mathbb{I}$ is the **unit operator**, $\mathbb{I}\,\psi = \psi$ for all ψ. The condition $\Gamma \leq \mathbb{I}$, which is an easy consequence of $\operatorname{Tr} \Gamma = 1$ and the condition $\Gamma \geq 0$, means that $(\psi, \Gamma \psi) \leq (\psi, \psi)$ for all ψ. Condition (3.1.20) is a far stronger statement than simply that Γ is a bounded operator. In fact (3.1.20) implies the **eigenfunction expansion**

$$\Gamma = \sum_{j=1}^{\infty} \lambda_j \Gamma_{\psi_j}, \tag{3.1.21}$$

where the functions ψ_j are an orthonormal basis for $L^2(\mathbb{R}^{3N}; \mathbb{C}^{q^N})$ and the real numbers λ_j, which are the eigenvalues of Γ, satisfy

$$\lambda_1 \geq \lambda_2 \geq \cdots \geq 0, \qquad \sum_{j=1}^{\infty} \lambda_j = 1.$$

Also, $\Gamma \psi_j = \lambda_j \psi_j$ for each j. In other words, Γ is a convex combination of pure state density matrices. The proof of (3.1.21) is left to the reader (or see, e.g., [168, Thm. 1.4]).

We say that Γ is an $\boldsymbol{H^1}$ **density matrix** if each ψ_j in (3.1.21) is in $H^1(\mathbb{R}^{3N};\mathbb{C}^{qN})$ and if

$$\sum_{j=1}^{\infty} \lambda_j \|\nabla\psi_j\|_2^2 < \infty,$$

with λ_j being the eigenvalues of Γ. Note that this is a stronger condition than merely assuming that Γ maps $L^2(\mathbb{R}^{3N};\mathbb{C}^{q^N})$ into $H^1(\mathbb{R}^{3N};\mathbb{C}^{q^N})$; this weaker condition is $\sum_{j=1}^{\infty}\lambda_j^2\|\nabla\psi_j\|_2^2 < \infty$. Analogous definitions can be made for $H^{1/2}$ density matrices.

Equation (3.1.21) also implies that Γ has a kernel function (which we also denote by Γ), i.e.,

$$(\Gamma\psi)(\underline{z}) = \int \Gamma(\underline{z},\underline{z}')\psi(\underline{z}')\mathrm{d}\underline{z}'. \tag{3.1.22}$$

The self-adjointness implies that $\Gamma(\underline{z},\underline{z}') = \overline{\Gamma(\underline{z}',\underline{z})}$. To calculate $\Gamma(\underline{z},\underline{z}')$ we first note that in the simple case of a pure state Γ_ψ, the kernel is obviously just $\psi(\underline{z})\overline{\psi}(\underline{z}')$. Then from (3.1.20) and an easy proof using the dominated convergence theorem to exchange limits and integrals, the following representation of $\Gamma(\underline{z},\underline{z}')$ can be obtained:

$$\Gamma(\underline{z},\underline{z}') = \sum_{j=1}^{\infty}\lambda_j\psi_j(\underline{z})\overline{\psi}_j(\underline{z}'). \tag{3.1.23}$$

There is a technical point that has to be noted. Equation (3.1.22) does *not* define the kernel function on the diagonal where $\underline{z}' = \underline{z}$, i.e., $\Gamma(\underline{z},\underline{z})$ is undefined for *all* $\underline{z} = (\boldsymbol{X},\underline{\sigma})$. The reason is that the set of points $(\boldsymbol{X},\boldsymbol{X})$ in $\mathbb{R}^{3N}\times\mathbb{R}^{3N}$ is merely a set of $6N$-dimensional Lebesgue measure zero. Therefore, the 'well known' equation

$$\operatorname{Tr}\Gamma = \int \Gamma(\underline{z},\underline{z})\mathrm{d}\underline{z} \tag{3.1.24}$$

does *not* follow from (3.1.22). However, if we agree to *define* $\Gamma(\underline{z},\underline{z})$ by (3.1.23), namely

$$\Gamma(\underline{z},\underline{z}) = \sum_{j=1}^{\infty}\lambda_j|\psi_j(\underline{z})|^2 \tag{3.1.25}$$

then (3.1.24) is true because

$$\operatorname{Tr}\Gamma = \sum_{j=1}^{\infty} \lambda_j$$

and each ψ_j is normalized. Note that (3.1.25) defines $\Gamma(\underline{z}, \underline{z})$ only for almost every $\underline{z}$, but that is sufficient for (3.1.24). The function $\Gamma(\underline{z}, \underline{z})$ on $(\mathbb{R}^3 \times \{1, \ldots, q\})^N$ defined using (3.1.25) is called the **diagonal part of the density matrix** Γ.

The eigenfunction expansion (3.1.21) is also important for defining energies and expectation values of certain operators. In Sections 3.2.1 and 3.2.2 we shall define, for any function $\psi \in H^1(\mathbb{R}^{3N}; \mathbb{C}^{q^N})$, an energy $\mathcal{E}(\psi)$ which is a quadratic form in ψ and to which is formally associated an operator H (the Hamiltonian), i.e., $\mathcal{E}(\psi) = (\psi, H\psi)$ in the sense discussed in (2.1.36). We now use (3.1.21) to *define the traces*

$$\boxed{\operatorname{Tr} H\Gamma = \operatorname{Tr}\Gamma H := \sum_{j=1}^{\infty} \lambda_j \mathcal{E}(\psi_j) =: \mathcal{E}(\Gamma).} \tag{3.1.26}$$

For unbounded operators H the expressions $\operatorname{Tr} H\Gamma$ and $\operatorname{Tr}\Gamma H$ are not well defined *a priori* but the right side of (3.1.26) is perfectly well defined (provided, of course, that $\sum_j \lambda_j |\mathcal{E}(\psi_j)| < \infty$.) The same notation (3.1.26) can be used to define $\operatorname{Tr} A\Gamma = \operatorname{Tr}\Gamma A$ for any operator, A, formally associated with a quadratic form $\mathcal{A}(\psi)$.

It is an obvious, but important consequence of (3.1.26) that

$$\inf\{\mathcal{E}(\psi) : (\psi, \psi) = 1\} = \inf\{\mathcal{E}(\Gamma) : \Gamma \text{ is a density matrix}\}.$$

One can speak of Γ belonging to a certain permutation symmetry type. It simply means that each ψ_j in (3.1.21) belongs to this type. In Section 3.1.3 it was stated that the only physically relevant permutation types are the totally symmetric (bosonic) or totally antisymmetric (fermionic) ones for each particle species, and it will be assumed henceforth, unless otherwise stated, that Γ belongs to a physical symmetry type.

3.1.5 Reduced Density Matrices

Assume, for simplicity, that the system under consideration contains only one species of particles, either fermions or bosons. The generalization of the following concepts to multiple species is obvious and left to the reader.

If Γ is a physical (bosonic or fermionic) N-particle density matrix with kernel $\Gamma(\underline{z}, \underline{z}')$ and if $1 \leq k < N$ we can define the **kernel of the k-particle reduced density matrix** to be

$$\gamma^{(k)}(z_1, \dots z_k; z_1', \dots, z_k') = \frac{N!}{(N-k)!} \int \Gamma(z_1, \dots, z_k, z_{k+1}, \dots, z_N; z_1', \dots, z_k', z_{k+1}, \dots, z_N) \mathrm{d}z_{k+1} \dots \mathrm{d}z_N. \tag{3.1.27}$$

Remarks 3.1. (1) The right side of (3.1.27) is formal (for the same reason that the right side of (3.1.24) is formal) but it acquires a unique meaning if the eigenfunction expansion (3.1.23) is used.

(2) Note that $\gamma^{(k)}$ is not normalized to have trace equal to one. Normalization conventions other than $N!/(N-k)!$ are often used in the literature.

(3) In (3.1.27) the variables $z_{k+1}, \dots, z_N$ are integrated out. Because Γ is bosonic or fermionic any other $N-k$ variables could have been chosen. Also, the ordering of $1, 2, \dots, k$ is arbitrary. Indeed, the correct definition of $\gamma^{(k)}$ when Γ is neither bosonic nor fermionic is to take *all* these possibilities and add them together with a weight 1 instead of $N!/(N-k)!$. When Γ is totally symmetric or totally antisymmetric all these choices give the same result, and that is why (3.1.27) is true in this *physical* case with the factor $N!/(N-k)!$. In the case of several species of particles it is obvious how to define the reduced density matrices for each species – or for a mixture of several species. In the interest of keeping the notation simple, we shall resist the temptation to write down the most general formula in the following treatment.

Example: For a determinantal wave function (3.1.16) the k-particle reduced density matrix is expressed in terms of the $u_1, \dots, u_N$ by

$$\gamma^{(k)}(z_1, \dots, z_k; z_1', \dots, z_k') = k! \sum_{\tau} \frac{1}{\sqrt{k!}} \det \left\{ u_{\tau_i}(z_j) \right\}_{i,j=1}^{k} \frac{1}{\sqrt{k!}} \det \{ \overline{u_{\tau_i}(z_j')} \}_{i,j=1}^{k}. \tag{3.1.28}$$

The sum runs over the $\binom{N}{k}$ choices, denoted by τ, of k functions from among the N given ones. In particular, the one-particle density matrix equals

$$\gamma^{(1)}(z, z') = \sum_{i=1}^{N} u_i(z) \overline{u_i(z')}.$$

For use later, we shall also write down the diagonal part of the two-particle density matrix, also called the two-particle density, of a determinantal wave function. It is given by

$$\gamma^{(2)}(z_1, z_2; z_1, z_2) = \gamma^{(1)}(z_1, z_1)\gamma^{(1)}(z_2, z_2) - |\gamma^{(1)}(z_1, z_2)|^2. \tag{3.1.29}$$

Associated with the kernel $\gamma^{(k)}$ is the operator $\gamma^{(k)}$, which acts on functions of k variables in the obvious way by integration, i.e.,

$$\left(\gamma^{(k)}\psi\right)(z_1, \ldots, z_k) = \int \gamma^{(k)}(z_1, \ldots, z_k; z'_1, \ldots, z'_k)\psi(z'_1, \ldots, z'_k)\mathrm{d}z'_1 \cdots \mathrm{d}z'_k.$$

It is called **the k-particle reduced density matrix**. It can also be written in terms of the **partial trace**, $\mathrm{Tr}^{(N-k)}$, as

$$\gamma^{(k)} = \frac{N!}{(N-k)!}\mathrm{Tr}^{(N-k)}\Gamma.$$

For our purposes, $\mathrm{Tr}^{(N-k)}$ can be thought of simply as a mnemonic device for (3.1.27), with the relation between kernel and operator given by (3.1.22).

As an operator, $\gamma^{(k)}$ has all the properties of a k-particle density matrix except for the normalization

$$\mathrm{Tr}\,\gamma^{(k)} = \frac{N!}{(N-k)!}.$$

To justify this statement it is necessary to verify that $\gamma^{(k)}$ is positive semidefinite. For any $\phi \in L^2(\mathbb{R}^{3k}; \mathbb{C}^{q^k})$, and with $\underline{z}^{(k)} := (z_1, \ldots, z_k)$, we have

$$\begin{aligned}(\phi, \gamma^{(k)}\phi) &= \int \overline{\phi^{(k)}(\underline{z}^{(k)'})}\phi^{(k)}(\underline{z}^{(k)})\gamma^{(k)}(\underline{z}^{(k)}, \underline{z}^{(k)'})\mathrm{d}\underline{z}^{(k)}\mathrm{d}\underline{z}^{(k)'} \\ &= \sum_{j=1}^{\infty}\lambda_j \int \left| \int \overline{\phi}(\underline{z}^{(k)})\psi_j(\underline{z}^{(k)}, \underline{z}^{(N-k)})\mathrm{d}\underline{z}^{(k)}\right|^2 \mathrm{d}\underline{z}^{(N-k)},\end{aligned} \tag{3.1.30}$$

and this is non-negative. (The reader is invited to prove, using the Schwarz inequality, that these integrands are summable and that the interchange of the order of summation and various integrations is justified by Fubini's theorem.)

As a consequence of being self adjoint and trace class, the kernel $\gamma^{(k)}$ also has an eigenfunction expansion

$$\gamma^{(k)}(\underline{z}^{(k)}, \underline{z}^{(k)'}) = \sum_{j=1}^{\infty}\lambda_j^{(k)} f_j(\underline{z}^{(k)})\overline{f_j(\underline{z}^{(k)'})}, \tag{3.1.31}$$

with $\lambda_j^{(k)} \geq 0$, $f_j \in L^2(\mathbb{R}^{3k}; \mathbb{C}^{q^k})$ orthonormal, and

$$\sum_{j=1}^{\infty} \lambda_j^{(k)} = \frac{N!}{(N-k)!}.$$

We also note that $\frac{(N-k)!}{N!}\gamma^{(k)}$ is an H^1 density matrix if Γ is one.

The expansion (3.1.31) permits us to define $\gamma^{(k)}$ on the diagonal, i.e., $\gamma^{(k)}(\underline{z}^{(k)}, \underline{z}^{(k)})$ in analogy with (3.1.25). In particular, it should be noted that the one-particle density ϱ_ψ defined in (3.1.4) is the diagonal part of $\gamma^{(1)}$.

An important application of partial traces is the definition of the **spin-summed density matrix**. Consider the case of the one-particle density matrix $\gamma^{(1)}$, which is a positive trace class operator on the one-particle space $L^2(\mathbb{R}^3; \mathbb{C}^q) \cong L^2(\mathbb{R}^3) \otimes \mathbb{C}^q$. By taking the partial trace over the $\mathbb{C}^q$ part, one obtains a positive trace class operator on $L^2(\mathbb{R}^3)$, which we shall denote as $\mathring{\gamma}^{(1)}$. Explicitly, in terms of kernels,

$$\mathring{\gamma}^{(1)}(\boldsymbol{x}, \boldsymbol{x}') = \sum_{\sigma=1}^{q} \gamma^{(1)}(\boldsymbol{x}, \sigma; \boldsymbol{x}', \sigma). \tag{3.1.32}$$

The trace of $\mathring{\gamma}^{(1)}$ on $L^2(\mathbb{R}^3)$ is the same as the trace of $\gamma^{(1)}$ on $L^2(\mathbb{R}^3; \mathbb{C}^q)$, namely $\int_{\mathbb{R}^3} \mathring{\gamma}^{(1)}(\boldsymbol{x}, \boldsymbol{x})\mathrm{d}\boldsymbol{x}$. The largest eigenvalue of $\mathring{\gamma}^{(1)}$ may be larger than the largest eigenvalue of $\gamma^{(1)}$; in fact, it is at most q times as big. We denote the largest eigenvalue of $\gamma^{(1)}$ and $\mathring{\gamma}^{(1)}$ by $\|\gamma^{(1)}\|_\infty$ and $\|\mathring{\gamma}^{(1)}\|_\infty$, respectively. We thus have

$$\|\mathring{\gamma}^{(1)}\|_\infty \leq q \|\gamma^{(1)}\|_\infty, \tag{3.1.33}$$

a fact that will be useful later.

For the purpose of this book the most important of the $\gamma^{(k)}$ are $\gamma^{(1)}$ and, to a much lesser extent, $\gamma^{(2)}$. The Hamiltonians H, or energy functionals $\mathcal{E}(\cdot)$ we shall consider in this book will have only one- or two-body terms in them. That is, formally

$$H = \sum_{i=1}^{N} h_i + \sum_{1 \leq i < j \leq N} W_{ij}.$$

This notation is meant to suggest that h is an operator (or quadratic form) on $L^2(\mathbb{R}^3; \mathbb{C}^q)$, e.g., $h = -\Delta + V(\boldsymbol{x})$, and $\sum_i h_i$ is simply $\sum_i [-\Delta_{\boldsymbol{x}_i} + V(\boldsymbol{x}_i)]$. For the energy $\mathcal{E}(\cdot)$, this translates into $\int (\sum_i |\nabla_{\boldsymbol{x}_i} \psi|^2 + \sum_i V(\boldsymbol{x}_i)|\psi|^2)\mathrm{d}\underline{z}$. Similarly, W is an operator on $L^2(\mathbb{R}^3; \mathbb{C}^q) \otimes L^2(\mathbb{R}^3; \mathbb{C}^q)$. We then have the *important formula*

$$\mathcal{E}(\Gamma) = \operatorname{Tr} H\Gamma = \operatorname{Tr} h\gamma^{(1)} + \frac{1}{2}\operatorname{Tr} W\gamma^{(2)}. \tag{3.1.34}$$

Again, the traces in (3.1.34) are defined by using the eigenfunction expansions (3.1.31) for $\gamma^{(1)}$ and $\gamma^{(2)}$.[4]

Formula (3.1.34) raises a tantalizing possibility about computing the minimum (or infimum) of $\mathcal{E}(\Gamma)$. Instead of considering all possible N-particle density matrices Γ, it suffices instead to consider all possible two-particle density matrices $\gamma^{(2)}$. Trivially $\gamma^{(1)}$ can be obtained from $\gamma^{(2)}$ by

$$\gamma^{(1)} = \frac{1}{N-1}\operatorname{Tr}^{(1)}\gamma^{(2)},$$

[4] Some readers might be concerned about the following technical point. We have defined $\operatorname{Tr} H\Gamma$ via the eigenfunction expansion of Γ, and similarly we define $\operatorname{Tr} h\gamma^{(1)}$ and $\operatorname{Tr} W\gamma^{(2)}$ via the eigenfunction expansions of $\gamma^{(1)}$ and $\gamma^{(2)}$. It remains to check that these different expansions give the same answer, and that (3.1.34) holds. For example, if $\gamma^{(1)} = \sum_i \kappa_i \gamma_{u_i}$ is the one-particle density matrix of $\Gamma = \sum_\alpha \lambda_\alpha \Gamma_{\Psi_\alpha}$, is it true that

$$\sum_i \kappa_i |u_i(z)|^2 = \sum_\alpha \lambda_\alpha \int |\Psi_\alpha(z, z_2, \ldots, z_N)|^2 \mathrm{d}z_2 \cdots \mathrm{d}z_N$$

for almost every $z = (\boldsymbol{x}, \sigma) \in \mathbb{R}^3 \times \mathbb{C}^q$? The answer is yes, and can easily be seen as follows. If V is a bounded, measurable function on $\mathbb{R}^3 \times \mathbb{C}^q$,

$$\begin{aligned}&(u_i, V\gamma^{(1)} u_i)\\ &= \sum_\alpha \lambda_\alpha \int \overline{u_i(z)} u_i(z') V(z) \overline{\Psi_\alpha(z, z_2, \ldots, z_N)} \Psi_\alpha(z', z_2, \ldots, z_N)\, \mathrm{d}z\, \mathrm{d}z'\, \mathrm{d}z_2 \cdots \mathrm{d}z_N.\end{aligned}$$

For almost every $z_2, \ldots, z_N$, we have

$$\begin{aligned}&\sum_i \int \overline{u_i(z)} u_i(z') V(z) \overline{\Psi_\alpha(z, z_2, \ldots, z_N)} \Psi_\alpha(z', z_2, \ldots, z_N)\, \mathrm{d}z\, \mathrm{d}z'\\ &= \int V(z) |\Psi_\alpha(z, z_2, \ldots, z_N)|^2 \mathrm{d}z\end{aligned}$$

and hence

$$\sum_i \kappa_i \int V(z)|u_i(z)|^2 \mathrm{d}z = \sum_i (u_i, V\gamma^{(1)} u_i) = \sum_\alpha \lambda_\alpha \int V(z) |\Psi_\alpha(z, z_2, \ldots, z_N)|^2\, \mathrm{d}z\, \mathrm{d}z_2 \cdots \mathrm{d}z_N.$$

Fubini's theorem justifies the interchange of integrals and sums. Since this is true for every (bounded) V we arrive at the statement above. A similar argument applies to the two-particle density and to the kinetic energy in (3.1.34).

so that $\operatorname{Tr} H\Gamma$ really only depends on $\gamma^{(2)}$. The difficulty is that we do not know how to characterize the set of two-particle density matrices $\gamma^{(2)}$ that arise as the reduction of some Γ. If we did, then the N-body problem would effectively be reduced to a two-body problem. It is definitely not the case that every self-adjoint positive semidefinite operator, γ, on $L^2(\mathbb{R}^3;\mathbb{C}^q)\otimes L^2(\mathbb{R}^3;\mathbb{C}^q)$ that satisfies $\operatorname{Tr}\gamma = N(N-1)$ arises from some Γ with the right symmetry. Some useful necessary conditions are known (see, e.g. [78, 147, 31]) but a non-trivial sufficient condition is not known. The quest for necessary and sufficient conditions has been a topic of research for many years – called the **N-representability problem**. We shall not mention it further here.

The *only* k for which the allowed k-particle reduced density matrices $\gamma^{(k)}$ can be characterized simply is $k = 1$. A positive semidefinite operator γ on the one-particle space $L^2(\mathbb{R}^3;\mathbb{C}^q)$ with $\operatorname{Tr}\gamma = N$ will be called **admissible** if it is the one-particle reduction of some density matrix Γ on the N-particle space. The boson and fermion cases are very different and we give them separately. In the boson case all such operators γ are admissible, but not all γ are admissible in the fermion case.

Theorem 3.1 (Admissible One-Body Density Matrices for Bosons). *Let γ be a self adjoint, positive semidefinite operator on $L^2(\mathbb{R}^3;\mathbb{C}^q)$ with finite trace*

$$\operatorname{Tr}\gamma = N$$

for some integer $N \geq 1$. Then there is a bosonic N-particle density matrix Γ such that $\gamma = N\operatorname{Tr}^{(N-1)}\Gamma$. Moreover, if $N \geq 2$, Γ can be chosen to be a pure state, i.e., $\Gamma = \Gamma_\psi$ for some ψ.

Proof. If $N = 1$ we simply take $\Gamma = \gamma$, so suppose $N \geq 2$. Let $\{f_j, \lambda_j\}_{j=1}^{\infty}$ be the orthonormal eigenfunctions and eigenvalues of γ. With the choice

$$\psi(\underline{z}) := N^{-1/2}\sum_{j=1}^{\infty}\lambda_j^{1/2}\prod_{i=1}^{N} f_j(z_i)$$

one easily checks that ψ is normalized and that $\gamma = N\operatorname{Tr}^{(N-1)}\Gamma_\psi$. ■

The following is the analogue of Theorem 3.1 for fermions. It is due to Coleman [31]. The conclusion (3.1.35) will be very important for us.

Theorem 3.2 (Admissible One-Body Density Matrices for Fermions). *Let γ satisfy the hypotheses of Theorem 3.1. Then there is a fermionic N-particle density matrix Γ such that $\gamma = N\operatorname{Tr}^{(N-1)}\Gamma$ if and only if γ satisfies the additional* *condition*

$$\boxed{\gamma \le \mathbb{I}.} \tag{3.1.35}$$

It is not true, in general, that Γ can be chosen to be a pure state. In particular, whenever γ has $N-1$ eigenvalues equal to 1 and at least $N+1$ positive eigenvalues then Γ cannot be pure.

Proof. We first prove that (3.1.35) suffices to insure the existence of a Γ. Let $\lambda_1 \ge \lambda_2 \ge \dots$ denote the eigenvalues of γ in decreasing order. By assumption, $0 \le \lambda_j \le 1$ and $\sum_{j=1}^{\infty} \lambda_j = N$. We can assume that $\lambda_{N+1} > 0$, otherwise a simple Slater determinant will satisfy the requirements. Note also that $\lambda_{N+1} \le N/(N+1) < 1$, since $N \ge \sum_{j=1}^{N+1} \lambda_j \ge (N+1)\lambda_{N+1}$.

Let $\chi_{[1,N]}$ be the characteristic function of $[1, N]$, i.e., $\chi_{[1,N]}(j) = 1$ for $1 \le j \le N$ and $\chi_{[1,N]}(j) = 0$ for $j > N$. With $\varepsilon = \min\{\lambda_N, 1 - \lambda_{N+1}\}$, we can write

$$\lambda_j = \varepsilon \chi_{[1,N]}(j) + (1-\varepsilon) f_j$$

with

$$f_j = \frac{1}{1-\varepsilon}\left(\lambda_j - \varepsilon \chi_{[1,N]}(j)\right).$$

Then $0 \le f_j \le 1$, and $\sum_{j=1}^{\infty} f_j = N$.

We can now rearrange the sequence f_j in decreasing order, and repeat the construction. After M iterations, we see that λ_j can be written as

$$\lambda_j = \sum_{k=1}^{M} c_k \chi_k(j) + R_j^M,$$

where χ_k is the characteristic function of some subset of N points in $\{1, 2, \dots\}$, and $\sum_{k=1}^{M} c_k \le 1$. In fact, if ε_k denotes the value of ε in the kth iteration, we have

$$c_k = \varepsilon_k \prod_{j=1}^{k-1} (1 - \varepsilon_j).$$

We can now distinguish two cases. If $\sum_{k=1}^{\infty} c_k = 1$, then $\lim_{M\to\infty} \sum_{j=1}^{\infty} R_j^M = 0$ and hence

$$\lambda_j = \sum_{k=1}^{\infty} c_k \chi_k(j).$$

In particular, γ can be written as a convex combination of rank N projections. If we take Γ to be the same convex combination of the rank 1 projections onto the Slater determinants corresponding to the rank N projections, Γ will satisfy all the requirements.

It remains to consider the case $\sum_{k=1}^{\infty} c_k =: d < 1$. The sequence R_j^M converges *strongly* to some sequence R_j as $M \to \infty$, with the property that $0 \leq R_j \leq 1 - d$ and $\sum_{j=1}^{\infty} R_j = N(1-d)$. That is, $\lim_{M\to\infty} \sum_{j=1}^{\infty} |R_j^M - R_j| = 0$. This follows easily from the dominated convergence theorem. For the following reason one readily concludes that $\lim_{k\to\infty} \varepsilon_k > 0$. Let us reorder the sequence $R_j^M/(1-\sum_{k=1}^{M} c_k)$ in decreasing order and call the resulting sequence r_j^M. Since $0 \leq r_j^M \leq 1$ and $\sum_j r_j^M = N$, we conclude that $r_{N+1}^M \leq N/(N+1)$, as we noted earlier. Then $1 - r_{N+1}^M \geq 1/(N+1)$, which implies that ε_k can only go to zero if r_N^M goes to zero. But r_j^M converges, as $M \to \infty$, to r_j, which equals the sequence $R_j/(1-d)$ rearranged in decreasing order. The numbers r_j are ordered, lie between 0 and 1, and sum to N. Hence $r_N > 0$. This implies that $\lim_{k\to\infty} \varepsilon_k > 0$.

For arbitrary numbers $0 \leq \varepsilon_j \leq 1$ and $M \in \mathbb{N} \cup \{\infty\}$,

$$\sum_{k=1}^{M} \varepsilon_k \prod_{j=1}^{k-1} (1-\varepsilon_j) = 1 - \prod_{j=1}^{M} (1-\varepsilon_k). \tag{3.1.36}$$

A simple proof of this identity (due to Ya. Sinai) is to rewrite it as a telescopic sum of the form $\sum_{k=1}^{M}(\alpha_{k-1} - \alpha_k) = \alpha_0 - \alpha_M$, where $\alpha_k = \prod_{j=1}^{k}(1-\varepsilon_j)$ for $k \geq 1$ and $\alpha_0 = 1$. We apply this to our situation above and note that the left side equals d when $M = \infty$. The right side equals 1, however, if the ε_k are not summable which, in particular, is the case if $\lim_{k\to\infty} \varepsilon_k > 0$. This contradicts the assumption that $d < 1$, and proves that (3.1.35) is indeed sufficient for N-representability.

Remark 3.2. What we have really just shown is that every sequence of numbers λ_j with $0 \leq \lambda_j \leq 1$ and $\sum_{j=1}^{\infty} \lambda_j = N$ can be written as a convex combination of characteristic functions of N elements. Since the latter are the extreme points in this convex set, this fact follows from a general result known as Choquet's theorem. In finite dimensions, it goes back to Caratheodory and Minkowski [151, Sect. 17]. For the infinite dimensional version, we refer the interested reader to [166, Sect. I.5]. Instead of using this abstract result, however, we chose to give the elementary explicit construction above.

We shall now prove the converse, namely that $\gamma = N\mathrm{Tr}^{(N-1)}\Gamma$ implies (3.1.35), i.e. $(\phi, \gamma\phi) \leq (\phi, \phi)$ for all $\phi \in L^2(\mathbb{R}^3; \mathbb{C}^q)$.

First, let us assume that Γ is pure, i.e., $\Gamma = \Gamma_\psi$ with $(\psi, \psi) = 1$. Define the **annihilation operator** $C_{N,\phi} : \mathcal{H}_N = \bigwedge^N L^2(\mathbb{R}^3; \mathbb{C}^q) \to \mathcal{H}_{N-1} = \bigwedge^{N-1} L^2(\mathbb{R}^3; \mathbb{C}^q)$ by

$$(C_{N,\phi}\psi)(z_1, \ldots, z_{N-1}) = N^{1/2} \int \psi(z_1, \ldots, z_N)\overline{\phi}(z_N)\mathrm{d}z_N. \tag{3.1.37}$$

Clearly,

$$(C_{N,\phi}\psi, C_{N,\phi}\psi)_{\mathcal{H}_{N-1}} = (\phi, \gamma\phi). \tag{3.1.38}$$

If χ is any function in $\mathcal{H}_{N-1}$ then one easily checks that

$$(\chi, C_{N,\phi}\psi)_{\mathcal{H}_{N-1}} = (C^\dagger_{N,\phi}\chi, \psi)_{\mathcal{H}_N}$$

with $C^\dagger_{N,\phi} : \mathcal{H}_{N-1} \to \mathcal{H}_N$ being the **creation operator** given by

$$(C^\dagger_{N,\phi}\chi)(z_1, \ldots, z_N) = N^{-1/2}\mathcal{A}\{\chi(z_1, \ldots, z_{N-1})\phi(z_N)\}. \tag{3.1.39}$$

Here, $\mathcal{A}$ is the antisymmetrizer (3.1.17). A simple algebraic exercise shows that

$$C_{N+1,\phi}C^\dagger_{N+1,\phi} + C^\dagger_{N,\phi}C_{N,\phi} = (\phi, \phi)\mathbb{I}_N \tag{3.1.40}$$

where $\mathbb{I}_N$ is the identity on $\mathcal{H}_N$. Thus, returning to (3.1.38),

$$\begin{aligned}(\phi, \gamma\phi) = (\psi, C^\dagger_{N,\phi}C_{N,\phi}\psi)_{\mathcal{H}_N} &= (\phi, \phi)(\psi, \psi)_{\mathcal{H}_N} - (\psi, C_{N+1,\phi}C^\dagger_{N+1,\phi}\psi)_{\mathcal{H}_N} \\ &= (\phi, \phi) - (C^\dagger_{N+1,\phi}\psi, C^\dagger_{N+1,\phi}\psi) \le (\phi, \phi).\end{aligned} \tag{3.1.41}$$

This is the desired goal. If Γ is not pure, we can apply (3.1.41) to each term in (3.1.21) and then use $\sum_j \lambda_j = 1$.

To prove the last sentence of the theorem it suffices to find a γ satisfying (3.1.35) but such that no pure Γ_ψ exists for which $\gamma = N\mathrm{Tr}^{(N-1)}\Gamma_\psi$. First, suppose $N = 2$ and that $\gamma^{(1)}$ has three non-zero eigenvalues and eigenfunctions $(1, f)$, (μ, g), $((1-\mu), h)$ with $0 < \mu < 1$ and with f, g, h orthonormal functions in $L^2(\mathbb{R}^3; \mathbb{C}^q)$. All other eigenvalues are zero. Assume now that $\gamma^{(1)}$ is the reduction of some Γ_ψ. If we take $\phi = f$ in (3.1.41) then, since $(f, \gamma^{(1)} f) = (f, f) = 1$, it follows from (3.1.41) that $C^\dagger_{3,f}\psi = 0$. By (3.1.40),

$$C^\dagger_{2,f}C_{2,f}\psi = \psi. \tag{3.1.42}$$

The following explicit form of (3.1.42) is a consequence of the definitions (3.1.37) and (3.1.39):

$$\mathcal{A}\left\{ f(z_2) \int \psi(z_1, z_2') \overline{f(z_2')} \mathrm{d}z_2 \right\} = \psi(z_1, z_2). \tag{3.1.43}$$

Let $2^{-1/2} f_2(z_1)$ denote the integral in (3.1.43). Since $(f, f) = 1$, it follows that $(f_2, f_2) = 1$. Then (3.1.43) reads

$$\psi(z_1, z_2) = 2^{-1/2} \det\{f_i(z_j)\}_{i,j=1}^2 \tag{3.1.44}$$

with $f_1 = f$. Moreover,

$$(f_1, f_2) = \int \psi(z_1, z_2) \overline{f}(z_1) \overline{f}(z_2) \mathrm{d}z_1 \mathrm{d}z_2,$$

and this vanishes because $f(z_1) f(z_2)$ is symmetric and ψ is antisymmetric; thus $(f_1, f_2) = 0$ and (3.1.44) is correctly normalized. From (3.1.44), we find that $\gamma^{(1)}$ has only two non-zero eigenvalues and eigenfunctions $(1, f_1)$ and $(1, f_2)$, which contradicts the initial assumption of eigenvalues $1, \mu, 1 - \mu$.

The same idea can easily be generalized to show that for any N, a pure state Γ_ψ cannot give rise to a $\gamma^{(1)}$ that has $(N - 1)$ eigenvalues 1 and all the remaining eigenvalues strictly less than 1. ∎

3.2 Many-Body Hamiltonians

In this section, we will describe in detail some of the many-body Hamiltonians whose ground state energy will be studied in this book. We also define what is meant by stability of the second kind. We start with the case of static nuclei, and consider the case of dynamic nuclei in Subsect. 3.2.2.

3.2.1 Many-Body Hamiltonians and Stability: Models with Static Nuclei

The **Coulomb Hamiltonian** for N non-relativistic electrons interacting through electrostatic forces with M static nuclei *fixed* at positions $\boldsymbol{R}_1, \ldots, \boldsymbol{R}_M \in \mathbb{R}^3$ is

$$H_{N,M} = -\frac{1}{2} \sum_{i=1}^{N} \Delta_i + \alpha V_C(\underline{\boldsymbol{X}}, \underline{\boldsymbol{R}}). \tag{3.2.1}$$

The function $V_C(\underline{X}, \underline{R})$ is the total Coulomb energy defined in Section 2.1.4, Eqs. (2.1.21)–(2.1.24) and $\alpha := e^2/\hbar c \approx 1/137$ is the **fine-structure constant**. The units are chosen as explained in Section 2.1.7. Note that the only free parameters besides N and M are the nuclear charge numbers $Z_1, \ldots, Z_M$. Although they are integers in nature, we shall *not* assume that here.

A simple but important observation is that (3.2.1) is symmetric under permutation of the electron labels. Hence it makes sense to restrict it to antisymmetric wave functions, as appropriate for electrons. For $\psi \in \bigwedge_{i=1}^N L^2(\mathbb{R}^3; \mathbb{C}^q)$ a totally antisymmetric wave function of space-spin variables, we write the quadratic form associated with $H_{N,M}$ as

$$\mathcal{E}_N(\psi) = (\psi, H_{N,M}\psi) = \sum_{i=1}^N T_\psi^i + \alpha V_\psi \tag{3.2.2}$$

(where T_ψ^i and V_ψ are defined in (3.1.10) and (3.1.12), respectively). Recall that $z_i = (\boldsymbol{x}_i, \sigma_i)$ where $\boldsymbol{x}_i \in \mathbb{R}^3$ and the spin index σ_i takes values in $\{1, \ldots, q\}$. In the case of electrons, $q = 2$ of course. If electrons were bosons we would then consider (3.2.2) with $\psi(z_1, \ldots, z_N)$ symmetric. The minimization problem associated with (3.2.2) is now

$$E_N(\underline{Z}, \underline{\boldsymbol{R}}) := \inf\{\mathcal{E}_N(\psi) : \psi \text{ bosonic or fermionic}, \|\psi\|_2 = 1, \psi \in H^1(\mathbb{R}^{3N})\} \tag{3.2.3}$$

with the obvious notation $\underline{Z} = (Z_1, \ldots, Z_M)$ and $\underline{\boldsymbol{R}} = (\boldsymbol{R}_1, \ldots, \boldsymbol{R}_M)$. $E_N(\underline{Z}, \underline{\boldsymbol{R}})$ is called the **ground state energy**. Note that for this problem the positions of the nuclei are fixed parameters and therefore the repulsion U is unimportant for the calculation of $E_N(\underline{Z}, \underline{\boldsymbol{R}})$, although it is crucial for questions such as binding and stability of the second kind.

The condition $E_N(\underline{Z}, \underline{\boldsymbol{R}}) > -\infty$ for all distinct values of $\boldsymbol{R}_1, \ldots, \boldsymbol{R}_M$ is called **stability of the first kind**. For non-relativistic Coulomb systems this kind of stability is not very hard to prove. To see this, simply omit the positive repulsive parts. The attractive parts can be studied one particle at a time; they satisfy the stability conditions in Section 2.2.1, and hence stability of the first kind follows. The estimate for $E_N(\underline{Z}, \underline{\boldsymbol{R}})$ obtained by this simple argument will depend on the nuclear coordinates $\underline{\boldsymbol{R}}$. Our goal is to eliminate this dependence by defining the **absolute ground state energy** to be

$$\boxed{E_{N,M}(\underline{Z}) := \inf\left\{E_N(\underline{Z}, \underline{\boldsymbol{R}}) : \underline{\boldsymbol{R}} \in \mathbb{R}^{3N}\right\}.} \tag{3.2.4}$$

With it we can define **stability of the second kind**, which says that

$$\boxed{E_{N,M}(\underline{Z}) \geq -\Xi(Z)(N+M)} \tag{3.2.5}$$

for some number $\Xi(Z)$ that depends only on $Z := \max(Z_1, \ldots, Z_M)$. Obviously, (3.2.5) is much harder to prove than stability of the first kind, for $U(\underline{\boldsymbol{R}})$ now plays a decisive role. (Without $U(\underline{\boldsymbol{R}})$, the minimum energy would occur at $\boldsymbol{R}_k = 0$ for all k, and would hence be proportional to $-N(\sum_k Z_k)^2 \sim -NM^2$.) As we shall see, *stability of the second kind holds only for fermions.*

As we mentioned in the prologue, the linear lower bound (3.2.5) is essential for our understanding of ordinary matter, where a full glass of water has twice the energy of a half-filled glass. Were the energy to grow by a larger power of the number of particles, one could extract a huge amount of energy simply by pouring one half-filled glass of water into another.

The relativistic analogue of (3.2.1) is given formally by

$$H_{N,M}^{\mathrm{rel}} = \sum_{i=1}^{N} \left(\sqrt{-\Delta_i + 1} - 1 \right) + \alpha V_C(\underline{\boldsymbol{X}}, \underline{\boldsymbol{R}}). \tag{3.2.6}$$

The symbol $\sqrt{-\Delta_i + 1} - 1$ is defined by considering the quadratic form associated with (3.2.6):

$$\mathcal{E}_N^{\mathrm{rel}}(\psi) = (\psi, H_{N,M}^{\mathrm{rel}} \psi) = \sum_{i=1}^{N} T_\psi^i + \alpha V_\psi,$$

where T_ψ^i is now defined in (3.1.11). Analogously to (3.2.3) we can define the ground state energy of $\mathcal{E}_N^{\mathrm{rel}}(\psi)$ to be

$$E_N^{\mathrm{rel}}(\underline{Z}, \underline{\boldsymbol{R}}) = \inf\{\mathcal{E}_N^{\mathrm{rel}}(\psi) : \psi \text{ bosonic or fermionic}, \ \|\psi\|_2 = 1, \ \psi \in H^{1/2}(\mathbb{R}^{3N})\}.$$

It is by no means clear that $E_N^{\mathrm{rel}}(\underline{Z}, \underline{\boldsymbol{R}})$ is finite. In fact, as we shall see in Chapter 8, even for the one electron, one nucleus case, $E_1^{\mathrm{rel}}(Z)$ is only finite if $Z\alpha \leq 2/\pi$. From this we would expect that, provided $Z_i\alpha \leq 2/\pi$ for all $i = 1, \ldots, M$ and provided that the nuclei stay fixed and apart, stability of the first kind should hold. In fact this is true (see Lemma 8.3 in Chapter 8) independent of the statistics, Fermi or Bose. Stability of the second kind is however a fairly deep result, relativistically. In addition it will turn out that *a bound on the fine structure constant α is also necessary to insure stability of the* *second kind.* Stability requires an understanding of the problem in the extreme relativistic limit where the mass of the electron no longer plays a role and can be set equal to zero.

The reason that the extreme relativistic limit is relevant is the pair of inequalities

$$|\boldsymbol{p}| - 1 \leq \sqrt{\boldsymbol{p}^2 + 1} - 1 \leq |\boldsymbol{p}|,$$

which are easily verified. Using the definition of the relativistic kinetic energy via the Fourier transform (cf. (3.1.11)) one obtains the inequalities

$$\widetilde{\mathcal{E}}_N^{\mathrm{rel}}(\psi) - N \leq \mathcal{E}_N^{\mathrm{rel}}(\psi) \leq \widetilde{\mathcal{E}}_N^{\mathrm{rel}}(\psi),$$

where

$$\widetilde{\mathcal{E}}_N^{\mathrm{rel}}(\psi) = \sum_{i=1}^{N} (\psi, |\boldsymbol{p}_i|\psi) + \alpha V_\psi.$$

Thus the stability of the second kind of $\mathcal{E}_N^{\mathrm{rel}}(\psi)$ and $\widetilde{\mathcal{E}}_N^{\mathrm{rel}}(\psi)$ are equivalent. Now we note that by scaling $\psi \mapsto \psi_\lambda$ given by

$$\psi_\lambda(\boldsymbol{x}_1, \sigma_1; \ldots; \boldsymbol{x}_N, \sigma_N) = \lambda^{3N/2}\psi(\lambda\boldsymbol{x}_1, \sigma_1; \ldots; \lambda\boldsymbol{x}_N, \sigma_N),$$

and also $\underline{\boldsymbol{R}} \mapsto \lambda\underline{\boldsymbol{R}}$,

$$\widetilde{\mathcal{E}}_N^{\mathrm{rel}}(\psi_\lambda) = \lambda\widetilde{\mathcal{E}}_N^{\mathrm{rel}}(\psi). \tag{3.2.7}$$

Thus,

$$\widetilde{E}_N^{\mathrm{rel}}(\underline{Z}) = \inf_{\psi, \underline{\boldsymbol{R}}} \{\widetilde{\mathcal{E}}_N^{\mathrm{rel}}(\psi) : \psi \in H^{1/2}(\mathbb{R}^{3N}),\ \|\psi\|_2 = 1\}$$

is either $-\infty$ or zero. By (3.2.7), if $\widetilde{\mathcal{E}}_N^{\mathrm{rel}}(\psi) < 0$ for some ψ then $\widetilde{E}_N^{\mathrm{rel}}(\underline{Z}) = -\infty$, while $\widetilde{E}_N^{\mathrm{rel}}(\underline{Z}) = 0$ if and only if $\widetilde{\mathcal{E}}_N^{\mathrm{rel}}(\psi) \geq 0$ for all ψ. Thus, *the stability of the first kind and of the second kind for $\mathcal{E}_N^{\mathrm{rel}}(\psi)$ are equivalent, and both are equivalent to the statement that $\widetilde{\mathcal{E}}_N^{\mathrm{rel}}(\psi) \geq 0$ for all ψ and all $\underline{\boldsymbol{R}}$.*

In Chapter 10 we consider the effect of external magnetic fields

$$\boldsymbol{B}(\boldsymbol{x}) = \operatorname{curl} \boldsymbol{A}(\boldsymbol{x}).$$

For spinless particles this entails replacing $\boldsymbol{p}$ by $\boldsymbol{p} + \sqrt{\alpha}\boldsymbol{A}(\boldsymbol{x})$, i.e.

$$-\Delta \to \left(-i\nabla + \sqrt{\alpha}\boldsymbol{A}(\boldsymbol{x})\right)^2 \tag{3.2.8}$$

in both the non-relativistic and relativistic cases. With the replacement (3.2.8) the Hamiltonians (3.2.1) and (3.2.6) become $H_{N,M}(\boldsymbol{A})$ or $H_{N,M}^{\mathrm{rel}}(\boldsymbol{A})$. It will turn out that the replacement (3.2.8) does not significantly affect the stability of matter question. In the non-relativistic case, this is a consequence of the diamagnetic inequality discussed in Chapter 4. The same holds in the relativistic case but the situation is much more complicated and will be discussed in detail in Chapter 8. We can generalize the notion of stability of the second kind (3.2.5) by generalizing (3.2.4) to include $\boldsymbol{B}$, i.e.,

$$E_{N,M}(\underline{Z}) = \inf_{\boldsymbol{B},\underline{\boldsymbol{R}}}\{E_N(\underline{Z}, \underline{\boldsymbol{R}}, \boldsymbol{B}) : \underline{\boldsymbol{R}} \in \mathbb{R}^{3M}\},$$

where $E_N(\underline{Z}, \underline{\boldsymbol{R}}, \boldsymbol{B})$ is the ground state energy with field $\boldsymbol{B}$. (By gauge invariance, the ground state energy depends on $\boldsymbol{A}$ only through $\boldsymbol{B}$.) Stability of the second kind continues to hold for fermions, i.e., $E_{N,M}(\underline{Z}) \geq -\Xi(Z)(N+M)$.

The situation is drastically changed if we allow the fermions to interact with the magnetic field through their spin, i.e.,

$$H_{N,M} \to H_{N,M}(\boldsymbol{A}) - \frac{\sqrt{\alpha}}{4} g \sum_{i=1}^{N} \boldsymbol{\sigma}_i \cdot \boldsymbol{B}(\boldsymbol{x}_i), \tag{3.2.9}$$

where g is the **gyromagnetic ratio** of the electrons – usually taken to be 2 – and where $\boldsymbol{\sigma} = (\sigma^1, \sigma^2, \sigma^3)$ denotes the vector of Pauli spin matrices (see Chapters 2 and 10). Now, for any fixed $\underline{\boldsymbol{R}}$ and $\underline{Z}$, the last term in (3.2.9) allows us to drive the energy to $-\infty$ by letting $\boldsymbol{B} \to \infty$ in a suitable way. To correct this, and restore stability of the second kind, we add one more term to the energy – the **magnetic field energy** defined in (2.1.16)

$$\mathcal{E}_{\text{mag}}(\boldsymbol{B}) = \frac{1}{8\pi} \int_{\mathbb{R}^3} |\boldsymbol{B}(\boldsymbol{x})|^2 \mathrm{d}\boldsymbol{x}.$$

Its role is similar to that of the nuclear repulsion U. Its inclusion prevents $\boldsymbol{B}$ from becoming too large – *provided* $Z\alpha^2$ *and* α *are both not too large, and* $0 \leq g \leq 2$. This is similar to the requirements for stability in the relativistic case.

3.2.2 Many-Body Hamiltonians: Models without Static Particles

For all the results in the following chapters where stability of the second kind for fermions is proved, the nuclei can be considered to be static. Their kinetic energy is not needed for stability.

As we shall see, a Coulomb system of *bosons* with static nuclei is not stable of the second kind. In the non-relativistic case the energy $E_{N,M}(\underline{Z})$ grows as $-[\min(N, M)]^{5/3}$. One might think that by taking the kinetic energy of the nuclei into account, the instability can be ameliorated. This is in fact so if the nuclei are fermions, but if they are also bosons the system is still unstable, but not as badly. The energy only grows as $-[\min(N, M)]^{7/5}$. This will be discussed in Chapter 7.

If both the positive and negative particles are dynamic, the wave functions are now functions of $\underline{\boldsymbol{X}}$ and $\underline{\boldsymbol{R}}$ and the corresponding spin variables. The energy of a non-relativistic system becomes

$$\mathcal{E}_N(\psi) \to \mathcal{E}_N(\psi) + \sum_{j=1}^{M} T_\psi^j$$

where $\mathcal{E}_N(\psi)$ is given in (3.2.2), and where $T_\psi^j = (2\mu)^{-1} \int |\nabla_{\boldsymbol{R}_j} \psi|^2$ is the kinetic energy of the positive particles with coordinates $\boldsymbol{R}_1, \ldots, \boldsymbol{R}_M$. Their mass is denoted by $\mu > 0$. Of course ψ is now in $L^2(\mathbb{R}^{3(N+M)})$. For simplicity neglecting the spin variables in the notation here, we have $\psi = \psi(\boldsymbol{x}_1, \ldots, \boldsymbol{x}_N; \boldsymbol{R}_1, \ldots, \boldsymbol{R}_M)$ and it satisfies appropriate symmetry requirements separately for permutations of the coordinates $\boldsymbol{x}$ and $\boldsymbol{R}$. (Cf. Section 3.1.3.) That is, it is antisymmetric in the electron coordinates $\boldsymbol{x}_i$ and either symmetric or antisymmetric in the nuclear positions $\boldsymbol{R}_i$ depending on the statistics of the nuclei.

Another model without static particles considered in this book is a relativistic gravitating system. It consists of neutral particles of mass m_n, e.g., neutrons, which interact only via the gravitational attraction. A more thorough discussion of the feasibility of such a model for neutron stars is given in Chapter 13. In any case the Hamiltonian of this model is then given by

$$\sum_{i=1}^{N} \left(\sqrt{-\Delta_i + m_n^2} - m_n \right) - G m_n^2 \sum_{1 \le i < j \le N} \frac{1}{|\boldsymbol{x}_i - \boldsymbol{x}_j|},$$

where G is the **gravitational constant**, which is

$$G = 6.674 \times 10^{-8} \text{ grams}^{-1} \text{ cm}^3 \text{ sec}^{-2}, \tag{3.2.10}$$

as determined by Cavendish in 1798 [28].

In dimensionless terms (in units where the neutron mass is one) this expression takes the form

$$\sum_{i=1}^{N} \left(\sqrt{-\Delta_i + 1} - 1 \right) - \kappa \sum_{1 \le i < j \le N} \frac{1}{|\boldsymbol{x}_i - \boldsymbol{x}_j|} \tag{3.2.11}$$

where $\kappa = Gm_n^2/\hbar c$ is a *dimensionless* constant. If m_n = mass of a neutron then

$$\kappa \approx 7 \times 10^{-37}.$$

The smallness of κ is the reason why gravity can often be neglected, except in very large systems.

The energy functional associated with (3.2.11) is given by

$$\mathcal{E}_N^{\text{grav}}(\psi) = \sum_{i=1}^{N} T_\psi^i - \kappa \left(\psi, \sum_{1 \le i < j \le N} \frac{1}{|\boldsymbol{x}_i - \boldsymbol{x}_j|} \psi \right)$$

with T_ψ^i defined in (3.1.11). It cannot be expected that this functional is bounded from below, in general. It has to be expected that beyond a critical N it becomes unbounded from below. As it will turn out, this critical number, N_c, is the famous **Chandrasekhar mass limit**; it will depend on the statistics and it will be smaller for bosons than for fermions. The reader is referred to Chapter 13.

One might consider a similar model for white dwarfs but then one has to take the electrostatic forces into account. The functional for a system of N electrons and M nuclei (usually helium nuclei) of mass m_n in units of the electron mass and nuclear charge Z is given by

$$\begin{aligned} &\sum_{i=1}^{N} T_\psi^{i,e} + \sum_{j=1}^{M} T_\psi^{j,n} - (\alpha Z + \kappa m_n^{-1}) \left(\psi, \sum_{i=1}^{N} \sum_{j=1}^{M} \frac{1}{|\boldsymbol{x}_i - \boldsymbol{R}_j|} \psi \right) \\ &\quad + (\alpha Z^2 - \kappa) \left(\psi, \sum_{1 \le i < j \le M} \frac{1}{|\boldsymbol{R}_i - \boldsymbol{R}_j|} \psi \right) \\ &\quad + (\alpha Z - \kappa m_n^{-2}) \left(\psi, \sum_{1 \le i < j \le N} \frac{1}{|\boldsymbol{x}_i - \boldsymbol{x}_j|} \psi \right). \end{aligned} \tag{3.2.12}$$

Here, e and n refer to electrons and nuclei respectively. The operator $T^{i,e}$ corresponds to $\sqrt{-\Delta_{\boldsymbol{x}_i} + 1} - 1$ while $T^{j,n}$ corresponds to $\sqrt{-\Delta_{\boldsymbol{R}_j} + m_n^2} - m_n$.

The conclusion about stability in Chapter 13 will be largely the same as for the neutron star model. There are still some open problems, however, which will be discussed in Chapter 13.

3.2.3 Monotonicity in the Nuclear Charges

We are interested in finding the ground state energy for a given number N of electrons and a certain number M of nuclei with charges $\underline{Z} = (Z_1, \ldots, Z_M)$. These nuclei might either be fixed at locations $\underline{\boldsymbol{R}} = (\boldsymbol{R}_1, \ldots, \boldsymbol{R}_M)$ or dynamic, in which case there are no fixed locations.

Recall that $E_N(\underline{Z}, \underline{\boldsymbol{R}})$ denotes the ground state energy in the fixed nuclear case, and $E_{N,M}(\underline{Z})$ is the ground state energy in the case of dynamic nuclei. In what follows, the form of the kinetic energy of the electrons and nuclei is not important. What is important is the fact that the Hamiltonian depends linearly on the nuclear charges. (Hence we have to exclude the case where the dynamic nuclei are coupled to a magnetic field, since in this case the nuclear charge appears in the kinetic energy in a non-linear way.)

Proposition 3.1 (Monotonicity in the Nuclear Charges). *Assume that $Z_k \leq Z$ for all $k = 1, \ldots, M$. Then*

$$E_N(\underline{Z}, \underline{\boldsymbol{R}}) \geq \min_{\underline{\widetilde{\boldsymbol{R}}} \subset \underline{\boldsymbol{R}}} E_N((Z, Z, \ldots, Z), \underline{\widetilde{\boldsymbol{R}}}). \tag{3.2.13}$$

Moreover, if $Z_k \leq \widetilde{Z}_k$ for all $k = 1, \ldots, M$, then

$$E_{N,M}(\underline{Z}) \geq E_{N,M}(\underline{\widetilde{Z}}). \tag{3.2.14}$$

In other words, in the case of fixed nuclear locations $\underline{\boldsymbol{R}} = (\boldsymbol{R}_1, \ldots, \boldsymbol{R}_M)$, a lower bound is obtained by replacing each nuclear charge Z_k by either 0 or the common upper bound Z, and taking the minimum over all such choices. In the dynamic case, even more is true, namely the energy is monotone non-increasing in each nuclear charge. The usefulness of this proposition lies in the fact that in order to prove stability of matter (of the first or second kind) it is enough to consider the case of equal nuclear charges. This observation was first made in [38, Lemma 2.3 *et seq.*]. We shall utilize this fact in the chapters to come.

Proof. Each Z_k appears linearly in the Hamiltonian, i.e., $H_{N,M}$ is an affine (operator valued) function of Z_k for fixed Z_j, $j \neq k$. Its ground state energy is the infimum over $(\psi, H_{N,M}\, \psi)$ and hence is concave separately in Z_k. (Note that this does *not* mean that it is a jointly concave function.) For fixed values of Z_j, $j \neq k$, the minimum of $E_N(\underline{Z}, \underline{\boldsymbol{R}})$ over $Z_k \in [0, Z]$ is attained either at $Z_k = 0$ or $Z_k = Z$. By applying this argument to all the nuclear charges separately, we end up with the statement (3.2.13).

In the dynamic case, the same concavity argument applies. However, for fixed Z_j, $j \neq k$, the energy $E_{N,M}(\underline{Z})$ can not be increasing for small Z_k. If it were, the energy could be lowered by setting $Z_k = 0$ which, however, has the same effect as moving the corresponding nucleus to infinity (which can be accomplished with infinitesimal kinetic energy cost for the nucleus). Hence $E_{N,M}$ is less than or equal to its value at $Z_k = 0$, and hence must be monotone decreasing by concavity. This proves (3.2.14). ■

3.2.4 Unrestricted Minimizers are Bosonic

For fermionic systems, the imposition of antisymmetry has a major impact on the stability question. For bosons, however, the imposition of symmetry often plays no role. One might as well minimize over all wave functions, irrespective of symmetry. That this is so (in the absence of magnetic fields) is the content of Corollary 3.1. When magnetic fields are present, the symmetry requirement is important and not automatic, although one does not expect it to affect the question of stability significantly.

The main result, Corollary 3.1, is a consequence of the following abstract theorem.

Theorem 3.3 (Symmetry of Minimizers). *Let $\mathcal{E}(\psi)$ be a (not necessarily bounded) quadratic form on $L^2(\mathbb{R}^{dN})$ for some d (e.g., $d = 3$) with the properties that*

(a) $\mathcal{E}(\psi) \geq c(\psi, \psi)$ for some constant c,
(b) $\mathcal{E}(\psi) \geq \mathcal{E}(|\psi|)$, where $|\psi|(\underline{X}) = |\psi(\underline{X})|$,
(c) $\mathcal{E}(\psi_\pi) = \mathcal{E}(\psi)$ for all permutations $\pi \in S_N$ permuting the N particles, i.e., $\psi_\pi(\boldsymbol{x}_1, \ldots, \boldsymbol{x}_N) = \psi(\boldsymbol{x}_{\pi(1)}, \ldots, \boldsymbol{x}_{\pi(N)})$.

Then

$$E_0 := \inf\{\mathcal{E}(\psi) : (\psi, \psi) = 1\} = \inf\{\mathcal{E}(\psi) : (\psi, \psi) = 1,\ \psi_\pi = \psi \text{ for all } \pi\}.$$

Proof. By assumption (a), $E_0 = \inf\{\mathcal{E}(\psi) : (\psi, \psi) = 1\}$ is finite. By assumption (b) and the fact that $(\psi, \psi) = (|\psi|, |\psi|)$, we can assume that $\psi = |\psi|$ without loss of generality when computing the infimum of $\mathcal{E}(\psi)$. The same is true when computing the infimum over symmetric ψ, because $|\psi|$ is symmetric if ψ is symmetric. For given non-negative ψ, we write $\psi = \psi_s + \psi_r$, where

$\psi_s = (1/N!) \sum_\pi \psi_\pi$ denotes the symmetric part of ψ and ψ_r is the remainder. We note that ψ_s is also non-negative. Using assumption (c) we have[5]

$$\mathcal{E}(\psi) = \mathcal{E}(\psi_s) + \mathcal{E}(\psi_r) \tag{3.2.15}$$

and, similarly,

$$(\psi, \psi) = (\psi_s, \psi_s) + (\psi_r, \psi_r). \tag{3.2.16}$$

Note that since $(\psi_\pi, \psi_{\pi'}) \geq 0$,

$$(\psi_s, \psi_s) \geq \frac{1}{N!}.$$

In particular, ψ_s does not vanish identically. Since $\mathcal{E}(\psi_s) \geq (\psi_s, \psi_s)E_0$ and $\mathcal{E}(\psi_r) \geq (\psi_r, \psi_r)E_0$, we conclude that if ψ is a minimizer for $\mathcal{E}$ so is $\psi_s/\|\psi_s\|$.

If there is no minimizer, we can apply the same argument to a minimizing sequence. That is, if $\psi^{(n)}$ is a sequence of non-negative normalized functions with $\lim_{n\to\infty} \mathcal{E}(\psi^{(n)}) = E_0$, then also $\lim_{n\to\infty} \mathcal{E}(\psi_s^{(n)})/(\psi_s^{(n)}, \psi_s^{(n)}) = E_0$. This proves the theorem. ■

Corollary 3.1 (Unrestricted Minimizers for Relativistic and Non-Relativistic Systems are Positive and Bosonic). *Consider an energy functional*

$$\mathcal{E}(\psi) = T_\psi + W_\psi,$$

corresponding to a Hamiltonian

$$H = \sum_{i=1}^{N} T_i + W(\underline{\boldsymbol{X}})$$

with $\mathcal{E}(\psi) = (\psi, H\psi)$. Here T_i is either $(2m)^{-1}\boldsymbol{p}_i^2$, $\psi \in H^1(\mathbb{R}^{dN})$ (non relativistic) or $T_i = \sqrt{\boldsymbol{p}_i^2 + m^2} - m$, $\psi \in H^{1/2}(\mathbb{R}^{dN})$ (relativistic). (Note that all

[5] To see the equality in (3.2.15), consider first the sesquilinear form associated with $\mathcal{E}$, given by $\widetilde{\mathcal{E}}(\phi, \psi) = (4i)^{-1}\mathcal{E}(\phi + i\psi) - (4i)^{-1}\mathcal{E}(\phi - i\psi) + (1/4)\mathcal{E}(\phi + \psi) - (1/4)\mathcal{E}(\phi + \psi)$. We can write $\widetilde{\mathcal{E}}(\psi_s, \psi_r) = (N!)^{-2} \sum_\pi \sum_\sigma \widetilde{\mathcal{E}}(\psi_\pi, \psi - \psi_\sigma)$. Since $\widetilde{\mathcal{E}}(\psi_\pi, \psi - \psi_\sigma) = \mathcal{E}(\psi, \psi_{\pi^{-1}} - \psi_{\pi^{-1}\sigma})$, this can be further written as $\widetilde{\mathcal{E}}(\psi_s, \psi_r) = (N!)^{-2} \sum_\pi \sum_\sigma \widetilde{\mathcal{E}}(\psi, \psi_{\pi^{-1}} - \psi_{\pi^{-1}\sigma}) = 0$. Hence $\mathcal{E}(\psi_s + \psi_r) = \widetilde{\mathcal{E}}(\psi_s + \psi_r, \psi_s + \psi_r) = \widetilde{\mathcal{E}}(\psi_s, \psi_s) + \widetilde{\mathcal{E}}(\psi_r, \psi_r) = \mathcal{E}(\psi_s) + \mathcal{E}(\psi_r)$. In other words, the cross-terms vanish.

the particles have the same mass.) The measurable function W is such that

$$T_\psi + W_\psi \geq c(\psi, \psi)$$

for some constant c and all ψ in H^1 or $H^{1/2}$. Furthermore, $W(\underline{X})$ is assumed to be symmetric, i.e.,

$$W(\boldsymbol{x}_1, \ldots, \boldsymbol{x}_N) = W(\boldsymbol{x}_{\pi(1)}, \ldots, \boldsymbol{x}_{\pi(N)})$$

for all permutations $\pi \in S_N$.

Then $\mathcal{E}(\psi)$ satisfies all the assumptions of Theorem 3.3. In particular the unrestricted infimum of $\mathcal{E}$ is the same as the infimum of $\mathcal{E}$ over non-negative totally symmetric functions ψ.

Proof. First notice that $T_\psi \geq T_{|\psi|}$. In the non-relativistic case, this simply follows from the fact that

$$|\nabla f(\boldsymbol{x})| \geq |\nabla |f(\boldsymbol{x})||^2$$

for almost every x, which holds for any function in $H^1(\mathbb{R}^d)$ [118, Thm. 7.8]. In the relativistic case, one uses the integral formula [118, Sects. 7.11 & 7.12]

$$(f, \sqrt{|\boldsymbol{p}|^2 + m^2}\, f) = \iint\limits_{\mathbb{R}^d \times \mathbb{R}^d} \frac{|f(\boldsymbol{x}) - f(\boldsymbol{y})|^2}{|\boldsymbol{x} - \boldsymbol{y}|^{(d+1)/2}} \left(\frac{m}{2\pi}\right)^{(d+1)/2} K_{(d+1)/2}(m|\boldsymbol{x} - \boldsymbol{y}|) \mathrm{d}\boldsymbol{x}\mathrm{d}\boldsymbol{y} \tag{3.2.17}$$

(with K the modified Bessel function of the third kind, which is non-negative) and the fact that $|f(\boldsymbol{x}) - f(\boldsymbol{y})| \geq ||f(\boldsymbol{x})| - |f(\boldsymbol{y})||$. A proof of this formula for $m = 0$ will be given in Lemma 8.1 in Chapter 8.

Moreover, $W_\psi = W_{|\psi|}$. That $\mathcal{E}(\psi_\pi) = \mathcal{E}(\psi)$ is obvious. Hence the last statement follows from Theorem 3.3. ∎

Remark 3.3. The proof above would not work in the case of magnetic fields, since the energy functional in this case does not decrease by taking the absolute value of the wave function. In fact, a minimizer will not be positive in this case, or even real. Moreover, the conclusion of the corollary itself does not hold, in general, if magnetic fields are present. Minimizers need not be bosonic. In fact, the bosonic symmetry requirement is essential for understanding vortex formation in dilute Bose gases subject to homogeneous magnetic fields [162, 126].

Remark 3.4. The corollary obviously generalizes to the case in which there are several species of particles and T and W are separately symmetric in the variables corresponding to each species. Then the ground state energy without symmetry restrictions coincides with the energy in which all the species are bosonic.

CHAPTER 4

Lieb–Thirring and Related Inequalities

4.1 LT Inequalities: Formulation

The non-relativistic versions of the inequalities discussed in this chapter were invented in 1975 [134, 135] in order to give an alternative, simpler proof of the stability of non-relativistic matter, first proved by Dyson and Lenard [44]. The basic idea, as we said in Section 1.1, was to relate the stability of quantum-mechanical Coulomb systems to the simpler Thomas–Fermi theory of Coulomb systems, which was known to have the required stability properties.

These inequalities have many extensions and a sizable literature, and are collectively known as Lieb–Thirring (LT) inequalities. While we do not follow the Thomas–Fermi route to stability in this book, for reasons explained earlier, the inequalities will, nevertheless, play an essential role in various aspects of the relativistic and non-relativistic theories.

In this introduction we shall explain the various inequalities. The next section gives their connection with kinetic energy inequalities, which was the original 1975 motivation. The rest of the chapter is devoted to proofs and can be skipped by anyone interested only in applications. The proofs require a theorem about traces of operators, which is given in an appendix to this chapter. The proofs given here follow closely the discussion in [118, Chapter 12], except for the theorem in the appendix to this chapter, which is only stated but not proved in [118].

The LT inequalities concern the Schrödinger operator on $L^2(\mathbb{R}^d)$, the space of square-integrable functions of d variables. For our purposes, $d = 3$ is the relevant case, but we may as well discuss all $d \geq 1$. The inequalities depend importantly on d. They are concerned with the sum of the γth power of the negative eigenvalues. For the application in later chapters, the sum of the negative eigenvalues (that is, $\gamma = 1$) will be the most relevant, but we shall also have use for other values of γ (in particular, $\gamma = 0$ and $\gamma = 1/2$).

Quite a few mathematicians have devoted a lot of time to explore generalizations of the inequalities and bounds on the optimal constants. For instance, there are generalizations to Riemannian manifolds. Only the Euclidean case will be discussed here.

The Schrödinger operator on $L^2(\mathbb{R}^d)$ is[1]

$$H = -\Delta + V(\boldsymbol{x}) \tag{4.1.1}$$

and the eigenvalue equation is

$$-\Delta\psi(\boldsymbol{x}) + V(\boldsymbol{x})\psi(\boldsymbol{x}) = E\psi(\boldsymbol{x}) \tag{4.1.2}$$

with $\psi \in L^2(\mathbb{R}^d)$. We assume that V is such that H is self-adjoint and bounded from below. For this purpose is suffices that V satisfies the requirements in (2.2.14), as explained in Section 2.2.1. The non-positive (i.e., negative or zero) eigenvalues, *if any*, are labeled $E_0 \leq E_1 \leq E_2 \leq \cdots \leq 0$, repeated according to multiplicity. There can be none, finitely many or infinitely many. They can be defined via the variational principle, see, e.g., [118, Sect. 11.2 *et seq.*].

The LT inequalities bound power sums of the *negative* eigenvalues by integrals of V_-, the negative part of V. Recall that any function V can be written as

$$V(\boldsymbol{x}) = V_+(\boldsymbol{x}) - V_-(\boldsymbol{x}) \tag{4.1.3}$$

with

$$V_+(\boldsymbol{x}) = \max\{V(\boldsymbol{x}), 0\}, \qquad V_-(\boldsymbol{x}) = \max\{-V(\boldsymbol{x}), 0\}. \tag{4.1.4}$$

With this convention, $V_-(\boldsymbol{x}) \geq 0$. We assume that V_- vanishes at infinity, i.e., the measure of the set where $V_-(\boldsymbol{x}) \geq t$ is finite for all $t > 0$.

4.1.1 The Semiclassical Approximation

Before presenting the inequalities let us discuss their 'semiclassical' interpretation, which will make them more transparent and natural. According to the semiclassical approach, which goes back to the earliest days of quantum mechanics, each volume $(2\pi)^d$ in $2d$-dimensional phase space (consisting of pairs of points $(\boldsymbol{p}, \boldsymbol{x})$ with $\boldsymbol{p} \in \mathbb{R}^d$ and $\boldsymbol{x} \in \mathbb{R}^d$) can support one quantum state. (If we restore Planck's constant h then the volume is $(2\pi\hbar)^d = h^d$.) This prescription

[1] In this chapter, there is no factor $1/2$ in front of the Laplacian, as there is elsewhere in the book. This is the standard convention used by spectral theorists.

can be quantified by using it to 'calculate' the *number* of negative (actually, non-positive) eigenvalues by integration, as follows:

$$\sum_{j\geq 0} |E_j|^0 \simeq (2\pi)^{-d} \iint\limits_{\mathbb{R}^d\times\mathbb{R}^d} \Theta(-\boldsymbol{p}^2 - V(\boldsymbol{x}))\mathrm{d}\boldsymbol{p}\mathrm{d}\boldsymbol{x}, \qquad (4.1.5)$$

where $\Theta(s) = 1$ if $s \geq 0$ and $= 0$ if $s < 0$. Our notation is that $|E_j|^0 = 1$ even if E_j is a zero eigenvalue. The integral in (4.1.5) can easily be evaluated by first doing the $\boldsymbol{p}$ integration over the ball in $\mathbb{R}^d$ of radius $\sqrt{V_-(\boldsymbol{x})}$. This integral is well known to be (see, e.g. [118, p.6]) $[\pi^{d/2}/\,\Gamma(d/2+1)]\ V_-(\boldsymbol{x})^{d/2}$. From this we conclude that the semiclassical approximation to the number of negative eigenvalues is $[(4\pi)^{-d/2}/\,\Gamma(d/2+1)]\ \int_{\mathbb{R}^d} V_-(\boldsymbol{x})^{d/2}\mathrm{d}\boldsymbol{x}$. Note that this quantity does not depend on V_+.

We can go further and 'calculate' power sums of the negative eigenvalues, $\sum_j |E_j|^\gamma$, for all $\gamma \geq 0$, by integrating the function $(2\pi)^{-d}|\boldsymbol{p}^2 + V(\boldsymbol{x})|^\gamma$ over the subset of phase-space in which $\boldsymbol{p}^2 + V(\boldsymbol{x}) < 0$. This is an easy exercise, similar to the one just mentioned. One does the $\boldsymbol{p}$ integration first and finds the **semiclassical approximation**

$$\sum_{j\geq 0} |E_j|^\gamma \simeq L^{\mathrm{cl}}_{\gamma,d} \int\limits_{\mathbb{R}^d} V_-(\boldsymbol{x})^{\gamma+d/2}\mathrm{d}\boldsymbol{x} \qquad \textit{(non-relativistic)}, \qquad (4.1.6)$$

where $L^{\mathrm{cl}}_{\gamma,d}$ in (4.1.6) is the **non-relativistic 'classical' constant**

$$\begin{aligned} L^{\mathrm{cl}}_{\gamma,d} &= (2\pi)^{-d} \int\limits_{\mathbb{R}^d:|\boldsymbol{p}|\leq 1} (1-\boldsymbol{p}^2)^\gamma \mathrm{d}\boldsymbol{p} \\ &= \frac{\Gamma(\gamma+1)}{(4\pi)^{d/2}\,\Gamma(\gamma+1+d/2)} \qquad \textit{(non-relativistic)}. \qquad (4.1.7) \end{aligned}$$

In a parallel fashion one can derive a semiclassical estimate for the negative eigenvalues of the operator

$$H = (-\Delta)^s + V(\boldsymbol{x})$$

for any $s > 0$. The only other case that will be of interest in this book is $s = 1/2$, which corresponds to the **'ultra-relativistic' Schrödinger operator** discussed in Section 2.2.1. Similarly to the non-relativistic case, we integrate $(2\pi)^{-d}||\boldsymbol{p}| + V(\boldsymbol{x})|^\gamma$ over the phase-space subset in which $|\boldsymbol{p}| + V(\boldsymbol{x}) \leq 0$. The result is the semiclassical approximation

$$\sum_{j\geq 0} |E_j|^\gamma \simeq L^{\mathrm{cl}}_{\gamma,d} \int_{\mathbb{R}^d} V_-(\boldsymbol{x})^{\gamma+d} \mathrm{d}\boldsymbol{x} \qquad (\textit{relativistic}), \tag{4.1.8}$$

where $L^{\mathrm{cl}}_{\gamma,d}$ is now the **relativistic 'classical' constant**

$$L^{\mathrm{cl}}_{\gamma,d} = (2\pi)^{-d} \int_{\mathbb{R}^d:|\boldsymbol{p}|\leq 1} (1-|\boldsymbol{p}|)^\gamma \mathrm{d}\boldsymbol{p} = \frac{\Gamma(\gamma+1)\,\Gamma((d+1)/2)}{\pi^{(d+1)/2}\Gamma(\gamma+1+d)} \qquad (\textit{relativistic}). \tag{4.1.9}$$

There are many theorems, which we shall not explore here, that state (under some smoothness conditions on V) that the formulas (4.1.6)–(4.1.9) are asymptotically exact for the potential λV as $\lambda \to +\infty$. The question we address here is whether (4.1.6) and (4.1.8) can be turned into *inequalities* for *arbitrary* V, with only the property that V_- is in the appropriate $L^p(\mathbb{R}^d)$ space and without $\lambda \to +\infty$. Naturally, we might expect that such inequalities will only hold with the constants $L^{\mathrm{cl}}_{\gamma,d}$ on the right side replaced by some other constants $L_{\gamma,d}$. The fact that $L^{\mathrm{cl}}_{\gamma,d}$ is valid asymptotically implies that $L_{\gamma,d}$ *can never be smaller than* $L^{\mathrm{cl}}_{\gamma,d}$.

An extension that will be important for us is the inclusion of a *magnetic field.* That is, the replacement[2]

$$-\Delta \longrightarrow (-i\nabla + \boldsymbol{A}(\boldsymbol{x}))^2\,, \tag{4.1.10}$$

where $\boldsymbol{A}$ is some vector field (with the physical interpretation that curl $\boldsymbol{A}(\boldsymbol{x})$ equals the magnetic field $\boldsymbol{B}(\boldsymbol{x})$). From the semiclassical perspective, the introduction of $\boldsymbol{A}$ makes no difference; we replace $\boldsymbol{p}^2$ by $(\boldsymbol{p}+\boldsymbol{A}(\boldsymbol{x}))^2$ but then we can make the change of variables $\boldsymbol{p} \to \boldsymbol{p} - \boldsymbol{A}(\boldsymbol{x})$ in the integration leading to (4.1.6) and end up with the same result as before, namely $L^{\mathrm{cl}}_{\gamma,d} \int V_-^{\gamma+d/2}$. This independence of $\boldsymbol{A}(\boldsymbol{x})$ also appears in classical statistical mechanics and leads to the observation of Van Leeuwen [182], in the early days of quantum mechanics, that there is no orbital diamagnetism, classically, and, therefore, orbital diamagnetism must be a quantum-mechanical phenomenon. It might be supposed that in the inequalities we seek an allowance might be necessary for a small increase of $L_{\gamma,d}$ if one wants to have a value of $L_{\gamma,d}$ that is *independent* of $\boldsymbol{A}$. Nevertheless, it is a fact that all presently known values of $L_{\gamma,d}$, both sharp ones and non-sharp ones, hold for arbitrary $\boldsymbol{A}$ fields. The unknown sharp constants might have to

[2] For notational convenience, we shall absorb the fine-structure constant α into $\boldsymbol{A}$ in this chapter, and write $\boldsymbol{p}+\boldsymbol{A}$ instead of $\boldsymbol{p}+\sqrt{\alpha}\boldsymbol{A}$.

be modified to allow for an $\boldsymbol{A}$ field, but this will have to be decided by future developments. The reason for this is that all known proofs use the resolvent of $-\Delta$ in one form or another, and it is well known that this resolvent satisfies a **diamagnetic inequality**. We shall discuss this further in Section 4.4.

It turns out that the desired inequalities can be achieved if and only if γ and d satisfy certain conditions (stated in the theorems below). These conditions are optimal, meaning that otherwise no inequality of the desired form can hold with any fixed, finite value of $L_{\gamma,d}$. The necessity of $\gamma > 0$ for $d = 1$ and $d = 2$ in the non-relativistic case, for example, comes from the well known fact that when $d \leq 2$ any *arbitrarily small*, negative V *always* has at least one negative eigenvalue. For $d = 1$, this negative eigenvalue is of the order λ^2 if V is replaced by λV for some small parameter λ, and hence an inequality of the desired form can only hold if $2\gamma \geq \gamma + 1/2$, or $\gamma \geq 1/2$.

4.1.2 The LT Inequalities; Non-Relativistic Case

We shall first discuss the LT inequalities in the non-relativistic case.

Theorem 4.1 (Non-Relativistic LT Inequality). *Fix $\gamma \geq 0$ and assume that the negative part of the potential V_- satisfies the condition $V_- \in L^{\gamma+d/2}(\mathbb{R}^d)$. Assume that $\boldsymbol{A} \in L^2_{\mathrm{loc}}(\mathbb{R}^d;\mathbb{R}^d)$.*[3] *Let $E_0 \leq E_1 \leq E_2 \leq \cdots$ be the non-positive eigenvalues, if any, of $(-i\nabla + \boldsymbol{A}(\boldsymbol{x}))^2 + V(\boldsymbol{x})$ in $\mathbb{R}^d$. Then, for suitable d, defined below, there is a finite constant $L_{\gamma,d}$, which is independent of V and $\boldsymbol{A}$, such that*

$$\boxed{\sum_{j\geq 0} |E_j|^\gamma \leq L_{\gamma,d} \int_{\mathbb{R}^d} V_-(\boldsymbol{x})^{\gamma+d/2}\,\mathrm{d}\boldsymbol{x}.} \tag{4.1.11}$$

This holds in the following cases:

$$\begin{aligned} \gamma &\geq \frac{1}{2} \quad \text{for } d = 1,\\ \gamma &> 0 \quad \text{for } d = 2,\\ \gamma &\geq 0 \quad \text{for } d \geq 3. \end{aligned} \tag{4.1.12}$$

[3] The space $L^2_{\mathrm{loc}}(\mathbb{R}^d)$ consists of functions that are square integrable in any ball in $\mathbb{R}^d$, but not necessarily in the whole of $\mathbb{R}^d$. The notation $L^2(\mathbb{R}^d;\mathbb{R}^d)$ means real-vector-valued functions such that each component is in $L^2(\mathbb{R}^d)$.

Otherwise, for any finite choice of $L_{\gamma,d}$ there is a V that violates (4.1.11). *We can take*

$$L_{\gamma,d} = (4\pi)^{-d/2} 2^{\gamma} \gamma \begin{cases} \frac{(d+\gamma)\Gamma(\gamma/2)^2}{2\,\Gamma(\gamma+1+d/2)} & \text{if } d > 1, \gamma > 0 \text{ or } d = 1, \gamma \geq 1, \\ \sqrt{\pi}/(\gamma^2 - 1/4) & \text{if } d = 1, \gamma > 1/2. \end{cases} \tag{4.1.13}$$

Remark 4.1. The fact that $E_0 > -\infty$ under the assumption on V stated in Theorem 4.1 already follows from the discussion in Section 2.2.1, Eq. (2.2.14). The necessity of the conditions (4.1.12) was explained at the end of Subsection 4.1.1.

The bounds (4.1.13) presented in this theorem are certainly not the best available, and they can be improved. We list them since they can be obtained relatively easily with the methods first used in 1975 [135], as shall be explained below. These methods work only for $\gamma > 1/2$ in case $d = 1$, and $\gamma > 0$ in case $d \geq 2$, however. The extension to $\gamma = 1/2$ in the case $d = 1$ was done by Weidl [185], and the sharp constant was obtained by Hundertmark, Lieb and Thomas [94]. The $\gamma = 0$ case for $d \geq 3$ was independently proved by Cwikel [35], Lieb [112] and Rosenblum [152] and is now known as the **CLR bound**. Their proofs are very different from each other and will not be presented here. Other proofs were given later by Li and Yau [107], Fefferman [56] and by Conlon [33]. A bound that is closely related to the CLR bound is in Corollary 4.2 on page 80.

The sharp (= optimal or best) value of $L_{\gamma,d}$ is known in some cases (but not in the physically most interesting case, $\gamma = 1$, where an upper bound on $L_{\gamma,d}$ will have to suffice for the present). In some cases $L_{\gamma,d} = L^{\mathrm{cl}}_{\gamma,d}$ and in others $L_{\gamma,d} > L^{\mathrm{cl}}_{\gamma,d}$. It is conjectured [135] that $L_{1,3} = L^{\mathrm{cl}}_{1,3}$. The best we have at present (due to Dolbeault, Laptev and Loss [41]) is

$$L_{1,3} \leq \frac{\pi}{\sqrt{3}} L^{\mathrm{cl}}_{1,3} \approx 1.814\, L^{\mathrm{cl}}_{1,3} \approx 0.0123. \tag{4.1.14}$$

In fact, it was shown in [41] that

$$\boxed{L_{\gamma,d} \leq \frac{\pi}{\sqrt{3}} L^{\mathrm{cl}}_{\gamma,d} \quad \text{for all } \gamma \geq 1 \text{ and for all } d \geq 1.} \tag{4.1.15}$$

We emphasize that the LT inequalities with these constants hold for arbitrary magnetic vector potentials $\boldsymbol{A}$. Moreover, [102, 2, 15]

$$L_{\gamma,d} = L^{\mathrm{cl}}_{\gamma,d} \quad \text{for all } \gamma \geq 3/2 \text{ and for all } d \geq 1. \tag{4.1.16}$$

It is also known that $L_{\gamma,d} > L^{\rm cl}_{\gamma,n}$ if $\gamma < 1$ [88]. For some review articles see [93, 102].

For $\gamma = 0$, the currently best known constant for $d = 3$, as obtained in [112], is

$$L_{0,3} \leq 6.87\, L^{\rm cl}_{0,3} = 6.87\, \frac{1}{6\pi^2} = 0.116. \tag{4.1.17}$$

As we shall now explain, this implies that

$$\boxed{L_{\gamma,3} \leq 6.87\, L^{\rm cl}_{\gamma,3}} \tag{4.1.18}$$

for any $\gamma \geq 0$.

We note that, in general, the ratio $L_{\gamma,d}/L^{\rm cl}_{\gamma,d}$ is decreasing in γ [2], i.e.,

$$\frac{L_{\gamma,d}}{L^{\rm cl}_{\gamma,d}} \geq \frac{L_{\gamma+\delta,d}}{L^{\rm cl}_{\gamma+\delta,d}} \tag{4.1.19}$$

for any $\delta \geq 0$. To see this, one simply notes that for $e \leq 0$ and $\delta > 0$

$$|e|^{\gamma+\delta} = c_{\gamma,\delta} \int_0^{-e} \mathrm{d}\lambda\, \lambda^{\delta-1} (e+\lambda)^\gamma$$

for some constant $c_{\gamma,\delta}$. Hence

$$\frac{\sum_{j\geq 0} |E_j|^{\gamma+\delta}}{\iint\limits_{p^2+V(x)\leq 0} |p^2+V(x)|^{\gamma+\delta} \mathrm{d}x\mathrm{d}p} = \frac{\int_0^\infty \mathrm{d}\lambda\, \lambda^{\delta-1} \sum_{j\geq 0,\, E_j+\lambda\leq 0} |E_j+\lambda|^\gamma}{\int_0^\infty \mathrm{d}\lambda\, \lambda^{\delta-1} \iint\limits_{p^2+V(x)+\lambda\leq 0} |p^2+V(x)+\lambda|^\gamma \mathrm{d}x\mathrm{d}p}.$$

To obtain an upper bound on the right side, one can certainly take the supremum over λ of the ratios of the integrands, which is less then $L_{\gamma,d}/L^{\rm cl}_{\gamma,d}$ by definition. On the other hand, $L_{\gamma+\delta,d}/L^{\rm cl}_{\gamma+\delta,d}$ is the supremum of the left side over all V, and hence $L_{\gamma+\delta,d}/L^{\rm cl}_{\gamma+\delta,d} \leq L_{\gamma,d}/L^{\rm cl}_{\gamma,d}$.

4.1.3 The LT Inequalities; Relativistic Case

We shall now discuss the relativistic case. Here, the Schrödinger operator is

$$H = |-i\nabla + A(x)| + V(x), \tag{4.1.20}$$

with corresponding eigenvalues.

Theorem 4.2 (Relativistic LT Inequality). *Fix $\gamma \geq 0$ and assume that the negative part of the potential V_- satisfies the condition $V_- \in L^{\gamma+d}(\mathbb{R}^d)$. Assume that $\boldsymbol{A} \in L^2_{\text{loc}}(\mathbb{R}^d;\mathbb{R}^d)$. Let $E_0 \leq E_1 \leq E_2 \leq \cdots$ be the non-positive eigenvalues, if any, of $|-i\nabla + \boldsymbol{A}(\boldsymbol{x})| + V$ in $\mathbb{R}^d$. Then, for $\gamma > 0$ in case $d = 1$, and $\gamma \geq 0$ in case $d \geq 2$, there is a finite constant $L_{\gamma,d}$, which is independent of V and $\boldsymbol{A}$, such that*

$$\boxed{\sum_{j\geq 0} |E_j|^\gamma \leq L_{\gamma,d} \int_{\mathbb{R}^d} V_-^{\gamma+d}(\boldsymbol{x})\,\mathrm{d}\boldsymbol{x}.} \tag{4.1.21}$$

We can take

$$L_{\gamma,d} = \gamma\, 2^\gamma \left(d + \frac{\gamma}{2}\right) \pi^{-d/2-1/2} \frac{\Gamma(\frac{\gamma}{2})^2 \Gamma(\frac{d+1}{2})}{\Gamma(1+d+\gamma)}. \tag{4.1.22}$$

Using a different method from the one presented in the proof below, following the method of [112], Daubechies [37] has shown that

$$L_{0,3} \leq 6.08\, L^{\text{cl}}_{0,3}$$

in the relativistic case. In combination with the argument above this actually implies that $L_{\gamma,3} \leq 6.08\, L^{\text{cl}}_{\gamma,3}$ for all $\gamma \geq 0$. In particular,

$$\boxed{L_{1,3} \leq 6.08\, L^{\text{cl}}_{1,3} = \frac{6.08}{24\pi^2} = 0.0257.} \tag{4.1.23}$$

In contrast, our bound (4.1.22) for $\gamma = 1$ and $d = 3$ states that $L_{1,3} \leq 7\pi\, L^{\text{cl}}_{1,3}$.

There is an LT inequality not only for $-\Delta$ and $\sqrt{-\Delta}$ but for all powers $(-\Delta)^s$ with $s > 0$. Since we need only $s = 1$ and $s = 1/2$ we shall not discuss the other cases here.

It will be noted that for relativistic mechanics it is necessary to consider the operator $T_\mu(\boldsymbol{p}) = \sqrt{\boldsymbol{p}^2 + \mu^2} - \mu$ for $\mu > 0$. To handle this we can use the simple facts that

$$\sqrt{-\Delta} - \mu \;\leq\; \sqrt{-\Delta + \mu^2} - \mu \;\leq\; \sqrt{-\Delta}. \tag{4.1.24}$$

Since we are only interested in showing that the total energy is bounded above and below by a constant times the particle number N, the presence of a term μN is irrelevant. Nevertheless, one can ask if Theorem 4.2 can be improved to take explicit account of $T_\mu(\boldsymbol{p}) = \sqrt{\boldsymbol{p}^2 + \mu^2} - \mu$. The answer is yes if the right side is replaced by the appropriate semiclassical expression

$$(2\pi)^{-d} \iint\limits_{T_\mu(\boldsymbol{p})+V(\boldsymbol{x})\le 0} \left|T_\mu(\boldsymbol{p}) + V(\boldsymbol{x})\right|^\gamma \mathrm{d}\boldsymbol{x}\mathrm{d}\boldsymbol{p}. \tag{4.1.25}$$

Note that for $\mu > 0$ this expression is finite if and only if $V \in L^{\gamma+d}(\mathbb{R}^d) \cap L^{\gamma+d/2}(\mathbb{R}^d)$. Moreover, the corresponding Lieb–Thirring estimates hold only in the non-relativistic range, i.e., for $\gamma \ge 1/2$ at $d = 1$, $\gamma > 0$ at $d = 2$, and $\gamma \ge 0$ at $d \ge 3$. The reason for this is that $T_\mu(\boldsymbol{p}) \approx \boldsymbol{p}^2/(2\mu)$ for small $|\boldsymbol{p}|$, and hence the low energy eigenvalues will be approximately equal to the non-relativistic ones. Only at large values of $|\boldsymbol{p}|$ do relativistic effects become important. The effects of positive mass μ were analyzed by Daubechies in [37]. For $d = 3$, the best known bounds for the number of eigenvalues of $T_\mu(\boldsymbol{p}) + V$ is 10.33 times the semiclassical expression ((4.1.25) with $\gamma = 0$) [70] and for the sum of negative eigenvalues it is 9.62 times the semiclassical expression ((4.1.25) with $\gamma = 1$) [37].

4.2 Kinetic Energy Inequalities

Let us focus on the case $\gamma = 1$ of the previous Section 4.1. The common setting is that we have an operator of the form $H = T + V$ and we know that the sum of the negative eigenvalues of H is bounded below, as follows.

$$\sum_{j\ge 0} |E_j| \le L \int\limits_{\mathbb{R}^d} V_-(\boldsymbol{x})^p \mathrm{d}\boldsymbol{x}. \tag{4.2.1}$$

In the cases of interest T is $-\Delta$ $(p = 1 + d/2)$ or $\sqrt{-\Delta}$ $(p = 1 + d)$, but it could be anything as far as the present section is concerned. Our goal is to use (4.2.1) to bound T instead of the E_j. This was the bound used in [134, 135] to relate the Schrödinger energy to the Thomas–Fermi energy.

Let $\psi(\boldsymbol{x}_1, \sigma_1, \boldsymbol{x}_2, \sigma_2, \ldots, \boldsymbol{x}_N, \sigma_N)$ be *any* N-particle wave function of space-spin ($\boldsymbol{x} \in \mathbb{R}^d$, $\sigma \in \{1, 2, \ldots, q\}$). It is not necessary for our purpose here to assume any symmetry properties. We are interested in the 'kinetic energy'

$$T_\psi := \left(\psi, \sum_{i=1}^{N} T_i\, \psi\right). \tag{4.2.2}$$

We assume that ψ is normalized, i.e., $(\psi, \psi) = 1$, and our goal is to bound T_ψ from below in terms of the one-body (spatial) density at $\boldsymbol{x} \in \mathbb{R}^d$ given by

$$\varrho_\psi(\boldsymbol{x}) := \sum_{i=1}^{N} \sum_{\underline{\sigma}} \int\limits_{\mathbb{R}^{(N-1)d}} |\psi(\boldsymbol{x}_1, \sigma_1, \ldots, \boldsymbol{x}_{i-1}, \sigma_{i-1}, \boldsymbol{x}, \sigma_i, \ldots, \boldsymbol{x}_N, \sigma_N)|^2 \times \mathrm{d}\boldsymbol{x}_1 \cdots \widehat{\mathrm{d}\boldsymbol{x}_i} \cdots \mathrm{d}\boldsymbol{x}_N. \tag{4.2.3}$$

Here, we use the same notation as in Chapter 3, Eq. (3.1.5), with $\widehat{\mathrm{d}\boldsymbol{x}_i}$ meaning that integration is over all variables but $\boldsymbol{x}_i$. Note that ψ is arbitrary and T_ψ has nothing to do with any V; the latter is introduced only as an aid in proving the following bound.

For the following theorem it is necessary to introduce the largest eigenvalue of the spin-summed one-particle density matrix of ψ. Recall from Chapter 3, Section 3.1.5, that the one-particle density matrix $\gamma_\psi^{(1)}$ of ψ is given as

$$\gamma_\psi^{(1)}(z, z') = \sum_{i=1}^{N} \int \psi(z_1, \ldots, z_{i-1}, z, \ldots, z_N) \times \overline{\psi}(z_1, \ldots, z_{i-1}, z', \ldots, z_N) \mathrm{d}z_1 \cdots \widehat{\mathrm{d}z_i} \cdots \mathrm{d}z_N. \tag{4.2.4}$$

The spin-summed one-particle density matrix $\mathring{\gamma}_\psi^{(1)}$ is then

$$\mathring{\gamma}_\psi^{(1)}(\boldsymbol{x}, \boldsymbol{x}') = \sum_{\sigma=1}^{q} \gamma^{(1)}(\boldsymbol{x}, \sigma, \boldsymbol{x}', \sigma). \tag{4.2.5}$$

Its largest eigenvalue will be denoted as $\|\mathring{\gamma}^{(1)}\|_\infty$. The one-particle density in (4.2.3) is just $\varrho_\psi(\boldsymbol{x}) = \mathring{\gamma}_\psi^{(1)}(\boldsymbol{x}, \boldsymbol{x})$.

Theorem 4.3 (Fundamental Kinetic Energy Inequality). *With the kinetic energy and the density as defined above, and assuming the inequality* (4.2.1), *the kinetic energy of a normalized ψ is bounded as*

$$\boxed{T_\psi \geq \frac{K}{\|\mathring{\gamma}_\psi^{(1)}\|_\infty^{p'/p}} \int\limits_{\mathbb{R}^d} \varrho_\psi(\boldsymbol{x})^{p'} \, \mathrm{d}\boldsymbol{x}} \tag{4.2.6}$$

where $p' = p/(p-1)$ *and*

$$\boxed{(pL)^{p'}(p'K)^{p} = 1.} \tag{4.2.7}$$

Moreover, inequality (4.2.1) *and inequality* (4.2.6) *(with* (4.2.7)*) are equivalent. The truth of* (4.2.6) *(with* (4.2.7)*) for all* ψ *implies the truth of* (4.2.1) *for all* V.

Remark 4.2. For antisymmetric functions ψ, we have seen in Chapter 3, Section 3.1.5, that $\|\mathring{\gamma}^{(1)}\|_\infty \le q$, and this fact will be crucial for our application in the proof of stability of matter! We shall use (4.2.6) with $q^{p'/p}$ in the denominator. In general, $\|\mathring{\gamma}^{(1)}\|_\infty \le N$ for any ψ because $\mathrm{Tr}\,[\mathring{\gamma}^{(1)}] = \mathrm{Tr}\,[\gamma^{(1)}] = N$. For bosonic wave functions, this inequality can be saturated for simple product wave functions, $\psi(\boldsymbol{x}_1, \ldots, \boldsymbol{x}_N) = \prod_{i=1}^N \phi(\boldsymbol{x}_i)$.

Remark 4.3. A slightly more general version of Theorem 4.3 also holds, in which one considers more general density matrices Γ and not just rank one density matrices as here; that is, one can replace $\psi(\underline{z})\overline{\psi}(\underline{z}')$ by $\Gamma(\underline{z}, \underline{z}')$. (Cf. Section 3.1.4 in Chapter 3 for a discussion of density matrices.) We leave this simple generalization to the reader.

Remark 4.4. If we take $q = 1$ and take ψ to be any **determinantal function**

$$\psi(\boldsymbol{x}_1, \boldsymbol{x}_2, \ldots, \boldsymbol{x}_N) = (N!)^{-1/2}\det\{\phi_i(\boldsymbol{x}_j)\}_{i,j=1}^N \tag{4.2.8}$$

where $\{\phi_i\}_{i=1}^N$ is any collection of N orthonormal functions in $L^2(\mathbb{R}^d)$, then we find that

$$\varrho_\psi(\boldsymbol{x}) = \sum_{i=1}^N |\phi_i(\boldsymbol{x})|^2 \qquad \text{and} \qquad T_\psi = \sum_{i=1}^N (\phi_i, T\phi_i)\,. \tag{4.2.9}$$

In this case, $\|\mathring{\gamma}_\psi^{(1)}\|_\infty = 1$.

Let us be very explicit and give the known value of $p,\ p'$ for $d = 3$ in the non-relativistic and relativistic case, as well as best known value of K. The following corollary follows from Theorems 4.1, 4.2 and 4.3, together with the bounds (4.1.14) and (4.1.23) on the optimal constants in the LT inequalities.

Corollary 4.1 (Kinetic Energy Inequalities). *For $d = 3$ and $T = (-i\nabla + \boldsymbol{A}(\boldsymbol{x}))^2$, there is a K independent of N such that*

$$\boxed{T_\psi \geq \frac{K}{\|\mathring{\gamma}_\psi^{(1)}\|_\infty^{2/3}} \int_{\mathbb{R}^3} \varrho_\psi(\boldsymbol{x})^{5/3} \mathrm{d}\boldsymbol{x}} \tag{4.2.10}$$

for any $\psi \in H^1_{\boldsymbol{A}}(\mathbb{R}^{3N})$. In fact, $K \geq \left(\frac{3}{\pi^2}\right)^{1/3} K^{\mathrm{cl}}$ and $K^{\mathrm{cl}} = \frac{3}{5}(6\pi^2)^{2/3}$.

For $d = 3$ and $T = |-i\nabla + \boldsymbol{A}(\boldsymbol{x})|$, we have

$$\boxed{T_\psi \geq \frac{K}{\|\mathring{\gamma}_\psi^{(1)}\|_\infty^{1/3}} \int_{\mathbb{R}^3} \varrho_\psi(\boldsymbol{x})^{4/3} \mathrm{d}\boldsymbol{x}} \tag{4.2.11}$$

for any $\psi \in H^{1/2}_{\boldsymbol{A}}(\mathbb{R}^{3N})$, with $K \geq (6.08)^{-1/3} K^{\mathrm{cl}}$ and $K^{\mathrm{cl}} = \frac{3}{4}\left(6\pi^2\right)^{1/3}$.

Proof of Theorem 4.3. Given ψ, we consider a one-body operator $H = T + V$, as above. We note that both T and V act only on the spatial part of a function in $L^2(\mathbb{R}^d; \mathbb{C}^q)$ and not on the spin variables. We then consider the N-body operator

$$K_N = \sum_{i=1}^{N} H_i,$$

acting on N-particle wave functions of space-spin. With the aid of the one-particle density matrix $\gamma_\psi^{(1)}$, we can write the expectation value of K_N as (see Eq. (3.1.34) in Chapter 3)

$$(\psi, K_N \psi) = \mathrm{Tr}\,[H\gamma_\psi^{(1)}] = \mathrm{Tr}\,[H\mathring{\gamma}_\psi^{(1)}].$$

Here, we abuse the notation slightly, since the symbol Tr for the trace stands for the trace over $L^2(\mathbb{R}^d; \mathbb{C}^q) \cong L^2(\mathbb{R}^d) \otimes \mathbb{C}^q$ in the second term, but for the trace only over $L^2(\mathbb{R}^d)$ in the third term. The fact that H is independent of spin implies that the $\mathbb{C}^q$ trace affects only $\gamma_\psi^{(1)}$.

The minimum of $\mathrm{Tr}\,[H\mathring{\gamma}]$ over all positive trace-class operators[4] $\mathring{\gamma}$ with $\|\mathring{\gamma}\|_\infty \leq \|\mathring{\gamma}_\psi^{(1)}\|_\infty$ is clearly given by the sum of the negative eigenvalues of H

[4] Operators A such that $\mathrm{Tr}\,|A| = \mathrm{Tr}\,\sqrt{A^\dagger A} < \infty$.

times $\|\mathring{\gamma}_\psi^{(1)}\|_\infty$. That is, the optimal choice for $\mathring{\gamma}$ is $\|\mathring{\gamma}_\psi^{(1)}\|_\infty$ times the projection onto the negative spectral subspace[5] of H. Hence

$$(\psi, K_N\psi) \geq \|\mathring{\gamma}_\psi^{(1)}\|_\infty \sum_{j\geq 0} E_j. \tag{4.2.12}$$

A consequence of (4.2.12), together with (4.2.1), is that

$$\begin{aligned}(\psi,\ K_N\,\psi) &= T_\psi + \int_{\mathbb{R}^d} V(\boldsymbol{x})\varrho_\psi(\boldsymbol{x})\mathrm{d}\boldsymbol{x} \\ &\geq \|\mathring{\gamma}_\psi^{(1)}\|_\infty \sum_{j\geq 0} E_j \\ &\geq -\|\mathring{\gamma}_\psi^{(1)}\|_\infty\, L \int_{\mathbb{R}^d} V_-(\boldsymbol{x})^p \mathrm{d}\boldsymbol{x}. \end{aligned} \tag{4.2.13}$$

This holds for any normalized ψ. Now we make a special choice for V, which will depend on the ψ being considered. We choose

$$V(\boldsymbol{x}) = -C\varrho_\psi(\boldsymbol{x})^{1/(p-1)}, \tag{4.2.14}$$

where $C > 0$ is some constant to be determined appropriately. From (4.2.13) and (4.2.14) we have that

$$T_\psi \geq C \int_{\mathbb{R}^d} \varrho_\psi(\boldsymbol{x})^{p'}\mathrm{d}\boldsymbol{x} - \|\mathring{\gamma}_\psi^{(1)}\|_\infty L C^p \int_{\mathbb{R}^d} \varrho_\psi(\boldsymbol{x})^{p'}\mathrm{d}\boldsymbol{x}. \tag{4.2.15}$$

Obviously, we choose C to make the right side of (4.2.15) as large as possible,

$$C = (p\,\|\mathring{\gamma}_\psi^{(1)}\|_\infty\, L)^{-p'/p}, \tag{4.2.16}$$

from which (4.2.6) and (4.2.7) follow.

Finally, to prove that (4.2.6) (with (4.2.7)) for all ψ implies the truth of (4.2.1) take $q = 1$ and take ψ to be the determinant formed from all the eigenfunctions corresponding to the negative eigenvalues of $H = T + V$ (if there are infinitely many negative eigenvalues just choose any N of them and let $N \to \infty$ at the end). For this ψ, $\|\mathring{\gamma}_\psi^{(1)}\|_\infty = 1$. We then have (using (4.2.6) and $K = (pL)^{-p'/p}/p'$

[5] For our H, this subspace consists of linear combinations of eigenfunctions with negative or zero eigenvalues.

from (4.2.7))

$$\begin{aligned}\sum_{j\geq 0} E_j &= T_\psi + \int_{\mathbb{R}^d} V(\boldsymbol{x})\varrho_\psi(\boldsymbol{x})\mathrm{d}\boldsymbol{x} \\ &\geq K \int_{\mathbb{R}^d} \varrho_\psi(\boldsymbol{x})^{p'}\,\mathrm{d}\boldsymbol{x} - \int_{\mathbb{R}^d} [(pL)^{1/p} V_-(\boldsymbol{x})][(pL)^{-1/p}\varrho_\psi(\boldsymbol{x})]\mathrm{d}\boldsymbol{x} \\ &\geq K \int_{\mathbb{R}^d} \varrho_\psi(\boldsymbol{x})^{p'}\,\mathrm{d}\boldsymbol{x} - \frac{1}{p'}\int_{\mathbb{R}^d} [(pL)^{-1/p}\varrho_\psi(\boldsymbol{x})]^{p'}\mathrm{d}\boldsymbol{x} - \frac{1}{p}\int_{\mathbb{R}^d} [(pL)^{1/p} V_-(\boldsymbol{x})]^p\mathrm{d}\boldsymbol{x} \\ &= -L \int_{\mathbb{R}^d} V_-(\boldsymbol{x})^p\mathrm{d}\boldsymbol{x}.\end{aligned}$$

The last inequality is Hölder's inequality $\int fg \leq (\int f^p)^{1/p}(\int g^{p'})^{1/p'}$ followed by $ab \leq a^p/p + b^{p'}/p'$. ■

4.3 The Birman–Schwinger Principle and LT Inequalities

Our goal here is to prove (4.1.11) and (4.1.21) and, more generally, the analogous inequality for $H = (-\Delta)^s + V(\boldsymbol{x})$. We will only consider the case $\boldsymbol{A} = 0$, and explain the generalization to arbitrary $\boldsymbol{A}$ in Section 4.4.

The first step is to note that we may as well replace $V = V_+ - V_-$ by the non-positive potential $-V_-$. The reason is that if we compare two operators H and $\widetilde{H} = (-\Delta)^s - V_-(\boldsymbol{x})$ we see that $H \geq \widetilde{H}$ (in the sense that $(\phi, H\phi) \geq (\phi, \widetilde{H}\phi)$ for all functions ϕ). This fact implies, by the variational min-max principle [118, Sect 12.1], that the corresponding eigenvalues are related by $E_j \geq \widetilde{E}_j$ for all j. Therefore, since we are interested in bounding the eigenvalues from below by $-\int V_-^p$, which is independent of V_+, we may as well omit the V_+ term in V. Henceforth, $H = (-\Delta)^s - V_-(\boldsymbol{x})$.

4.3.1 The Birman–Schwinger Formulation of the Schrödinger Equation

The next step is to rewrite the Schrödinger equation for a *negative* eigenvalue $-e$ as $((-\Delta)^s + e)\,\psi(\boldsymbol{x}) = V_-(\boldsymbol{x})\psi(\boldsymbol{x})$. If we now define $\phi(\boldsymbol{x}) = \sqrt{V_-(\boldsymbol{x})}\,\psi(\boldsymbol{x})$ we obtain $((-\Delta)^s + e)\,\psi(\boldsymbol{x}) = \sqrt{V_-(\boldsymbol{x})}\,\phi(\boldsymbol{x})$, which implies that

$\psi = ((-\Delta)^s + e)^{-1} \sqrt{V_-(\boldsymbol{x})}\, \phi(\boldsymbol{x})$ or, equivalently,

$$\phi = K_e \phi \tag{4.3.1}$$

where K_e (called the **Birman–Schwinger kernel** [18, 157]) is the integral kernel given by

$$K_e(\boldsymbol{x}, \boldsymbol{y}) = \sqrt{V_-(\boldsymbol{x})}\, G_e(\boldsymbol{x} - \boldsymbol{y}) \sqrt{V_-(\boldsymbol{y})}, \tag{4.3.2}$$

where $G_e(\boldsymbol{x} - \boldsymbol{y})$ is the integral kernel (or **Green's function**) for the inverse of the positive operator $(-\Delta)^s + e$. Explicitly, $(K_e \phi)(\boldsymbol{x}) = \int_{\mathbb{R}^d} K_e(\boldsymbol{x}, \boldsymbol{y}) \phi(\boldsymbol{y}) \mathrm{d}\boldsymbol{y}$ and similarly for G_e.

The kernel $G_e(\boldsymbol{x} - \boldsymbol{y})$ is well known and is given by the inverse Fourier transform

$$G_e(\boldsymbol{x} - \boldsymbol{y}) = \int_{\mathbb{R}^d} \frac{1}{|2\pi \boldsymbol{k}|^{2s} + e} \exp\left(2\pi i \boldsymbol{k} \cdot (\boldsymbol{x} - \boldsymbol{y})\right) \mathrm{d}\boldsymbol{k}. \tag{4.3.3}$$

While we will not need an explicit formula for G_e let us record the function $G_e(\boldsymbol{x})$ for the usual Laplacian ($s = 1$) and dimensions 1 and 3:

$$G_e(\boldsymbol{x}) = \frac{1}{2\sqrt{e}} \exp\left(-\sqrt{e}\, |\boldsymbol{x}|\right), \quad d = 1,\ s = 1$$

$$G_e(\boldsymbol{x}) = \frac{1}{4\pi |\boldsymbol{x}|} \exp\left(-\sqrt{e}\, |\boldsymbol{x}|\right), \quad d = 3,\ s = 1. \tag{4.3.4}$$

Equation (4.3.1) says that when $-e$ is a negative eigenvalue of H then 1 is an eigenvalue of K_e. Although it is not *a priori* clear that $\phi = \sqrt{V_-}\psi$ is in $L^2(\mathbb{R}^d)$ we claim that it is; moreover, we claim that there is a one-to-one correspondence between an eigenvalue $-e$ of H and an eigenvalue 1 of K_e. This is important for us because it enables us to reformulate the problem in terms of traces of powers of K_e on $L^2(\mathbb{R}^d)$.

From now on we will assume that $s = 1$, and comment on the generalization to $s < 1$ at the end of the proof. If $\psi \in H^1(\mathbb{R}^d)$ then $\phi = \sqrt{V_-}\psi$ is in $L^2(\mathbb{R}^d)$ by our assumptions on V, as explained in Section 2.2.1. If ψ satisfies $-\Delta\psi - V_-\psi = -e\psi$ then clearly $\phi \neq 0$ since otherwise $-\Delta\psi = -e\psi$ which is impossible for an H^1 function. Moreover, K_e maps $L^2(\mathbb{R}^d)$ into $L^2(\mathbb{R}^d)$ by the Hardy–Littlewood–Sobolev inequality [118, Thm. 4.3]. So ϕ is an $L^2(\mathbb{R}^d)$ eigenfunction of an $L^2(\mathbb{R}^d) \to L^2(\mathbb{R}^d)$ operator. Conversely, if $\phi \in L^2(\mathbb{R}^d)$ satisfies $K_e\phi = \phi$ then, if we define $\psi = (-\Delta + e)^{-1}\sqrt{V_-}\phi$, we have that $(-\Delta + e)\psi = \sqrt{V_-}\phi = \sqrt{V_-}K_e\phi = V_-\psi$. The fact that $\psi \in L^2(\mathbb{R}^d)$ follows from $(\psi, \psi) = (\sqrt{V_-}\phi, (-\Delta + e)^{-2}\sqrt{V_-}\phi) < e^{-1}(\sqrt{V_-}\phi, (-\Delta + e)^{-1}\sqrt{V_-}\phi) = e^{-1}(\phi, \phi)$. Moreover, $\psi \neq 0$ since otherwise $\sqrt{V_-}\phi$ would be zero and hence $K_e\phi = 0$. This one-to-one correspondence between ψ and ϕ implies that the multiplicities of the eigenvalue $-e$ of $-\Delta - V_-$ and the eigenvalue 1 of K_e are the same.

There is another important observation about (4.3.1) and (4.3.2). For any given $e > 0$ the operator K_e has a spectrum of eigenvalues, which are all non-negative. (K_e is compact and, moreover, Tr $(K_e)^m < \infty$, for suitable $m > 0$, as we shall see.) The observation is that K_e, as an operator, is monotone decreasing in e. That is, if $e < e'$ then $K_e - K_{e'} > 0$, as an operator inequality. This monotonicity is true for any operator of the form $(A + e)^{-1}$ with $A \geq 0$ since the difference is just $(e' - e)(A + e)^{-1}(A + e')^{-1}$, which is clearly positive. Hence all the eigenvalues of K_e are monotone decreasing in e. They are also easily seen to be continuous.[6]

These facts stated above are shown schematically in Figure 4.1 in which the eigenvalues λ_j of K_e are plotted as functions of e. This figure immediately leads to the following conclusion: For each number $e > 0$ we can define

$$N_e := \text{number of eigenvalues of } H \text{ less than or equal to } -e,$$

including multiplicity, as usual. We can also define B_e to be the number of eigenvalues of K_e that are ≥ 1. Then

$$\boxed{N_e = B_e.} \tag{4.3.5}$$

4.3.2 Derivation of the LT Inequalities

Let $\gamma > 0$. To exploit (4.3.5) we write

$$\sum_{j\geq 0} |E_j|^\gamma = \sum_{j\geq 0} e_j^\gamma = \gamma \int_0^\infty e^{\gamma-1} N_e \, \mathrm{d}e \tag{4.3.6}$$

which is easily verified by integration by parts while noting that the derivative of N_e is just a sum of unit delta-functions at the numbers e_j.

[6] The simplest way to see continuity is to note that $0 \leq K_e - K_{e'} \leq [(e' - e)/e']K_e$ for $0 < e \leq e'$. By the min-max principle [118, Thm. 12.1], the eigenvalues of K_e differ from the corresponding eigenvalues of $K_{e'}$ by at most $(e' - e)/e'$ times the norm (i.e., largest eigenvalue) of K_e.

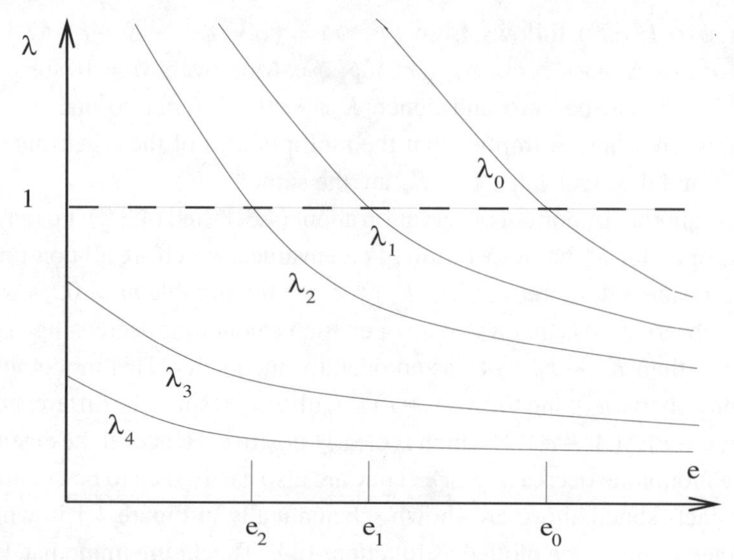

Figure 4.1: The eigenvalues $\lambda_0 \geq \lambda_1 \geq \cdots$ of K_e are schematically shown as a function of e. Only the first five are shown here. The eigenvalues of the Schrödinger equation are $E_i = -e_i$ and the e_i are the values of e for which $\lambda_i = 1$.

On the other hand, B_e is certainly $\leq \sum_{\lambda_j \geq 1} \lambda_j^m \leq \mathrm{Tr}\,(K_e)^m$ for any number $m > 0$. By Theorem 4.5 in the appendix to this chapter and (4.3.3),

$$\begin{aligned} N_e = B_e &\leq \mathrm{Tr}\,(K_e)^m \\ &\leq \mathrm{Tr}\,(V_-)^{m/2}(G_e)^m(V_-)^{m/2} \\ &= \int_{\mathbb{R}^d} V_-(\boldsymbol{x})^m G_e^m(\boldsymbol{0})\,\mathrm{d}\boldsymbol{x} \\ &= \left(\int_{\mathbb{R}^d} \frac{1}{((2\pi \boldsymbol{k})^2 + e)^m}\,\mathrm{d}\boldsymbol{k} \right) \int_{\mathbb{R}^d} V_-(\boldsymbol{x})^m\,\mathrm{d}\boldsymbol{x} \end{aligned} \tag{4.3.7}$$

as long as $m \geq 1$. The expression $(\int \ldots)$ on the right side is just $G_e^m(\boldsymbol{0})$, as we see from the analogue of (4.3.3) for the kernel $G_e^m(\boldsymbol{x} - \boldsymbol{y})$. This expression is finite if and only if $2m > d$, in which case it is, by scaling, $C_{d,m} e^{-m+d/2}$ with

$$C_{d,m} = \int_{\mathbb{R}^d} \frac{1}{((2\pi \boldsymbol{k})^2 + 1)^m}\,\mathrm{d}\boldsymbol{k} = (4\pi)^{-d/2} \frac{\Gamma(m - d/2)}{\Gamma(m)}. \tag{4.3.8}$$

Unfortunately, when $e^{-m+d/2}$ is inserted in (4.3.6) we obtain a divergent integral either at $e = 0$ or at $e = \infty$, for any m. To remedy the situation let us consider the potential $W_e(\boldsymbol{x}) = [V(\boldsymbol{x}) + e/2]_- = \max\{-V(\boldsymbol{x}) - e/2,\, 0\}$ (which depends on the value of e under consideration in (4.3.6)). Clearly, W_e is in $L^{\gamma+d/2}(\mathbb{R}^d)$ if V_- is. Moreover, with $N_e(V)$ denoting the number of eigenvalues of $-\Delta + V$ which are $\leq -e$, it is easy to see that

$$N_e(-V_-) = N_{e/2}(-V_- + e/2) \leq N_{e/2}(-W_e) \tag{4.3.9}$$

because $W_e \geq V_- - e/2$. Let us, therefore, replace e by $e/2$ and V_- by W_e in the right side of (4.3.7) and then insert the result in (4.3.6). We obtain

$$\begin{aligned}\sum_{j\geq 0} |E_j|^\gamma &\leq \gamma\, C_{d,m} \int\limits_{\mathbb{R}^d} \left(\int\limits_0^{2V_-(\boldsymbol{x})} e^{\gamma-1-m+d/2}\, 2^{m-d/2}\, (V_-(\boldsymbol{x}) - e/2)^m \mathrm{d}e \right) \mathrm{d}\boldsymbol{x} \\ &= (4\pi)^{-d/2} 2^\gamma\, \gamma\, m \frac{\Gamma(m - d/2)\Gamma(-m + \gamma + d/2)}{\Gamma(\gamma + 1 + d/2)} \int\limits_{\mathbb{R}^d} V_-(\boldsymbol{x})^{\gamma+d/2}\, \mathrm{d}\boldsymbol{x}.\end{aligned} \tag{4.3.10}$$

In order for (4.3.10) to be finite we have to require m to satisfy $d/2 < m < \gamma + d/2$, so we choose $m = (\gamma + d)/2$. For $d = 1$ this choice is only greater or equal to 1 if $\gamma \geq 1$. Hence, if $d = 1$ and $1/2 < \gamma < 1$, we choose $m = 1$. This leads to the value given in Theorem 4.1 and concludes the proof of that theorem (except for the case $d = 1,\ \gamma = 1/2$ and $d \geq 3,\ \gamma = 0$).

The proof for $s = 1/2$, or any other s for that matter, proceeds in exactly the same way and we leave the details for the reader. To obtain the constants in (4.1.22), one chooses $m = d + \gamma/2$ in the relativistic case. This proves Theorem 4.2 (except for the case $d \geq 2,\ \gamma = 0$).

Remark 4.5. For $\gamma = 1$ and $d = 3$, our choice of m is $m = (\gamma + d)/2 = 2$ in the non-relativistic case. In this special case it is very easy to prove inequality (4.3.7) without having to employ the general Theorem 4.5. In fact, using a simple Schwarz inequality and Plancherel's identity (2.1.40) we have

$$\begin{aligned}\mathrm{Tr}\,(K_e)^2 &= \iint\limits_{\mathbb{R}^3\times\mathbb{R}^3} V_-(\boldsymbol{x})V_0(\boldsymbol{y})G_e(\boldsymbol{x} - \boldsymbol{y})^2 \mathrm{d}\boldsymbol{x}\mathrm{d}\boldsymbol{y} \leq \int\limits_{\mathbb{R}^3} V_-(\boldsymbol{x})^2 \mathrm{d}\boldsymbol{x} \int\limits_{\mathbb{R}^3} G_e(\boldsymbol{y})^2 \mathrm{d}\boldsymbol{y} \\ &= \int\limits_{\mathbb{R}^3} V_-(\boldsymbol{x})^2 \mathrm{d}\boldsymbol{x} \int\limits_{\mathbb{R}^3} \frac{1}{\left(|2\pi \boldsymbol{k}|^2 + e\right)^2} \mathrm{d}\boldsymbol{k}.\end{aligned}$$

This case is of particular relevance in the proof of stability of non-relativistic matter discussed in Chapter 7.

4.3.3 Useful Corollaries

The special case $\gamma = 0$ in Theorem 4.1, the CLR bound, is not covered in the foregoing proof. It is a bound on N_0, the total number of non-positive eigenvalues of $-\Delta + V(\boldsymbol{x})$, and is valid only for $d \geq 3$.

There is, however, a bound on N_e, the number of eigenvalues below $-e$, which is valid in all dimensions when $e > 0$, and which will be useful later in this book. Its proof is actually contained in the proof of Theorem 4.1 in the previous subsection, and so we take linguistic liberties and call it a corollary.

Corollary 4.2 (Bound on N_e). *Let $d \geq 1$, $\gamma > 0$ and $e > 0$. Then N_e, the number of eigenvalues of $-\Delta + V(\boldsymbol{x})$ less than or equal to $-e$, is bounded by*[7]

$$N_e \leq \frac{2^\gamma}{e^\gamma} \frac{1}{(4\pi)^{d/2}} \frac{\Gamma(\gamma)}{\Gamma(d/2+\gamma)} \int_{\mathbb{R}^d} [V(\boldsymbol{x}) + e/2]_-^{d/2+\gamma} \, \mathrm{d}\boldsymbol{x}. \qquad (4.3.11)$$

Similarly, in the relativistic case, the number of eigenvalues of $\sqrt{-\Delta} + V(\boldsymbol{x})$ below $-e$ is bounded by

$$N_e \leq \frac{2^\gamma}{e^\gamma} \frac{1}{\pi^{(d+1)/2}} \frac{\Gamma(\frac{d+1}{2})\Gamma(\gamma)}{\Gamma(d+\gamma)} \int_{\mathbb{R}^d} [V(\boldsymbol{x}) + e/2]_-^{d+\gamma} \, \mathrm{d}\boldsymbol{x}. \qquad (4.3.12)$$

Corollary 4.2 is proved by combining (4.3.5), (4.3.7) and (4.3.9).

One more inequality that is easily derived from the proof of Theorem 4.1 (which even extends down to $e = 0$) is due to Birman [18] and Schwinger [157]. It is valid for $d = 3$ only.

Corollary 4.3 (Birman–Schwinger Bound). *For $e \geq 0$ and $d = 3$*

$$N_e \leq \frac{1}{(4\pi)^2} \iint_{\mathbb{R}^3 \times \mathbb{R}^3} \frac{[V(\boldsymbol{x}) + e]_- [V(\boldsymbol{y}) + e]_-}{|\boldsymbol{x} - \boldsymbol{y}|^2} \mathrm{d}\boldsymbol{x}\mathrm{d}\boldsymbol{y} \qquad (4.3.13)$$

$$\leq \frac{1}{4(4\pi)^{2/3}} \left(\int_{\mathbb{R}^3} [V(\boldsymbol{x}) + e]_-^{3/2} \mathrm{d}\boldsymbol{x} \right)^{4/3}. \qquad (4.3.14)$$

[7] We emphasize that e in (4.3.11) is the energy and *not* 2.718.

Proof. We first consider the case $e > 0$. For $0 < \varepsilon < e$, we use

$$N_e(V) \leq N_\varepsilon(-[V(\boldsymbol{x}) + e - \varepsilon]_-).$$

According to Eq. (4.3.5), this latter quantity equals the number of eigenvalues of

$$\sqrt{[V + e - \varepsilon]_-}\frac{1}{-\Delta + \varepsilon}\sqrt{[V + e - \varepsilon]_-}$$

that are greater or equal to 1. For $d = 3$, the integral kernel of this operator is given by (compare with (4.3.4))

$$\sqrt{[V(\boldsymbol{x}) + e - \varepsilon]_-}\,\sqrt{[V(\boldsymbol{y}) + e - \varepsilon]_-}\,\frac{\exp\left(-\sqrt{\varepsilon}|\boldsymbol{x} - \boldsymbol{y}|\right)}{4\pi\,|\boldsymbol{x} - \boldsymbol{y}|}.$$

The number of eigenvalues greater or equal to 1 is certainly less than the sum of the square of all eigenvalues. The latter is simply the square of the Hilbert–Schmidt norm, which equals

$$\frac{1}{(4\pi)^2}\iint_{\mathbb{R}^3\times\mathbb{R}^3}\frac{[V(\boldsymbol{x}) + e - \varepsilon]_-[V(\boldsymbol{y}) + e - \varepsilon]_-}{|\boldsymbol{x} - \boldsymbol{y}|^2}\exp\left(-2\sqrt{\varepsilon}|\boldsymbol{x} - \boldsymbol{y}|\right)\mathrm{d}\boldsymbol{x}\mathrm{d}\boldsymbol{y}.$$

The exponential factor can simply be bounded by one. The remaining integral is monotone in ε and converges to the right side of (4.3.13) as $\varepsilon \to 0$.

The reason we cannot apply this argument directly for $\varepsilon = 0$ is the fact that (4.3.5) does *not* hold, in general, for $e = 0$. To circumvent this problem and prove (4.3.13) for $e = 0$, consider a perturbation of V of the form $V(\boldsymbol{x}) \to V(\boldsymbol{x}) - \lambda W(\boldsymbol{x})$, where $W(\boldsymbol{x})$ is a smooth and rapidly decaying function that is strictly positive. By the variational principle, the number of strictly negative eigenvalues of the operator $-\Delta + V - \lambda W$ is greater or equal to $N_0(V)$. Hence

$$N_0(V) \leq \lim_{\lambda\to 0}\lim_{e\to 0} N_e(V - \lambda W).$$

Applying the bound for $N_e(V - \lambda W)$ above and using monotone convergence, we arrive at the desired result.

Finally, the inequality (4.3.14) follows from (4.3.13) by an application of the Hardy–Littlewood–Sobolev inequality, see [118, Thm. 4.3]. ■

Compared with the CLR bound (Eq. (4.1.11) for $\gamma = 0$) the inequality in (4.3.14) contains the 4/3 power of the dimensionless quantity $\int V_-^{3/2}$ instead of the first power. This might look like an innocent detail, but it can make a huge difference. In fact, if V is the sum of N widely separated pieces, the number of bound states will roughly be the sum of the number of bound states for each piece, and so will be proportional to N – not $N^{4/3}$! *The important point about the LT inequalities is that there is no exponent outside the integral*, for any $\gamma \geq 0$, and this confirms the intuition obtained from the semiclassical expressions. This fact is crucial for the understanding of stability of matter, as will be discussed in the chapters to follow.

4.4 Diamagnetic Inequalities

The proof of the Lieb–Thirring inequalities presented in the previous section carries over to the magnetic case essentially without change. In particular, Theorems 4.1 and 4.2 also hold, with the same constants, when $-\Delta$ is replaced by $(-i\nabla + \boldsymbol{A}(\boldsymbol{x}))^2$ for some vector potential $\boldsymbol{A}(\boldsymbol{x})$. In fact, all the best presently known constants in these inequalities are independent of $\boldsymbol{A}(\boldsymbol{x})$, since all the proofs use the **diamagnetic inequality** in one way or the other. We shall explain this inequality in the following.

Note, however, that the situation changes dramatically when spin is introduced (cf. Section 2.1.6 and Chapter 9), since there can be cancellations between the $(-i\nabla + \boldsymbol{A}(\boldsymbol{x}))^2$ term and the extra term $\boldsymbol{\sigma} \cdot \boldsymbol{B}$. LT-type inequalities in this case were given in [123, 47, 133, 25, 26, 169, 49, 50], but we shall not discuss them here.

One formulation of the diamagnetic inequality is the fact that the integral kernel of the resolvent of $(-i\nabla + \boldsymbol{A}(\boldsymbol{x}))^2$, i.e.,

$$\frac{1}{(-i\nabla + \boldsymbol{A}(\boldsymbol{x}))^2 + e}(\boldsymbol{x}, \boldsymbol{y})$$

for $e > 0$ is point-wise bounded above by the corresponding quantity for $\boldsymbol{A} = 0$, that is

$$\left| \frac{1}{(-i\nabla + \boldsymbol{A}(\boldsymbol{x}))^2 + e}(\boldsymbol{x}, \boldsymbol{y}) \right| \leq \frac{1}{-\Delta + e}(\boldsymbol{x}, \boldsymbol{y}) = G_e(\boldsymbol{x} - \boldsymbol{y}) \quad \text{for all } \boldsymbol{x}, \boldsymbol{y} \in \mathbb{R}^d. \tag{4.4.1}$$

Obviously, (4.4.1) is enough to establish the LT inequalities. Its proof is due to Simon [165]. We first consider the following Lemma.

Lemma 4.1 (Kato's Inequality). *Let* $\mathrm{sgn}(u) = \bar{u}/|u|$ *if* $u(\boldsymbol{x}) \neq 0$, *and* 0 *otherwise. Assume*[8] *that* $\boldsymbol{A} \in C^1(\mathbb{R}^d)$, $u \in L^1_{\mathrm{loc}}(\mathbb{R}^d)$ *such that* $(-i\nabla + \boldsymbol{A})^2 u \in L^2_{\mathrm{loc}}(\mathbb{R}^d)$. *Then*[9]

$$-\Delta|u| \leq \Re\, \mathrm{sgn}(u)(-i\nabla + \boldsymbol{A}(\boldsymbol{x}))^2 u. \tag{4.4.2}$$

For simplicity, we shall only prove this Lemma if $u \in C^2$ and $|u| \in C^2$. We refer to [150, Chapter X.4] for the extension.

Proof. Since $2|u|\nabla|u| = \nabla|u|^2 = 2\Re\, \bar{u}\nabla u$, we first note that

$$|u|\nabla|u| = \Re\, \bar{u}\nabla u = \Re\, \bar{u}(\nabla + i\boldsymbol{A})u.$$

Taking the divergence of this equation, we obtain

$$|\nabla|u||^2 + |u|\Delta|u| = |(\nabla + i\boldsymbol{A})u|^2 + \Re\, \bar{u}(\nabla + i\boldsymbol{A})^2 u.$$

By writing u as its absolute value times a phase, it is easy to see that

$$|(\nabla + i\boldsymbol{A})u|^2 \geq |\nabla|u||^2 \tag{4.4.3}$$

This implies the statement of the lemma. ∎

An immediate corollary of the inequality (4.4.3) is that

$$(\psi, (-i\nabla + \boldsymbol{A}(\boldsymbol{x}))^2\psi) \geq (|\psi|, -\Delta\,|\psi|) \tag{4.4.4}$$

for any $\psi \in H^1(\mathbb{R}^d)$. The analogue in the relativistic case is

$$(\psi, |-i\nabla + \boldsymbol{A}(\boldsymbol{x})|\psi) \geq (|\psi|, \sqrt{-\Delta}\,|\psi|),$$

which can be obtained from Eq. (4.4.8) in the following theorem by a suitable $t \to 0$ limit. These inequalities show that the ground state energy of single particle in a potential V is lowest when the magnetic field is absent.

With the aid of Kato's inequality, we can now prove

Theorem 4.4 (Diamagnetic Inequality). *Let* $\boldsymbol{A} \in L^2_{\mathrm{loc}}(\mathbb{R}^d;\mathbb{R}^d)$. *Then, for* $e > 0$ *and for almost every* $\boldsymbol{x}, \boldsymbol{y} \in \mathbb{R}^d$,

$$\left|\frac{1}{(-i\nabla + \boldsymbol{A}(\boldsymbol{x}))^2 + e}(\boldsymbol{x}, \boldsymbol{y})\right| \leq \frac{1}{-\Delta + e}(\boldsymbol{x}, \boldsymbol{y}). \tag{4.4.5}$$

[8] C^k consists of functions with continuous derivatives up to order k.

[9] $\Re\, z$ denotes the real part of a complex number z.

Moreover, in the relativistic case,

$$\left| \frac{1}{|-i\nabla + A(x)| + e}(x, y) \right| \le \frac{1}{\sqrt{-\Delta} + e}(x, y). \tag{4.4.6}$$

Similarly, for the heat kernel at $t > 0$,

$$\left| e^{-t(-i\nabla + A(x))^2}(x, y) \right| \le e^{t\Delta}(x, y) \tag{4.4.7}$$

and

$$\left| e^{-t|-i\nabla + A(x)|}(x, y) \right| \le e^{-t\sqrt{-\Delta}}(x, y). \tag{4.4.8}$$

In the case without magnetic fields, the heat kernel can be computed explicitly [118, Sects. 7.9 & 7.10]. In the non-relativistic case, it is given by

$$e^{t\Delta}(x, y) = \frac{1}{(4\pi t)^{d/2}} \exp\left(-|x - y|^2/(4t)\right),$$

whereas in the relativistic case it is

$$e^{-t\sqrt{-\Delta}}(x, y) = \Gamma\left(\tfrac{d+1}{4}\right) \pi^{-(d+1)/2} \frac{t}{\left(t^2 + |x - y|^2\right)^{(d+1)/2}}.$$

Proof. By an approximation argument [167, Theorem 15.4], it is enough to consider the case $A \in C^1$. Thus, we can apply Kato's inequality in Lemma 4.1.

The Green's function

$$G_e^A(x, y) = \frac{1}{(-i\nabla + A(x))^2 + e}(x, y)$$

satisfies the differential equation

$$(-i\nabla_x + A(x))^2 G_e^A(x, y) + e G_e^A(x, y) = \delta(x - y)$$

in the sense of distributions. If we multiply this equation by $\operatorname{sgn}(G_e^A)$ and use Kato's inequality (4.4.2), we obtain

$$(-\Delta_x + e)\, |G_e^A(x, y)| \le \delta(x - y) \Re \operatorname{sgn}(G_e^A(x, y)) \le \delta(x - y). \tag{4.4.9}$$

In particular, multiplying Eq. (4.4.9) by $G_e(y', x)$ and integrating over x, this yields the desired inequality

$$|G_e^A(y', y)| \le G_e(y', y).$$

In order to prove the corresponding result in the relativistic case, we first note that the diamagnetic inequality for the resolvent implies a corresponding result for the heat kernel. More precisely, since

$$e^{-B} = \lim_{n\to\infty} \left(\frac{n}{n+B}\right)^n$$

the inequality (4.4.5) implies (4.4.7) for any $t > 0$. This diamagnetic inequality for the heat kernel can then be extended to the relativistic case, using the fact that

$$e^{-t|p|} = \frac{1}{\sqrt{\pi}} \int_0^\infty \frac{\mathrm{d}\lambda}{\sqrt{\lambda}} e^{-\lambda - t^2|p|^2/(4\lambda)}. \tag{4.4.10}$$

This formula implies (4.4.8). From this bound one obtains the result (4.4.6) for the resolvent by integrating over t, using that

$$\frac{1}{Y} = \int_0^\infty \mathrm{d}t\, e^{-tY}$$

for $Y > 0$. We omit the technical details. ■

A natural way to view these domination questions is by means of functional integrals, which can be defined for $(-\Delta)^s$ for $0 < s \leq 1$. A very good discussion of this is in Simon's book [167].

4.5 Appendix: An Operator Trace Inequality

The following inequality about traces of operators is needed in order to bound the traces of the Birman–Schwinger kernel that appear in the proof of the LT inequalities. This inequality was originally proved in [135, Appendix B] with the aid of a sophisticated theorem of Epstein [46]. Earlier, Seiler and Simon [160] had proved the theorem using interpolation techniques when $\mathcal{H} = L^2(\mathbb{R}^d)$, A is a convolution operator and B is a multiplication operator, which is the case we actually need for application to the Birman–Schwinger kernel.

Subsequently, Araki [4] found a proof that was a bit longer but which used only well known facts and which, therefore, can easily be given in a self-contained fashion. We prove the theorem here using Araki's method and refer the reader to [4] for a generalization of the theorem.

Theorem 4.5 (Traces of Powers). *Let A and B be positive, self-adjoint operators on a separable Hilbert space $\mathcal{H}$. Then, for each real number $m \geq 1$,*

$$\boxed{\operatorname{Tr}\,(B^{1/2}AB^{1/2})^m \leq \operatorname{Tr}\, B^{m/2}A^m B^{m/2}.} \tag{4.5.1}$$

Proof. Step 1: We will need the following little proposition. Suppose that $\mathbf{a} = \{a_1 \geq a_2 \geq a_3 \geq \ldots\}$ and $\mathbf{b} = \{b_1 \geq b_2 \geq b_3 \geq \ldots\}$ are two infinite ordered sequences of real numbers satisfying the condition $\mathbf{a} \prec \mathbf{b}$, meaning that

$$\sum_{j=1}^{n} a_j \leq \sum_{j=1}^{n} b_j \qquad \text{for} \quad n = 1,\ 2,\ 3,\ \ldots. \tag{4.5.2}$$

Then

$$\sum_{j=1}^{n} e^{a_j} \leq \sum_{j=1}^{n} e^{b_j} \qquad \text{for} \quad n = 1,\ 2,\ 3,\ \ldots. \tag{4.5.3}$$

Remark 4.6. Inequality (4.5.3) can be rewritten in the following way: If $\boldsymbol{\lambda}$ and $\boldsymbol{\sigma}$ are two (decreasingly) ordered sequences of non-negative numbers then

$$\prod_{j=1}^{n} \sigma_j \leq \prod_{j=1}^{n} \lambda_j \text{ for all } n = 1,\ 2, 3, \ldots$$

$$\Longrightarrow \sum_{j=1}^{n} \sigma_j \leq \sum_{j=1}^{n} \lambda_j \text{ for all } n = 1,\ 2,\ 3, \ldots. \tag{4.5.4}$$

Remark 4.7. This proposition holds with e^x replaced by any monotone-increasing, convex function. It was first proved by Hardy, Littlewood and Polya in 1929 [85], but is often known as a theorem of Karamata, who proved it independently in 1932 [98]. A nice proof, using 'linearization' is in [11, Sec. 28–30]. Another useful reference is [90]. The proof given here also works for all monotone convex functions and is close to that in [98].

To prove (4.5.3) we write $e^x = \int_{\mathbb{R}} (x-y)_+ e^y \mathrm{d}y$, where $z_+ := \max\{z,\ 0\}$. We then see that (4.5.3) will follow for the function e^x if we can prove the analogue of (4.5.3) for the monotone-increasing, convex function $x \mapsto (x-y)_+$, for each fixed real number y. But this is easy to do. We note the obvious fact that either $\sum_{j=1}^{n}(a_j - y)_+ = \sum_{j=1}^{k}(a_j - y)$ for some $k \leq n$ or else $\sum_{j=1}^{n}(a_j - y)_+ = (a_1 - y)_+ = 0$. The case $\sum_{j=1}^{n}(a_j - y)_+ = 0$ is trivial and otherwise

$$\sum_{j=1}^{n}(a_j - y)_+ = \sum_{j=1}^{k}(a_j - y) \leq \sum_{j=1}^{k}(b_j - y) \leq \sum_{j=1}^{k}(b_j - y)_+ \leq \sum_{j=1}^{n}(b_j - y)_+, \tag{4.5.5}$$

which proves (4.5.3).

Step 2: Define the positive operators

$$X = B^{m/2} A^m B^{m/2} \quad \text{and} \quad Y = (B^{1/2} A B^{1/2})^m. \tag{4.5.6}$$

We let $\boldsymbol{\lambda}$ denote the eigenvalues of X in decreasing order (followed by an infinite string of zeros in case our matrices are finite dimensional). Similarly $\boldsymbol{\sigma}$ are the eigenvalues of Y. Clearly, inequality (4.5.4) will prove our theorem if we can show that $\prod_{j=1}^{n} \lambda_j \geq \prod_{j=1}^{n} \sigma_j$ for every $n \geq 1$. To do this, consider the n-fold *antisymmetric* tensor product $\bigwedge^n \mathcal{H}$, and the operators $\Lambda = X \otimes X \otimes \cdots \otimes X$ and $\Sigma = Y \otimes Y \otimes \cdots \otimes Y$ acting on this space. The *largest* eigenvalue of Λ is exactly what we are looking for, namely $\prod_{j=1}^{n} \lambda_j$, and that of Σ is $\prod_{j=1}^{n} \sigma_j$. To conclude our proof, therefore, we have to show that the largest eigenvalue of Λ is not less than that of Σ.

Fortunately, Λ and Σ can be written in another way. Define $\alpha = A \otimes A \otimes \cdots \otimes A$ and $\beta = B \otimes B \otimes \cdots \otimes B$. Then, since ordinary operator products and tensor products commute, we have that

$$\Lambda = \beta^{m/2} \alpha^m \beta^{m/2} \quad \text{and} \quad \Sigma = (\beta^{1/2} \alpha \beta^{1/2})^m, \tag{4.5.7}$$

which is the form of the operators in our proposed inequality about the largest eigenvalues for the particular case $n = 1$, except that the operators are different. That is to say, A is replaced by α, B is replaced by β and the Hilbert space is replaced by $\bigwedge^n \mathcal{H}$.

With Λ and Σ as given in (4.5.7), for unknown, but positive operators α and β, we only have to prove that the largest eigenvalues e_Λ and e_Σ satisfy $e_\Lambda \geq e_\Sigma$. Let us assume (after scaling) that $e_\Lambda = 1$. This means that, as an operator, $\Lambda \leq \mathbb{I}$. Our goal will be achieved if we can then infer that $\Sigma \leq \mathbb{I}$.

Now $\Lambda \leq \mathbb{I}$ implies that $\alpha^m \leq \beta^{-m}$. Taking the mth root (for $m \geq 1$) is *operator monotone*,[10] i.e., $\alpha^m \leq \beta^{-m}$ implies that $\alpha \leq \beta^{-1}$. We then have that

[10] Note that although taking a root is operator monotone, taking powers ≥ 1 is not, in general. The fact that $x \mapsto x^p$ is operator monotone for $0 < p < 1$ follows easily from the integral representation $x^p = \pi^{-1} \sin(p\pi) \int_0^\infty t^p (t^{-1} - (t + x)^{-1}) \mathrm{d}t$, together with the observation that

$\beta^{1/2}\alpha\beta^{1/2} \leq \beta^{1/2}\beta^{-1}\beta^{1/2} = \mathbb{I}$. Hence, $\Sigma = (\beta^{1/2}\alpha\beta^{1/2})^m \leq \mathbb{I}$, and our proof is complete. ■

Remark 4.8. We have actually proved more than (4.5.1), which refers to the sum of *all* the eigenvalues. We have proved that the inequality holds for the largest eigenvalue and for the sum of the n largest eigenvalues of each operator.

$x \mapsto -1/x$ is operator monotone. We refer to [17, 42] for more information on operator monotone functions.

CHAPTER 5

Electrostatic Inequalities

5.1 General Properties of the Coulomb Potential

The interaction energy of charged particles is described by the Coulomb potential, and the proof of stability of matter depends to a large extent on having good estimates for the size of this interaction for systems with many particles.

From this point on, until the end of the book, we will restrict our attention to the physical case of $\mathbb{R}^3$. Many of the results have a natural extension to different dimensions, but we shall not explore them here.

Charge distributions may be continuous or discrete and hence it is convenient to consider them as Borel measures.[1] The **potential function** Φ associated with a Borel measure μ is defined by

$$\Phi(\boldsymbol{x}) = \int_{\mathbb{R}^3} \frac{1}{|\boldsymbol{x} - \boldsymbol{y}|} \mu(\mathrm{d}\boldsymbol{y}). \tag{5.1.1}$$

For a positive measure μ this expression is always well defined in the sense that it might be $+\infty$. If μ is a signed Borel measure, i.e., $\mu = \sigma - \tau$ with σ, τ positive Borel measures, the potential may not be well defined because of cancellation problems. If, however, we assume that the positive measure $\nu = \sigma + \tau$ is finite in the sense that

$$\int_{\mathbb{R}^3} \frac{1}{1 + |\boldsymbol{x}|} \nu(\mathrm{d}\boldsymbol{x}) < \infty, \tag{5.1.2}$$

it is not hard to see that under such an assumption the potential $\Phi(\boldsymbol{x})$ is finite almost everywhere, in fact it is in $L^1_{\mathrm{loc}}(\mathbb{R}^3)$ (see [118, Chap. 9]). Note that any signed measure can be written as the difference of two positive measures

[1] For readers unfamiliar with measures, just think of $\mu(\mathrm{d}\boldsymbol{x})$ as $\varrho(\boldsymbol{x})\mathrm{d}\boldsymbol{x}$ where ϱ is some function in $L^1_{\mathrm{loc}}(\mathbb{R}^3)$. See [118, Chapter 1].

with disjoint support. We shall not use this fact, which is known as the 'Hahn decomposition', in any way.

The **Coulomb energy** $D(\mu, \mu)$ of a charge distribution μ is defined by

$$D(\mu, \mu) = \frac{1}{2} \iint_{\mathbb{R}^3 \times \mathbb{R}^3} \frac{1}{|\boldsymbol{x} - \boldsymbol{y}|} \mu(\mathrm{d}\boldsymbol{x}) \mu(\mathrm{d}\boldsymbol{y}). \tag{5.1.3}$$

Again, this expression is well defined in the case where μ is positive. *In the following a* **charge distribution** *will be a Borel measure* $\mu = \sigma - \tau$ *where both,* σ *and* τ *are positive measures satisfying the condition (5.1.2) and, moreover, both have a finite Coulomb energy.* We do not require that σ and τ have disjoint support. The following theorem shows that the expression (5.1.3) is well defined for charge distributions.

The measures we are mostly concerned with are absolutely continuous, that is, they are of the form $\mu(\mathrm{d}\boldsymbol{x}) = \varrho(\boldsymbol{x})\mathrm{d}\boldsymbol{x}$ for an integrable function ϱ. In this case, we shall slightly abuse notation and use the symbol

$$D(\varrho, \varrho) = \frac{1}{2} \iint_{\mathbb{R}^3 \times \mathbb{R}^3} \frac{\varrho(\boldsymbol{x})\varrho(\boldsymbol{y})}{|\boldsymbol{x} - \boldsymbol{y}|} \mathrm{d}\boldsymbol{x}\, \mathrm{d}\boldsymbol{y} \tag{5.1.4}$$

for the Coulomb energy of the corresponding measure.

Quite generally we define the **interaction energy** of two charge distributions μ, ν by

$$D(\mu, \nu) = \frac{1}{2} \iint_{\mathbb{R}^3 \times \mathbb{R}^3} \frac{1}{|\boldsymbol{x} - \boldsymbol{y}|} \mu(\mathrm{d}\boldsymbol{x}) \nu(\mathrm{d}\boldsymbol{y}). \tag{5.1.5}$$

Theorem 5.1 (Positive Definiteness of the Coulomb Potential). *Assume that μ is a charge distribution as defined above. Then $0 \leq D(\mu, \mu) < \infty$. Moreover, for any two charge distributions μ and ν*

$$D(\mu, \nu)^2 \leq D(\mu, \mu) D(\nu, \nu). \tag{5.1.6}$$

Proof. A calculation (see, e.g. [118, 5.10(3)]) shows that

$$\frac{1}{|\boldsymbol{x} - \boldsymbol{y}|} = \frac{1}{\pi^3} \int_{\mathbb{R}^3} \frac{1}{|\boldsymbol{x} - \boldsymbol{z}|^2} \frac{1}{|\boldsymbol{y} - \boldsymbol{z}|^2} \mathrm{d}\boldsymbol{z}. \tag{5.1.7}$$

(The fact that the two sides of (5.1.7) are proportional to each other follows from the observation that they are both functions only of $|\boldsymbol{x} - \boldsymbol{y}|$ and both are homogeneous of degree -1; the only thing that has to be worked out is the constant π^{-3}.) Hence, for any charge distributions μ,

$$D(\mu, \mu) = \frac{1}{2\pi^3} \int_{\mathbb{R}^3} \left[\int_{\mathbb{R}^3} \frac{1}{|x - z|^2} \mu(\mathrm{d}x) \right]^2 \mathrm{d}z \geq 0. \tag{5.1.8}$$

It is easy to see that $D(\mu, \mu) < \infty$, since $D(\sigma, \sigma)$ and $D(\tau, \tau)$ are finite by assumption. This fact, as well as inequality (5.1.6), follows from Schwarz's inequality and the representation

$$D(\mu, \nu) = \frac{1}{2\pi^3} \int_{\mathbb{R}^3} \left[\int_{\mathbb{R}^3} \frac{1}{|x - z|^2} \mu(\mathrm{d}x) \right] \left[\int_{\mathbb{R}^3} \frac{1}{|x - z|^2} \nu(\mathrm{d}x) \right] \mathrm{d}z. \tag{5.1.9}$$

■

Another fact about the Coulomb potential, which will be of great importance, is **Newton's theorem**. It states that *outside* a charge distribution that is rotationally symmetric about a point x_0, the associated potential looks as if the charge is concentrated at the point x_0. Recall that a measure μ is rotationally symmetric with respect to a point x_0 if for every measurable set A and any rotation $\mathcal{R}$ which leaves x_0 fixed,

$$\mu(\mathcal{R}A) = \mu(A). \tag{5.1.10}$$

Theorem 5.2 (Newton's Theorem, [144, Thm. XXXI]). *Let μ be a charge distribution that is rotationally symmetric with respect to the origin. Then*

$$\Phi(x) = \frac{1}{|x|} \int_{|y| \leq |x|} \mu(\mathrm{d}y) + \int_{|y| > |x|} \frac{1}{|y|} \mu(\mathrm{d}y). \tag{5.1.11}$$

Consequently, if, for some $R > 0$, $\mu(\{y : |y| > R\}) = 0$ then

$$\Phi(x) = \frac{1}{|x|} \mu(\mathbb{R}^3) \qquad \text{for } |x| > R. \tag{5.1.12}$$

Likewise, if $\mu(\{y : |y| < R\}) = 0$, then

$$\Phi(x) = \text{constant} = \int_{|y| > R} \frac{1}{|y|} \mu(\mathrm{d}y) \qquad \text{for } |x| < R. \tag{5.1.13}$$

Proof. Since the measure μ is rotationally symmetric with respect to the origin, so is the potential Φ, i.e., $\Phi(\boldsymbol{x}) = \Phi(\boldsymbol{y})$ whenever $|\boldsymbol{x}| = |\boldsymbol{y}|$. Thus, $\Phi(\boldsymbol{x}) = \langle\Phi\rangle(\boldsymbol{x})$ where $\langle\Phi\rangle(\boldsymbol{x})$ denotes the average of $\Phi(|\boldsymbol{x}|\boldsymbol{\omega})$ over the unit vector $\boldsymbol{\omega}$. Using (5.1.1) this requires that we calculate the integral

$$\int_{\mathbb{S}^2} \frac{1}{||\boldsymbol{x}|\boldsymbol{\omega} - \boldsymbol{y}|} \mathrm{d}\omega \tag{5.1.14}$$

where $\mathrm{d}\omega$ is the uniform normalized surface measure on the unit 2-sphere. By rotation invariance we can let the vector $\boldsymbol{y}$ point towards the north pole. Resorting to polar coordinates, one is led to the calculable integral

$$\frac{1}{2}\int_{-1}^{1} \frac{1}{(|\boldsymbol{x}|^2 + |\boldsymbol{y}|^2 - 2|\boldsymbol{x}||\boldsymbol{y}|s)^{1/2}} \mathrm{d}s = \min\left\{\frac{1}{|\boldsymbol{x}|}, \frac{1}{|\boldsymbol{y}|}\right\}. \tag{5.1.15}$$

Formula (5.1.11) now follows. ■

5.2 Basic Electrostatic Inequality

When estimating the Coulomb potential of a collection of nuclei and electrons one faces two essential difficulties. The first is that the potential is singular at the location of the nuclei. The second, more serious one, is that the potential can also be large away from the nuclei because there might be many nuclei. The following electrostatic inequality, proved in [138, Lemma 1], disentangles these two issues. It suffices for several of the proofs of stability in this book, but a more sophisticated version turns out to be necessary for some stability results concerned with the quantized electromagnetic field. That version, which uses this basic version as input, will be given in Section 5.4.

To describe the inequality in detail we need the notion of a **Voronoi cell** with respect to a collection of nuclei of *equal* charge Z located at points $\boldsymbol{R}_j$. Let $\boldsymbol{R}_1, \ldots, \boldsymbol{R}_M$ be M distinct points in $\mathbb{R}^3$. Define the cell $\Gamma_j \subset \mathbb{R}^3$ by

$$\Gamma_j = \{\boldsymbol{x} \in \mathbb{R}^3 : |\boldsymbol{x} - \boldsymbol{R}_j| < |\boldsymbol{x} - \boldsymbol{R}_i|, \quad \text{for all } i \neq j\}. \tag{5.2.1}$$

Clearly Γ_j is open and it is easily seen to be convex. Its boundary, $\partial\Gamma_j$, is a finite collection of segments of planes and, possibly, the point at infinity.

Denote (half) the **nearest neighbor distance** of the nucleus located at $\boldsymbol{R}_j \in \Gamma_j$ by

$$D_j := \frac{1}{2} \min_{i \neq j} |\boldsymbol{R}_i - \boldsymbol{R}_j|. \tag{5.2.2}$$

This is the distance from $\boldsymbol{R}_j$ to $\partial\Gamma_j$. The Coulomb potential felt by an electron at the point $\boldsymbol{x}$, generated by M nuclei of equal charge Z, is the negative of

$$W(\boldsymbol{x}) = Z \sum_k \frac{1}{|\boldsymbol{x} - \boldsymbol{R}_k|},$$

which is a **harmonic**[2] function on the complement of all the $\boldsymbol{R}_j$. In the following, it will be important that all the nuclear charges be equal. Concerning the question of stability of matter, this is not an important restriction, as noted in Section 3.2.3.

Define

$$\mathfrak{D}(\boldsymbol{x}) = \min\{|\boldsymbol{x} - \boldsymbol{R}_i| : 1 \leq i \leq M\}, \tag{5.2.3}$$

and set

$$\Phi(\boldsymbol{x}) = W(\boldsymbol{x}) - \frac{Z}{\mathfrak{D}(\boldsymbol{x})}. \tag{5.2.4}$$

In other words, for $\boldsymbol{x} \in \Gamma_j$, the number $-\Phi(\boldsymbol{x})$ is the value of the potential at the point $\boldsymbol{x}$ due to all the nuclei *outside the Voronoi cell* Γ_j. It is important to note that Φ is a continuous function on $\mathbb{R}^3$ (here we use the fact that all the nuclei have the same charge Z). Although Φ is harmonic in each Voronoi cell, it is not harmonic on the whole of $\mathbb{R}^3$ since Φ is not differentiable on the boundary of the Voronoi cells. Its derivative makes a jump as we go from one cell to a neighboring cell.

An obvious connection between Φ and the Coulomb repulsion energy of nuclei of equal charge Z is

$$Z^2 \sum_{k<l} \frac{1}{|\boldsymbol{R}_k - \boldsymbol{R}_l|} = \frac{1}{2} Z \sum_j \Phi(\boldsymbol{R}_j). \tag{5.2.5}$$

Our main concern is to estimate the potential energy of an electron, taking into account all the other electrons and all the nuclei – except for the nucleus in the cell in which the electron finds itself. The Coulomb repulsion among the nuclei is included. The following theorem gives such an estimate. We will utilize it later by essentially identifying the arbitrary measure μ that appears in the theorem with the (smoothed) charge density of all the electrons.

[2] A harmonic function φ is one satisfying $\Delta\varphi = 0$. Equivalently, the average of φ over any sphere equals the value of φ at the center of the sphere. See [118, Chapter 9].

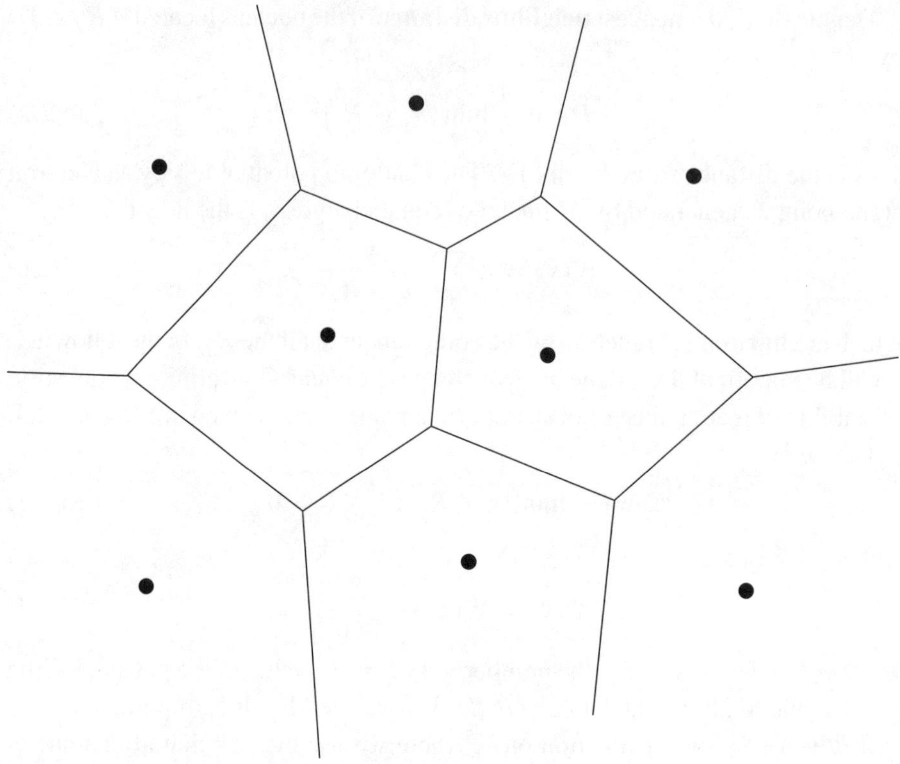

Figure 5.1: Example of Voronoi cells in two dimensions.

Theorem 5.3 (Basic Electrostatic Inequality). *For any charge distribution $\mu = \mu_+ - \mu_-$ with $D(\mu_+, \mu_+)$, $D(\mu_-, \mu_-) < \infty$, and for any set of distinct points $\boldsymbol{R}_1, \ldots, \boldsymbol{R}_M \in \mathbb{R}^3$,*

$$D(\mu, \mu) - \int \Phi(\boldsymbol{x})\mu(\mathrm{d}\boldsymbol{x}) + \sum_{k<l} \frac{Z^2}{|\boldsymbol{R}_k - \boldsymbol{R}_l|} \geq \frac{1}{8} \sum_j \frac{Z^2}{D_j}, \tag{5.2.6}$$

where Φ is defined in (5.2.4) and D_j in (5.2.2).

Proof. Our first goal is to find a charge distribution, ν, that generates the potential Φ. That is to say,

$$\Phi(\boldsymbol{x}) = \int_{\mathbb{R}^3} \frac{1}{|\boldsymbol{x} - \boldsymbol{y}|} \nu(\mathrm{d}\boldsymbol{y}) \tag{5.2.7}$$

or, equivalently,

$$-\Delta\Phi = 4\pi\nu. \tag{5.2.8}$$

To this end, pick an infinitely differentiable function f of compact support, and calculate

$$\begin{aligned}\int \Phi(\boldsymbol{x})\Delta f(\boldsymbol{x})\mathrm{d}\boldsymbol{x} &= \sum_j \int_{\Gamma_j} \Phi(\boldsymbol{x})\Delta f(\boldsymbol{x})\mathrm{d}\boldsymbol{x} \\ &= \sum_j \int_{\Gamma_j} \operatorname{div}(\Phi\nabla f)(\boldsymbol{x})\mathrm{d}\boldsymbol{x} - \sum_j \int_{\Gamma_j} \nabla\Phi(\boldsymbol{x})\cdot\nabla f(\boldsymbol{x})\mathrm{d}\boldsymbol{x} \\ &= \sum_j \int_{\partial\Gamma_j} \Phi(\boldsymbol{x})\hat{\boldsymbol{n}}_j\cdot\nabla f(\boldsymbol{x})\mathrm{d}S - \sum_j \int_{\Gamma_j} \nabla\Phi(\boldsymbol{x})\cdot\nabla f(\boldsymbol{x})\mathrm{d}\boldsymbol{x},\end{aligned}$$

where $\hat{\boldsymbol{n}}_j$ is the outward normal to $\partial\Gamma_j$ and $\mathrm{d}S$ denotes the two-dimensional Euclidean surface measure on $\partial\Gamma_j$. Since Φ and ∇f are continuous the sum of all of the boundary integrals is zero. Furthermore,

$$\begin{aligned}-\sum_j \int_{\Gamma_j} \nabla f(\boldsymbol{x})\cdot\nabla\Phi(\boldsymbol{x})\mathrm{d}\boldsymbol{x} &= -\sum_j \int_{\Gamma_j} \operatorname{div}(f\nabla\Phi)(\boldsymbol{x})\mathrm{d}\boldsymbol{x} + \sum_j \int_{\Gamma_j} f(\boldsymbol{x})\Delta\Phi(\boldsymbol{x})\mathrm{d}\boldsymbol{x} \\ &= -\sum_j \int_{\Gamma_j} \operatorname{div}(f\nabla\Phi)(\boldsymbol{x})\mathrm{d}\boldsymbol{x}\end{aligned} \tag{5.2.9}$$

since Φ is harmonic in Γ_j. Hence we have

$$\int_{\mathbb{R}^3} \Phi(\boldsymbol{x})\Delta f(\boldsymbol{x})\mathrm{d}\boldsymbol{x} = -\sum_j \int_{\partial\Gamma_j} f(\boldsymbol{x})\hat{\boldsymbol{n}}_j\cdot\nabla\Phi(\boldsymbol{x})\mathrm{d}S. \tag{5.2.10}$$

Note that this expression does not vanish since $\nabla\Phi$ is not a continuous function.

The boundary $\partial\Gamma_j$ consists of two-dimensional planar segments each separating Γ_j from some other Voronoi cell Γ_i (and possibly the point at infinity, which we can ignore for the present since f has compact support and hence all integrands have compact support). Note that on the right side of (5.2.10) each boundary segment appears *twice*, once as the boundary of a Voronoi cell and once as the boundary of its neighbor. The outward normals, however, point in *opposite* directions. Therefore, if we insert (5.2.4) in the right side of (5.2.10) we see that the contribution of $W(\boldsymbol{x})$ for each segment cancels since $W(\boldsymbol{x})$ is a differentiable function away from the $\boldsymbol{R}_j$, and hence is differentiable in a neighborhood of $\partial\Gamma_j$. Thus (5.2.10) becomes

$$\int_{\mathbb{R}^3} \Phi(\boldsymbol{x})\Delta f(\boldsymbol{x})\mathrm{d}\boldsymbol{x} = \sum_j \int_{\partial\Gamma_j} f(\boldsymbol{x})\hat{\boldsymbol{n}}_j \cdot \nabla \frac{Z}{\mathfrak{D}(\boldsymbol{x})}\mathrm{d}S. \tag{5.2.11}$$

Choose any two Voronoi cells Γ_j and Γ_k that share a planar segment. Since every point on this segment has the same distance to the point $\boldsymbol{R}_k$ and $\boldsymbol{R}_j$ we learn that the gradients of $\mathfrak{D}(\boldsymbol{x})^{-1}$, one taken from the interior of Γ_j and one taken from the interior of Γ_k and then evaluated on the segment, have the same magnitude but opposite orientation, i.e.,

$$\hat{\boldsymbol{n}}_j \cdot \nabla \frac{1}{|\boldsymbol{x} - \boldsymbol{R}_j|} = \hat{\boldsymbol{n}}_k \cdot \nabla \frac{1}{|\boldsymbol{x} - \boldsymbol{R}_k|}$$

for all $\boldsymbol{x}$ on the segment. This allows us to rewrite (5.2.11) as

$$\int_{\mathbb{R}^3} \Phi(\boldsymbol{x})\Delta f(\boldsymbol{x})\mathrm{d}\boldsymbol{x} = 2Z \int_{\cup_j\partial\Gamma_j} f(\boldsymbol{x})\hat{\boldsymbol{n}}_j \cdot \nabla \frac{1}{|\boldsymbol{x} - \boldsymbol{R}_j|}\mathrm{d}S. \tag{5.2.12}$$

It follows from this formula that the measure ν that generates $\Phi(\boldsymbol{x})$ in the sense of (5.2.7), (5.2.8) is a surface charge density concentrated on the planar part of $\cup_j\partial\Gamma_j$. On the segment joining the Voronoi cells Γ_j and Γ_k it has the magnitude

$$-\frac{Z}{2\pi}\hat{\boldsymbol{n}}_j \cdot \nabla \frac{1}{|\boldsymbol{x} - \boldsymbol{R}_j|} = -\frac{Z}{2\pi}\hat{\boldsymbol{n}}_k \cdot \nabla \frac{1}{|\boldsymbol{x} - \boldsymbol{R}_k|}.$$

Moreover, $\nu(\mathrm{d}\boldsymbol{x})$ is positive, which follows from the fact that the Voronoi cells are convex, and hence $\hat{\boldsymbol{n}}_j \cdot (\boldsymbol{x} - \boldsymbol{R}_j) \geq 0$ for all $\boldsymbol{x} \in \partial\Gamma_j$. Thus we have achieved our first goal, to identify the measure ν in (5.2.7), (5.2.8).

For any measure $\mu(\mathrm{d}\boldsymbol{x})$ we can write

$$\begin{aligned} D(\mu, \mu) - \int \Phi(\boldsymbol{x})\mu(\mathrm{d}\boldsymbol{x}) + \sum_{k<l} \frac{Z^2}{|\boldsymbol{R}_k - \boldsymbol{R}_l|} \\ = D(\mu - \nu, \mu - \nu) - D(\nu, \nu) + \sum_{k<l} \frac{Z^2}{|\boldsymbol{R}_k - \boldsymbol{R}_l|} \\ \geq -D(\nu, \nu) + \sum_{k<l} \frac{Z^2}{|\boldsymbol{R}_k - \boldsymbol{R}_l|} \end{aligned} \tag{5.2.13}$$

since $D(\mu - \nu, \mu - \nu) \geq 0$ by Theorem 5.1.

It remains to calculate $D(\nu, \nu)$. First, note that

$$W(\boldsymbol{x}) = Z \sum_j \int_{\mathbb{R}^3} \delta(\boldsymbol{y} - \boldsymbol{R}_j) \frac{1}{|\boldsymbol{y} - \boldsymbol{x}|} \mathrm{d}\boldsymbol{y},$$

where δ is the Dirac delta-function, so that (5.2.7) implies that $\int_{\mathbb{R}^3} W(\boldsymbol{x})\nu(\mathrm{d}\boldsymbol{x}) = Z \sum_j \Phi(\boldsymbol{R}_j)$. We use this fact in the following way. With the aid of (5.2.5),

$$\begin{aligned} D(\nu, \nu) &= \frac{1}{2} \int_{\mathbb{R}^3} \Phi(\boldsymbol{x})\nu(\mathrm{d}\boldsymbol{x}) = \frac{1}{2} \int_{\mathbb{R}^3} W(\boldsymbol{x})\nu(\mathrm{d}\boldsymbol{x}) - \frac{1}{2} \int_{\mathbb{R}^3} \frac{Z}{\mathfrak{D}(\boldsymbol{x})} \nu(\mathrm{d}\boldsymbol{x}) \\ &= \frac{1}{2} \sum_j Z\Phi(\boldsymbol{R}_j) - \frac{1}{2} \int_{\mathbb{R}^3} \frac{Z}{\mathfrak{D}(\boldsymbol{x})} \nu(\mathrm{d}\boldsymbol{x}) \\ &= \sum_{k<l} \frac{Z^2}{|\boldsymbol{R}_k - \boldsymbol{R}_l|} - \frac{1}{2} \int_{\mathbb{R}^3} \frac{Z}{\mathfrak{D}(\boldsymbol{x})} \nu(\mathrm{d}\boldsymbol{x}). \end{aligned}$$

Next,

$$\frac{1}{2} \int_{\mathbb{R}^3} \frac{Z}{\mathfrak{D}(\boldsymbol{x})} \nu(\mathrm{d}\boldsymbol{x}) = - \sum_j \frac{Z^2}{8\pi} \int_{\partial\Gamma_j} \frac{1}{|\boldsymbol{x} - \boldsymbol{R}_j|} \hat{\boldsymbol{n}}_j \cdot \nabla \frac{1}{|\boldsymbol{x} - \boldsymbol{R}_j|} \mathrm{d}S,$$

which, by a straightforward calculation, equals

$$- \sum_j \frac{Z^2}{16\pi} \int_{\partial\Gamma_j} \hat{\boldsymbol{n}}_j \cdot \nabla \frac{1}{|\boldsymbol{x} - \boldsymbol{R}_j|^2} \mathrm{d}S = \sum_j \frac{Z^2}{16\pi} \int_{\Gamma_j^c} \Delta \frac{1}{|\boldsymbol{x} - \boldsymbol{R}_j|^2} \mathrm{d}\boldsymbol{x},$$

where Γ_j^c is the *complement* of Γ_j. Note the change of sign! Since

$$\Delta \frac{1}{|\boldsymbol{x} - \boldsymbol{R}_j|^2} = \frac{2}{|\boldsymbol{x} - \boldsymbol{R}_j|^4},$$

we need only calculate a lower bound to the integral

$$\frac{Z^2}{8\pi} \int_{\Gamma_j^c} \frac{1}{|\boldsymbol{x} - \boldsymbol{R}_j|^4} \mathrm{d}\boldsymbol{x}. \tag{5.2.14}$$

This is an integral over the complement of the Voronoi cell Γ_j, which is a set containing the half-space whose boundary plane touches the ball of radius D_j centered at $\boldsymbol{R}_j$ (in fact the boundary of this half-space contains the planar segment separating Γ_j from its nearest neighbor). Thus we get a lower bound by just integrating (5.2.14) over that half-space. By shifting and rotating coordinates we may assume that $\boldsymbol{R}_j = 0$ and that the bounding half-plane is parallel to the y–z plane. Hence

$$\frac{Z^2}{8\pi}\int_{\Gamma_j^c} \frac{1}{|\boldsymbol{x}-\boldsymbol{R}_j|^4}\mathrm{d}\boldsymbol{x} \geq \frac{Z^2}{8\pi}\int_{-\infty}^{\infty}\mathrm{d}z\int_{-\infty}^{\infty}\mathrm{d}y\int_{D_j}^{\infty}\mathrm{d}x\frac{1}{(x^2+y^2+z^2)^2} = \frac{Z^2}{8D_j},$$

which proves the theorem. ■

5.3 Application: Baxter's Electrostatic Inequality

The basic inequality, Theorem 5.3, will now be applied to proving an inequality of Baxter [10] that was originally proved very differently using some sophisticated concepts of potential theory.[3] The derivation from Theorem 5.3 given here is due to Solovej (private communication).

In order to bound the Coulomb potential $V_C(\underline{X}, \underline{R})$ from below, we would like to apply Theorem 5.3 to a measure that is appropriate to N electrons, namely $\mu(\mathrm{d}\boldsymbol{x}) = \sum_{i=1}^{N}\delta(\boldsymbol{x}-\boldsymbol{x}_i)\mathrm{d}\boldsymbol{x}$. Inequality (5.2.6) is of no use in this case, however, because of the infinite self-energy of μ. What we are after here is an inequality of the form (5.2.6), *without* these infinite self-energy terms. This is the content of the following theorem.

Theorem 5.4 (Baxter's Inequality). *Let $V_C(\underline{X}, \underline{R})$ be the Coulomb potential of N electrons with coordinates $\boldsymbol{x}_i \in \mathbb{R}^3$ and M nuclei with common nuclear charge Z and with coordinates $\boldsymbol{R}_j \in \mathbb{R}^3$, as given in (2.1.21)–(2.1.24), namely*

$$V_C(\underline{X}, \underline{R}) = \sum_{1\leq i<j\leq N}\frac{1}{|\boldsymbol{x}_i-\boldsymbol{x}_j|} - \sum_{i=1}^{N}\sum_{j=1}^{M}\frac{Z}{|\boldsymbol{x}_i-\boldsymbol{R}_j|} + \sum_{k<l}\frac{Z^2}{|\boldsymbol{R}_k-\boldsymbol{R}_l|}. \tag{5.3.1}$$

As before let Γ_j denote the Voronoi cell belonging to $\boldsymbol{R}_j$ and let $\mathfrak{D}(\boldsymbol{x})$ denote the distance from $\boldsymbol{x}$ to the nearest nucleus; if $\boldsymbol{x} \in \Gamma_j$ then $\mathfrak{D}(\boldsymbol{x}) = |\boldsymbol{x}-\boldsymbol{R}_j|$. With

[3] Baxter's inequality in [10] did not have the positive term on the right side of (5.3.2).

D_j half the nearest neighbor distance defined in (5.2.2), there is the inequality

$$V_C(\underline{X}, \underline{R}) \geq -(2Z+1)\sum_{i=1}^{N} \frac{1}{\mathfrak{D}(x_i)} + \frac{Z^2}{8}\sum_{j=1}^{M} \frac{1}{D_j}. \tag{5.3.2}$$

Remark 5.1. This theorem says, in effect, that for a lower bound the electrostatic interactions can be thought of as canceling each other except for the residual interaction of each electron with its nearest nucleus. Note that (5.3.2) gives a lower bound on the many-body potential V_C in terms of a one-body potential for the electrons. The striking feature of this inequality is that $(N+M)^2$ terms on the left side are bounded by only $N+M$ terms on the right. A defect of this theorem is that the Coulomb singularity on the right side has a weight $2Z+1$ instead of the expected Z. For non-relativistic matter, where stability holds for all α and Z, this defect only amounts to a modified numerical constant in the energy estimate, but for relativistic matter, where there is a critical value of $\alpha Z \leq 2/\pi$ for stability, the factor $2Z+1$ will yield an incorrect value for the critical value of αZ. This defect will be remedied in the refined inequality in the next section.

Proof. For simplicity, denote $d_i = \mathfrak{D}(x_i)$. Let $\mu_i(\mathrm{d}x) = (d_i^2\pi)^{-1}\delta(|x - x_i| - d_i/2)\mathrm{d}x$ be the normalized uniform measure supported on a sphere of radius $d_i/2$ centered at x_i, and let $\mu = \sum_{i=1}^{N} \mu_i$. (Note that we use $d_i/2$ and not d_i.) If we replace the electron point charges by the smeared out spherical charges μ_i, the electrostatic interaction among the electrons is reduced because the interaction energy between two spheres is less than or equal to that between two points. This follows from Newton's Theorem, which also implies that the interaction between the smeared electrons and the nuclei is not changed, since the radius $d_i/2$ is less than the distance of x_i to any of the nuclei. Hence

$$\begin{aligned} V_C(\underline{X}, \underline{R}) &\geq \sum_{1\leq i<j\leq N} 2D(\mu_i, \mu_j) - \sum_{j=1}^{M} \int_{\mathbb{R}^3} \frac{Z}{|x - R_j|}\mu(\mathrm{d}x) + \sum_{k<l} \frac{Z^2}{|R_k - R_l|} \\ &= D(\mu, \mu) - \sum_{j=1}^{M} \int_{\mathbb{R}^3} \frac{Z}{|x - R_j|}\mu(\mathrm{d}x) + \sum_{k<l} \frac{Z^2}{|R_k - R_l|} - \sum_{i=1}^{N} \frac{1}{d_i}. \end{aligned} \tag{5.3.3}$$

We have used that $D(\mu_i, \mu_i) = 1/d_i$. We shall show that the right side of (5.3.3) is bounded from below by the right side of (5.3.2).

An application of Theorem 5.3 yields the lower bound

$$V_C(\underline{X}, \underline{R}) \geq -\sum_{i=1}^{N} \left(\int_{\mathbb{R}^3} \frac{Z}{\mathfrak{D}(\boldsymbol{x})} \mu_i(\mathrm{d}\boldsymbol{x}) + \frac{1}{d_i} \right) + \frac{Z^2}{8} \sum_{j=1}^{M} \frac{1}{D_j}. \tag{5.3.4}$$

To finish the proof it suffices to show that for $\boldsymbol{x}$ in the support of μ_i, we have $\mathfrak{D}(\boldsymbol{x}) \geq d_i/2$. This follows from the triangle inequality, which implies that for any k

$$|\boldsymbol{x} - \boldsymbol{R}_k| \geq |\boldsymbol{x}_i - \boldsymbol{R}_k| - |\boldsymbol{x}_i - \boldsymbol{x}| \geq d_i - d_i/2 = d_i/2. \tag{5.3.5}$$

In the last step, we used that $|\boldsymbol{x}_i - \boldsymbol{R}_k| \geq d_i$ by definition, and $|\boldsymbol{x}_i - \boldsymbol{x}| = d_i/2$ for $\boldsymbol{x}$ on the sphere centered at $\boldsymbol{x}_i$. ■

5.4 Refined Electrostatic Inequality

As remarked in the previous section, the lower bound in Theorem 5.4 has the virtue of giving a lower bound to the many-body Coulomb potential V_C in terms of a one-body potential for the electrons, but it has the defect that the Coulomb singularity has the wrong prefactor $2Z + 1$ on the right side, instead of the desired factor Z. This is of crucial importance when studying the stability of relativistic matter. The following theorem, which remedies this defect, was proved in [138, Thm. 6]. The price one has to pay is a slightly more negative one-body potential in the Voronoi cells Γ_j, but we emphasize the fact that the singularities at the nuclei are unchanged, namely $Z/|\boldsymbol{x} - \boldsymbol{R}_j|$.

Theorem 5.5 (Refined Electrostatic Inequality). *For $0 < \lambda < 1$ let W^λ be the function on $\mathbb{R}^3$ whose value in the Voronoi cell Γ_j is defined to be*

$$W^\lambda(\boldsymbol{x}) = \frac{Z}{|\boldsymbol{x} - \boldsymbol{R}_j|} + \begin{cases} \frac{1}{2} D_j^{-1}(1 - D_j^{-2}|\boldsymbol{x} - \boldsymbol{R}_j|^2)^{-1} & \text{for } |\boldsymbol{x} - \boldsymbol{R}_j| \leq \lambda D_j \\ \left(\sqrt{2Z} + \frac{1}{2}\right) |\boldsymbol{x} - \boldsymbol{R}_j|^{-1} & \text{for } |\boldsymbol{x} - \boldsymbol{R}_j| > \lambda D_j, \end{cases} \tag{5.4.1}$$

with D_j defined in (5.2.2). Then, with $\underline{X} = (\boldsymbol{x}_1, \dots, \boldsymbol{x}_N)$ and $\underline{R} = (\boldsymbol{R}_1, \dots, \boldsymbol{R}_M)$ as before,

$$\boxed{V_C(\underline{X}, \underline{R}) \geq -\sum_{i=1}^{N} W^{\lambda}(\boldsymbol{x}_i) + \frac{Z^2}{8} \sum_{j=1}^{M} \frac{1}{D_j}.} \tag{5.4.2}$$

Remark 5.2. The optimal choice of λ in (5.4.1) that makes W^{λ} as small as possible is independent of $\underline{X}$ and $\underline{R}$ and depends only on Z. To see this, note that the upper choice in (5.4.1) is increasing in $|\boldsymbol{x} - \boldsymbol{R}_j|$ and the lower choice is decreasing. Hence the optimal λ is determined by setting the two terms equal, that is,

$$\frac{1}{2(1-\lambda^2)} = \frac{\sqrt{2Z} + 1/2}{\lambda}. \tag{5.4.3}$$

In other words, one can take the minimum of the two choices in (5.4.1) and forget about λ. (When $|\boldsymbol{x} - \boldsymbol{R}_j| > D_j$, but still in the Voronoi cell Γ_j, the first choice has to be interpreted as $+\infty$ when taking the minimum.) When $|\boldsymbol{x} - \boldsymbol{R}_j| \geq D_j$, $W^{\lambda}(\boldsymbol{x}) = (\sqrt{Z} + 1/\sqrt{2})^2/|\boldsymbol{x} - \boldsymbol{R}_j|$.

Proof. In order to be able to employ Theorem 5.3, we shall first smear out the electron charges in a definite (possibly non-spherical) way. This will decrease the contribution of the electron–electron repulsion to the Coulomb energy, but increase the electron–nuclear attraction. The measure $\mu(\mathrm{d}\boldsymbol{x})$ will be taken to be the sum of all the smeared charges of all the electrons. The term $D(\mu, \mu)$ will underestimate the interaction of two electrons, which is fine, but it will also include a self-interaction term for each electron, which is not fine and which will have to be subtracted. These subtractions form part of the terms in (5.4.1). Another part of these terms arises from the fact that $\int \Phi(\boldsymbol{x})\mu(\mathrm{d}\boldsymbol{x})$ underestimates the electron–nuclear interaction.

More precisely, we shall replace the point charge at $\boldsymbol{x}_i$ by a charge distribution ν_i which is defined as follows. We distinguish two cases, depending on whether the distance of the electron $\boldsymbol{x}_i$ to the nearest nucleus at $\boldsymbol{R}_j$ is bigger or smaller than λD_j.

Case 1. If $|\boldsymbol{x}_i - \boldsymbol{R}_j| \leq \lambda D_j$, the measure ν_i is supported on S_i, the sphere of radius D_j around $\boldsymbol{R}_j$. Note that by definition S_i lies in the Voronoi cell Γ_j. The charge ν_i is chosen in such a way that the potential created by ν_i

equals $|x - x_i|^{-1}$ for all x with $|x - R_j| > D_j$, that is, for all x outside S_i. This property determines ν_i uniquely. Moreover, ν_i is a positive measure. This follows from the fact that the function $x \mapsto |x - x_i|^{-1} - \int_{S_i} |y - x|^{-1} \nu_i(\mathrm{d}y)$ is **superharmonic**[4] inside S_i and vanishes at S_i; hence it is a non-negative function whose normal derivative, which equals the surface charge by Gauss's law, has to be non-negative.[5]

The potential $\int_{S_i} |y - x|^{-1} \nu_i(\mathrm{d}y)$ generated by ν_i can be evaluated explicitly. For $|x - R_j| \geq D_j$ it is $|x - x_i|^{-1}$, while for $|x - R_j| < D_j$ it is given by

$$\frac{D_j}{|x_i - R_j|} \frac{1}{|x - x_i^*|} \tag{5.4.4}$$

where $x_i^* - R_j = (x_i - R_j) D_j^2 / |x_i - R_j|^2$. The point x_i^* is in fact obtained from x_i by spherical inversion. That is, the potential inside S_i is the same as the potential created by a point charge located at x_i^* outside S_i with charge $D_j / |x_i - R_j|$.

The self-energy of ν_i is then easily computed to be

$$D(\nu_i, \nu_i) = \frac{1}{2|x_i - x_i^*|} \frac{D_j}{|x_i - R_j|} = \frac{1}{2D_j} \frac{1}{1 - |x_i - R_j|^2 / D_j^2}. \tag{5.4.5}$$

Case 2. If $x_i \in \Gamma_j$ and $|x_i - R_j| > \lambda D_j$, we choose S_i to be the sphere centered at x_i with radius $|x_i - R_j| / \zeta$ for some $\zeta > 1$ to be determined later. Note that S_i does not necessarily lie entirely inside Γ_j. The measure ν_i is again supported on S_i, but this time with uniform surface charge distribution. Its self-energy therefore equals

$$D(\nu_i,\ \nu_i) = \frac{\zeta}{2|x_i - R_j|}. \tag{5.4.6}$$

In both cases, ν_i is a positive measure, and

$$\int_{S_i} \frac{\nu_i(\mathrm{d}x)}{|x - y|} \leq \frac{1}{|y - x_i|}$$

[4] A function f is superharmonic if $\Delta f \leq 0$ in the sense of distributions. It is a theorem [118, Thm. 9.3] that this is equivalent to the statement that for all x, $f(x)$ is greater than or equal to the average of f over any sphere centered at x.

[5] In physics this charge distribution is called the screening charge. It is the charge on a perfectly conducting grounded sphere that is induced by a unit charge located at a point x_i inside the conductor. In mathematics it is known as the harmonic measure corresponding to the point x_i.

for all $\boldsymbol{y} \in \mathbb{R}^3$. Using this twice,

$$2D(\nu_i,\ \nu_j) = \iint\limits_{S_i \times S_j} \frac{\nu_i(\mathrm{d}\boldsymbol{x})\nu_j(\mathrm{d}\boldsymbol{y})}{|\boldsymbol{x}-\boldsymbol{y}|} \leq \int\limits_{S_j} \frac{\nu_j(\mathrm{d}\boldsymbol{y})}{|\boldsymbol{x}_i-\boldsymbol{y}|} \leq \frac{1}{|\boldsymbol{x}_i-\boldsymbol{x}_j|}.$$

In particular,

$$\sum_{i<j} \frac{1}{|\boldsymbol{x}_i-\boldsymbol{x}_j|} \geq D\left(\textstyle\sum_i \nu_i,\ \sum_i \nu_i\right) - \sum_i D(\nu_i,\ \nu_i). \tag{5.4.7}$$

The self-energy terms are given in (5.4.5) in case 1 and in (5.4.6) in case 2, respectively.

We now apply Theorem 5.3, which states that

$$D\left(\textstyle\sum_i \nu_i,\ \sum_i \nu_i\right) + \sum_{k<l} \frac{Z^2}{|\boldsymbol{R}_k-\boldsymbol{R}_l|} \geq \sum_{i=1}^{N} \int\limits_{S_i} \Phi(\boldsymbol{x})\nu_i(\mathrm{d}\boldsymbol{x}) + \frac{Z^2}{8}\sum_{j=1}^{M}\frac{1}{D_j}$$

with Φ given in (5.2.4). In order to calculate $\int_{S_i} \Phi(\boldsymbol{x})\nu_i(\mathrm{d}\boldsymbol{x})$ we again have to distinguish the two cases given above.

Case 1. In this case, S_i is entirely contained in some Γ_j, and hence

$$\int\limits_{S_i} \Phi(\boldsymbol{x})\nu_i(\mathrm{d}\boldsymbol{x}) = \sum_{k,\,k\neq j} \int\limits_{S_i} \frac{Z}{|\boldsymbol{x}-\boldsymbol{R}_k|}\nu_i(\mathrm{d}\boldsymbol{x}) = \sum_{k,\,k\neq j} \frac{Z}{|\boldsymbol{x}_i-\boldsymbol{R}_k|},$$

since all the $\boldsymbol{R}_k$ for $k \neq j$ are located outside S_i.

Case 2. Writing $\Phi(\boldsymbol{x}) = W(\boldsymbol{x}) - Z/\mathfrak{D}(\boldsymbol{x})$ as in (5.2.4) and using the fact that all the $\boldsymbol{R}_k$ lie outside S_i by construction, we have

$$\int\limits_{S_i} \Phi(\boldsymbol{x})\nu_i(\mathrm{d}\boldsymbol{x}) = \sum_{k=1}^{M} \frac{Z}{|\boldsymbol{x}_i-\boldsymbol{R}_k|} - Z\int\limits_{S_i} \frac{\nu_i(\mathrm{d}\boldsymbol{x})}{\mathfrak{D}(\boldsymbol{x})}.$$

Recall that $\mathfrak{D}(\boldsymbol{x})$ is the distance to the nearest nucleus defined in (5.2.3). Assume that $\boldsymbol{x}_i \in \Gamma_j$. We claim that for all $\boldsymbol{x} \in S_i$ and all k

$$|\boldsymbol{R}_k - \boldsymbol{x}| \geq |\boldsymbol{R}_j - \boldsymbol{x}_i|\left(1 - \frac{1}{\zeta}\right). \tag{5.4.8}$$

This follows from the triangle inequality in (5.3.5), $|\boldsymbol{R}_k - \boldsymbol{x}| \geq |\boldsymbol{R}_k - \boldsymbol{x}_i| - |\boldsymbol{x} - \boldsymbol{x}_i| = |\boldsymbol{R}_k - \boldsymbol{x}_i| - |\boldsymbol{x}_i - \boldsymbol{R}_j|/\zeta$. The fact that $|\boldsymbol{R}_k - \boldsymbol{x}_i| \geq |\boldsymbol{R}_j - \boldsymbol{x}_i|$, by definition of $\boldsymbol{R}_j$ as the nearest nucleus, yields (5.4.8). In particular, $\mathfrak{D}(\boldsymbol{x}) \geq |\boldsymbol{R}_j - \boldsymbol{x}_i| \left(1 - \frac{1}{\zeta}\right)$, and hence

$$\int_{S_i} \Phi(\boldsymbol{x}) \nu_i(\mathrm{d}\boldsymbol{x}) \geq -\frac{Z}{|\boldsymbol{R}_j - \boldsymbol{x}_i|} \frac{1}{\zeta - 1} + \sum_{k,\, k \neq j} \frac{Z}{|\boldsymbol{x}_i - \boldsymbol{R}_k|}. \tag{5.4.9}$$

The last term is the desired interaction energy with all but the nearest nucleus, and the first expression is an error term.

Altogether, we have shown that

$$V_C(\underline{\boldsymbol{X}}, \underline{\boldsymbol{R}}) \geq -\sum_{i=1}^{N} \frac{Z}{\mathfrak{D}(\boldsymbol{x}_i)} + \frac{Z^2}{8} \sum_{j=1}^{M} \frac{1}{D_j} - \sum_{i=1}^{N} F^\lambda(\boldsymbol{x}_i), \tag{5.4.10}$$

where F^λ is given as follows. In case 1, i.e., when $\boldsymbol{x} \in \Gamma_j$ and $|\boldsymbol{x} - \boldsymbol{R}_j| \leq \lambda D_j$, we see from (5.4.5) that

$$F^\lambda(\boldsymbol{x}) = \frac{1}{2} D_j^{-1} (1 - D_j^{-2} |\boldsymbol{x} - \boldsymbol{R}_j|^2)^{-1}.$$

In case 2, when $\boldsymbol{x} \in \Gamma_j$ and $|\boldsymbol{x} - \boldsymbol{R}_j| > \lambda D_j$, it follows from (5.4.4) and (5.4.9) that

$$F^\lambda(\boldsymbol{x}) = \frac{1}{|\boldsymbol{x} - \boldsymbol{R}_j|} (\zeta/2 + Z/(\zeta - 1)).$$

In order to minimize this last term in case 2, we choose $\zeta = 1 + \sqrt{2Z}$. This yields (5.4.2). ■

CHAPTER 6

An Estimation of the Indirect Part of the Coulomb Energy

6.1 Introduction

In this chapter we shall bound the difference between the Coulomb energy of a system of many electrons in a state ψ and the electrostatic energy of the corresponding charge distribution ϱ_ψ. This difference is also known as the indirect part of the Coulomb energy. Such a bound is not only relevant for the discussion of stability of matter, but is of importance in other areas such as density functional theory in quantum chemistry [51]. While the results on this chapter will be used in one of our proofs of stability of non-relativistic matter in the next chapter, they are not strictly needed for the proof, and we shall give a different proof avoiding their use. A reader who is only interested in the quickest proof of stability can skip this chapter. The results will be of importance in the discussion of relativistic matter in Chapter 8, however.

As explained in Chapter 3, a quantum mechanical system of N particles has a state which, generally, is described by a density matrix. For our purposes here, only the (spin summed) diagonal part of this density matrix is relevant. Also, statistics does not play any role in this chapter; for our purposes here it does not matter whether we are dealing with Bose, Fermi or mixed statistics. In fact, the exchange energy defined below depends only on the N particle density (spin summed, in case of spin).

We denote the N particle density as $P_N(\boldsymbol{x}_1, \ldots, \boldsymbol{x}_N)$. It should be thought of as $|\psi(\boldsymbol{x}_1, \ldots, \boldsymbol{x}_N)|^2$ or, more generally, as $\sum_{\underline{\sigma}} |\psi(\boldsymbol{x}_1, \sigma_1, \ldots, \boldsymbol{x}_N, \sigma_N)|^2$ in the case of non-zero spin, but that does not matter. The function P_N is non-negative, and we assume it to be normalized, as usual, as

$$\int_{\mathbb{R}^{3N}} P_N(\boldsymbol{x}_1, \ldots, \boldsymbol{x}_N) \mathrm{d}\boldsymbol{x}_1 \cdots \mathrm{d}\boldsymbol{x}_N = 1. \tag{6.1.1}$$

For particles with charges e_i, $1 \leq i \leq N$, the electrostatic energy is defined by

$$I_P = \sum_{1\leq i<j\leq N} e_i e_j \int_{\mathbb{R}^{3N}} \frac{P_N(\boldsymbol{x}_1, \ldots, \boldsymbol{x}_N)}{|\boldsymbol{x}_i - \boldsymbol{x}_j|} \mathrm{d}\boldsymbol{x}_1 \cdots \mathrm{d}\boldsymbol{x}_N \qquad (6.1.2)$$

where e_i is the charge of particle i; it is *not* assumed that all the e_i are the same, or that P_N is a permutation symmetric function of the $\boldsymbol{x}_i$. For studying the question of stability of matter, it is desirable to have a lower bound on I_P in terms of the single particle *charge density*, which is defined for each $\boldsymbol{x}$ in $\mathbb{R}^3$ by

$$Q(\boldsymbol{x}) = \sum_{i=1}^{N} Q_i(\boldsymbol{x}) \qquad (6.1.3)$$

and where the *charge density* of particle i is given by

$$Q_i(\boldsymbol{x}) = e_i \int_{\mathbb{R}^{3(N-1)}} P_N(\boldsymbol{x}_1, \ldots, \boldsymbol{x}_{i-1}, \boldsymbol{x}, \boldsymbol{x}_{i+1}, \ldots, \boldsymbol{x}_N) \mathrm{d}\boldsymbol{x}_1 \cdots \mathrm{d}\widehat{\boldsymbol{x}}_i \cdots \mathrm{d}\boldsymbol{x}_N. \qquad (6.1.4)$$

As usual $\mathrm{d}\widehat{\boldsymbol{x}}_i$ means that the $\boldsymbol{x}_i$ integration is omitted. Observe that $Q_i(\boldsymbol{x})$ is a non-negative function with integral

$$\int_{\mathbb{R}^3} Q_i(\boldsymbol{x}) \mathrm{d}x = e_i \qquad (6.1.5)$$

(because of the normalization (6.1.1)). Hence the total charge is given by

$$\int_{\mathbb{R}^3} Q(\boldsymbol{x}) \mathrm{d}\boldsymbol{x} = \sum_{i=1}^{N} e_i. \qquad (6.1.6)$$

The electrostatic energy associated with the charge density $Q(x)$ is given by (compare with Eq. (5.1.3))

$$D(Q, Q) = \frac{1}{2} \iint_{\mathbb{R}^3 \times \mathbb{R}^3} \frac{Q(\boldsymbol{x}) Q(\boldsymbol{y})}{|\boldsymbol{x} - \boldsymbol{y}|} \mathrm{d}\boldsymbol{x} \mathrm{d}\boldsymbol{y} \qquad (6.1.7)$$

and is called the **direct part of the Coulomb energy**. It is the classical Coulomb energy associated with a 'fluid' of charge density Q. Since $Q(\boldsymbol{x}) \geq 0$, $D(Q, Q)$ is always well defined in the sense that it is either finite or $+\infty$. Accordingly, **the indirect part of the Coulomb energy**, denoted by E_P, is defined by the equation

$$I_P = D(Q, Q) + E_P. \tag{6.1.8}$$

Thus, E_P is the difference between the true energy I_P and the classical approximation $D(Q, Q)$. Sometimes it is called the **exchange plus correlation energy**. It is the aim of this chapter to give a lower bound on E_P in terms of Q. We emphasize again that our bound on E_P holds for *arbitrary* normalized N-particle densities P_N.

6.2 Examples

The first example comes from **Hartree's theory**. Consider N spinless particles (i.e. $q = 1$), each with the same charge e. Assume that they are not correlated, in which case their wave function is given by a simple product

$$\psi(\boldsymbol{x}_1, \ldots, \boldsymbol{x}_N) = f_1(\boldsymbol{x}_1) \cdots f_N(\boldsymbol{x}_N), \tag{6.2.1}$$

where each f_i is in $L^2(\mathbb{R}^3)$ and normalized. If all the f_i are the same, this would describe bosons, but this is not important. The N-particle density in this case would be

$$P_N(\boldsymbol{x}_1, \ldots, \boldsymbol{x}_N) = |\psi(\boldsymbol{x}_1, \ldots, \boldsymbol{x}_N)|^2 = |f_1(\boldsymbol{x}_1)|^2 \cdots |f_N(\boldsymbol{x}_N)|^2. \tag{6.2.2}$$

A simple computation yields

$$Q(\boldsymbol{x}) = e \sum_{i=1}^{N} |f_i(\boldsymbol{x})|^2 \tag{6.2.3}$$

and

$$I_P = D(Q, Q) - e^2 \sum_{i=1}^{N} D(|f_i|^2, |f_i|^2). \tag{6.2.4}$$

Hence E_P is the (negative) sum of the self-energies of the charge distributions $e|f_i(\boldsymbol{x})|^2$ in this case.

Another example is provided by a **Hartree–Fock** wave function. Again the charges are all taken to be equal to e. Then ψ is the antisymmetric function of space and spin $z = (\boldsymbol{x}, \sigma)$ given by a determinant

$$\psi(z_1, \ldots, z_N) = (N!)^{-1/2} \det(\phi_i(z_j)) \tag{6.2.5}$$

where the functions ϕ_i are orthonormal in $L^2(\mathbb{R}^3;\mathbb{C}^q)$. Using (3.1.28) and the equation (3.1.29) for the two-particle density, one sees that for the corresponding N-particle density

$$P_N(\boldsymbol{x}_1,\ldots,\boldsymbol{x}_N)=\sum_{\sigma}|\psi(z_1,\ldots,z_N)|^2$$

the indirect part of the Coulomb energy is

$$E_P=-\frac{1}{2}e^2\iint\frac{\left|\gamma^{(1)}(z,z')\right|^2}{|\boldsymbol{x}-\boldsymbol{x}'|}\mathrm{d}z\mathrm{d}z' \tag{6.2.6}$$

where

$$\gamma^{(1)}(z,z')=\sum_{i=1}^{N}\phi_i(z)\overline{\phi_i(z')} \tag{6.2.7}$$

denotes the one-particle density matrix of ψ. In this context E_P is called the **exchange term**.

An *approximation* to E_P in terms of Q was computed in [40] using perturbation theory, i.e. using the eigenfunctions of the kinetic energy operator (i.e. the Laplacian) in a large cubic box Λ. One chooses the $\phi_i(z)$ to be a product of spin and space wave function, i.e.,

$$\phi_{\alpha,\boldsymbol{k}}(z)=\chi_\alpha(\sigma)\frac{1}{\sqrt{|\Lambda|}}e^{2\pi i\boldsymbol{k}\cdot\boldsymbol{x}}, \tag{6.2.8}$$

where $\chi_\alpha(\sigma)$ with $\alpha=1,\ldots,q$ is an orthonormal basis in $\mathbb{C}^q$, $|\Lambda|$ denotes the volume of Λ, and the allowed values of $\boldsymbol{k}$ are

$$\boldsymbol{k}=\frac{\boldsymbol{n}}{|\Lambda|^{1/3}}\quad\text{with } \boldsymbol{n}\in\mathbb{Z}^3_+. \tag{6.2.9}$$

A dimensional argument immediately shows that for $|\Lambda|$ and N large, and with $\varrho=N/|\Lambda|$ fixed,

$$E_P=-Ce^2q^{-1/3}\varrho^{4/3}|\Lambda|. \tag{6.2.10}$$

Closer inspection shows that $C=0.93$.

Formula (6.2.10) suggests that in the general case, i.e., for N-particle densities coming from antisymmetric functions of space and spin, E_P given by (6.1.8) should be bounded below by

$$-Ce^{2/3}q^{-1/3}\int_{\mathbb{R}^3} Q(\boldsymbol{x})^{4/3}\mathrm{d}\boldsymbol{x} \tag{6.2.11}$$

for some suitable universal constant C. For the case where all the e_i equal e, (6.2.11) is correct *provided the factor* $q^{-1/3}$ *is omitted.* That the value of q and the Pauli principle play no role in the question of bounding E_P in terms of Q *alone* we shall show now.

Given any symmetric particle density $P_N(\boldsymbol{x}_1, \ldots, \boldsymbol{x}_N)$, it is easy to find functions $\psi_q \in \bigwedge^N L^2(\mathbb{R}^3; \mathbb{C}^q)$ for $q = 1, 2, 3, \ldots$ each of which is antisymmetric in the $z_i = (\boldsymbol{x}_i, \sigma_i)$ and such that all the functions ψ_q have the same N-particle density P_N. To see this, define a function $\theta(\boldsymbol{x}_1, \ldots, \boldsymbol{x}_N)$ to be antisymmetric in the $\boldsymbol{x}_i$ and independent of the σ_i, and to take only the values ± 1. Consider the function

$$\psi_q(z_1, \ldots z_N) = q^{-N/2}\sqrt{P_N(\boldsymbol{x}_1, \ldots, \boldsymbol{x}_N)}\theta(\boldsymbol{x}_1, \ldots, \boldsymbol{x}_N) \tag{6.2.12}$$

which is antisymmetric and independent of the σ_i. Obviously we have

$$\sum_{\underline{\sigma}} |\psi_q(z_1, \ldots, z_N)|^2 = P_N(\boldsymbol{x}_1, \ldots, \boldsymbol{x}_N) \tag{6.2.13}$$

for each $\boldsymbol{x}_1, \ldots, \boldsymbol{x}_N$, and for each value of q. Hence any symmetric N-particle density P_N can come from an antisymmetric wave function with arbitrary value of q.

Thus, the best general estimate we could aim for is

$$E_P \geq -Ce^{2/3}\int_{\mathbb{R}^3} Q(\boldsymbol{x})^{4/3}\mathrm{d}\boldsymbol{x} \tag{6.2.14}$$

with C being a universal constant. One might try to improve the constant in the estimate by *excluding* certain symmetries for wave functions for example symmetric (i.e. bosonic) functions. That this is impossible can be seen in a similar fashion as above. Thus the Coulomb repulsion is insensitive to the symmetry properties of a wave function and is therefore not able to see the spin. The reason q entered in (6.2.11) was that the kinetic energy operator was taken into account, i.e., changing q meant changing the ground state of the kinetic energy operator and hence it meant changing $Q(\boldsymbol{x})$. Our point is that when $Q(\boldsymbol{x})$ is the *only* available information then (6.2.14) is the best one can hope for. This is in contrast to the kinetic energy which is sensitive to the symmetry properties of the spatial part of the wave function. In fact, as was shown in Chapter 4, a spin dependent bound for the form

$$T_\psi \geq Cq^{-2/3}e^{-5/3} \int_{\mathbb{R}^3} Q(\boldsymbol{x})^{5/3} \mathrm{d}\boldsymbol{x} \tag{6.2.15}$$

can be obtained. The difference, i.e., the spin dependence of the kinetic energy estimate and the spin independence of the exchange estimate, can be (somewhat sloppily) rephrased by saying that only an off-diagonal operator (like the Laplacian) can 'see' the symmetry properties of a wave function and therefore the spin.

6.3 Exchange Estimate

The main theorem of this chapter is the following lower bound on E_P. The charges e_i do not have to be equal to each other, but it is important that they are all positive or all negative. Note that $e_i Q_i \geq 0$ by definition.

Theorem 6.1 (Exchange Estimate). *Let P_N be a normalized N-particle density. Assume that the e_i have the same sign for all $1 \leq i \leq N$. Then the indirect term E_P of the Coulomb energy given by (6.1.8) satisfies the estimate*

$$E_P \geq -C \left(\int_{\mathbb{R}^3} \left(\sum_{i=1}^{N} e_i Q_i(\boldsymbol{x}) \right)^{4/3} \mathrm{d}\boldsymbol{x} \right)^{1/2} \left(\int_{\mathbb{R}^3} |Q(\boldsymbol{x})|^{4/3} \mathrm{d}\boldsymbol{x} \right)^{1/2}, \tag{6.3.1}$$

where C is some constant satisfying

$$C \leq 1.68.$$

Here $Q(\boldsymbol{x})$ and $Q_i(\boldsymbol{x})$ are given by the expressions (6.1.3) and (6.1.4). In case all the e_i equal a common value e then

$$\boxed{E_P \geq -C|e|^{2/3} \int_{\mathbb{R}^3} |Q(\boldsymbol{x})|^{4/3} \mathrm{d}\boldsymbol{x}.} \tag{6.3.2}$$

Remarks 6.1.

(a) This theorem was originally proved in [111] with the aid of the Hardy-Littlewood maximal function, with a bound on the constant C given by 8.52

instead of 1.68. The improved constant was obtained by Lieb and Oxford in [125]. Moreover, by numerically optimizing the smearing function μ in our equation (6.6.9) below, it was shown in [29] that the Lieb–Oxford method presented here can yield $C < 1.636$.

(b) Since the number of particles is fixed one might expect that the sharp constant in (6.3.1) is N dependent. This is in fact true. In the case of one particle the constant C_1 can in principle be computed exactly. Since $I_P = 0$ in this case, $E_P = -D(Q, Q)$ and we have that

$$C_1 = \sup\left\{\frac{D(Q,Q)}{\int Q(\boldsymbol{x})^{4/3}\mathrm{d}\boldsymbol{x}} : Q(\boldsymbol{x}) \geq 0, \int Q(\boldsymbol{x})\mathrm{d}\boldsymbol{x} = 1\right\}. \tag{6.3.3}$$

It is not difficult to see that C_1 is finite (see item (f) below). The existence of a maximizer was shown in [125]. The corresponding variational equation is the Lane–Emden equation of order 3, which was studied in [179] and [76]. In particular, its solutions are tabulated. The result is

$$C_1 = 1.092. \tag{6.3.4}$$

This constant plays a role in the Chandrasekhar mass limit for gravitating bodies (see Chapter 13). The Lane–Emden equation goes back to Lane [101] in his study of gravitating gas spheres in the year 1870!

A lower bound for C_2 was computed in [125]:

$$C_2 \geq 1.234 > C_1 \tag{6.3.5}$$

In general it is not hard to see that

$$C_N \leq C_{N+1}, \tag{6.3.6}$$

and we refer to [125] for details. The constant C in (6.3.1), which is valid for *all* particle numbers, is the worst possible case, and we note (from the above bound on C_2) that the bound 1.68 cannot be improved very much.

(c) It is not clear if the (unknown) optimal constant C in the theorem is improved by making the assumption that all the e_i are equal. Our proof, however, does not get simpler or better in the equal charge case.

(d) A slightly unpleasant feature of the bound (6.3.1) is that it does not just depend on the total charge density $Q(\boldsymbol{x})$, but rather on the individual charge densities $Q_i(\boldsymbol{x})$.

(e) Note that there is no *upper* bound on E_P in terms of the one-particle density Q. Even for a nice, smooth Q, I_P (and hence E_P) can be $+\infty$. For example, if $P_2(\boldsymbol{x}_1, \boldsymbol{x}_2) = C \exp(-|\boldsymbol{x}_1|^2 - |\boldsymbol{x}_2|^2)/|\boldsymbol{x}_1 - \boldsymbol{x}_2|^2$ then this is the case. This example also shows why it is necessary that all charges have the same sign in Theorem 6.1 for, otherwise, E_P could be $-\infty$.

(f) This last remark is of a more technical nature. Since E_P is the difference of two positive quantities and since the only assumption on P_N is that it is normalized, the reader might worry that E_P is not well defined. Conceivably I_P and $D(Q, Q)$ could both be infinity and yet E_P, being the difference of the two, is somehow finite. This does not affect the validity of Theorem 6.1 as the following reasoning shows. We can assume that $\int |Q(\boldsymbol{x})|^{4/3} \mathrm{d}\boldsymbol{x} < \infty$ for otherwise there is nothing to prove. By the Hardy–Littlewood–Sobolev inequality (see [118, Thm. 4.3]) we have that

$$D(Q, Q) \leq C \|Q\|_{6/5}^2 \tag{6.3.7}$$

and by Hölder's inequality

$$\begin{aligned} \|Q\|_{6/5} &\leq \left(\int_{\mathbb{R}^3} |Q(\boldsymbol{x})| \mathrm{d}\boldsymbol{x} \right)^{1/3} \left(\int_{\mathbb{R}^3} |Q(\boldsymbol{x})|^{4/3} \mathrm{d}\boldsymbol{x} \right)^{1/2} \\ &= \left| \sum_{i=1}^N e_i \right|^{1/3} \left(\int_{\mathbb{R}^3} |Q(\boldsymbol{x})|^{4/3} \mathrm{d}\boldsymbol{x} \right)^{1/2}. \end{aligned} \tag{6.3.8}$$

Hence, whenever $Q \in L^{4/3}(\mathbb{R}^3)$ (so that the right side of (6.3.1) and (6.3.2) is finite) $D(Q, Q)$ is finite and E_P is well defined. It might be $+\infty$, but never $-\infty$.

The proof of Theorem 6.1 is contained in Sections 6.4–6.6. For simplicity, we shall consider the case that all $e_i > 0$ even though the main application of Theorem 6.1 is to electrons for which $e_i = -e < 0$. This is done to avoid writing absolute values everywhere, but it is of no consequence since the only relevant quantities are the products $e_k e_j > 0$.

6.4 Smearing Out Charges

The first step in the proof of Theorem 6.1 is a generalization of a lemma originally due to Onsager [145].

Lemma 6.1 (Onsager's Lemma). *Consider N positive charges e_i located at distinct points $\boldsymbol{x}_1, \ldots, \boldsymbol{x}_N$ in $\mathbb{R}^3$. For each $1 \leq i \leq N$, let $\mu_{\boldsymbol{x}_i}$ be a non-negative, bounded, function that is spherically symmetric about $\boldsymbol{x}_i$, with $\int \mu_{\boldsymbol{x}_i}(\boldsymbol{x})\mathrm{d}\boldsymbol{x} = 1$. Then for any non-negative, integrable, function ϱ there is the inequality*

$$\sum_{i<j} \frac{e_i e_j}{|\boldsymbol{x}_i - \boldsymbol{x}_j|} \geq -D(\varrho, \varrho) + 2\sum_{i=1}^{N} e_i D(\varrho, \mu_{\boldsymbol{x}_i}) - \sum_{i=1}^{N} e_i^2 D(\mu_{\boldsymbol{x}_i}, \mu_{\boldsymbol{x}_i}). \tag{6.4.1}$$

Proof. The functions $\mu_{\boldsymbol{x}_i}$ being bounded guarantees that $D(\varrho, \mu_{\boldsymbol{x}_i})$ and $D(\mu_{\boldsymbol{x}_i}, \mu_{\boldsymbol{x}_i})$ are finite. We can assume that $D(\varrho, \varrho)$ is not infinite, because if it were infinite the right side of (6.4.1) would be $-\infty$ and the lemma is trivial. We know from Theorem 5.1 that $D(\,\cdot\,,\,\cdot\,)$ is positive definite and hence

$$D\left(\varrho - \sum_{i=1}^{N} e_i \mu_{\boldsymbol{x}_i}, \varrho - \sum_{i=1}^{N} e_i \mu_{\boldsymbol{x}_i}\right) \geq 0 \tag{6.4.2}$$

which implies that

$$\sum_{i \neq j} e_i e_j D(\mu_{\boldsymbol{x}_i}, \mu_{\boldsymbol{x}_j}) \geq -D(\varrho, \varrho) + 2\sum_{i=1}^{N} e_i D(\varrho, \mu_{\boldsymbol{x}_i}) - \sum_{i=1}^{N} e_i^2 D(\mu_{\boldsymbol{x}_i}, \mu_{\boldsymbol{x}_i}). \tag{6.4.3}$$

Since $\mu_{\boldsymbol{x}_i}$ and $\mu_{\boldsymbol{x}_j}$ are non-negative and spherically symmetric around the centers $\boldsymbol{x}_i$ and $\boldsymbol{x}_j$, we know from Newton's Theorem (Theorem 5.2) that

$$D(\mu_{\boldsymbol{x}_i}, \mu_{\boldsymbol{x}_j}) \leq \frac{1}{2} \frac{1}{|\boldsymbol{x}_i - \boldsymbol{x}_j|} \tag{6.4.4}$$

which proves (6.4.1). ■

Remark 6.2. The point of the above lemma is that it estimates a quantity

$$\sum_{i<j} \frac{e_i e_j}{|\boldsymbol{x}_i - \boldsymbol{x}_j|} \tag{6.4.5}$$

in which correlations are important by another one where the correlations among the $\boldsymbol{x}_i$ are not important. That is, the right side of (6.4.1) depends only on how the points $\boldsymbol{x}_i$ are distributed relative to ϱ, but not relative to each other – as opposed to the left side of (6.4.1).

Lemma 6.1 immediately allows us to get a lower bound for E_P in terms of the one-particle densities $Q_i(\boldsymbol{x})$ and $Q(\boldsymbol{x}) = \sum_{i=1}^{N} Q_i(\boldsymbol{x})$. Choosing $\varrho = Q$, multiplying (6.4.1) by the N-particle density $P_N(\boldsymbol{x}_1, \ldots, \boldsymbol{x}_N)$ and integrating we arrive at

$$I_P \geq -D(Q,Q) + 2\sum_{i=1}^{N} \int_{\mathbb{R}^3} D(Q, \mu_{\boldsymbol{x}_i}) Q_i(\boldsymbol{x}_i) \mathrm{d}\boldsymbol{x}_i$$

$$- \sum_{i=1}^{N} e_i \int_{\mathbb{R}^3} D(\mu_{\boldsymbol{x}_i}, \mu_{\boldsymbol{x}_i}) Q_i(\boldsymbol{x}_i) \mathrm{d}\boldsymbol{x}_i. \tag{6.4.6}$$

The normalization $\int P_N(\boldsymbol{x}_1, \ldots, \boldsymbol{x}_N)\mathrm{d}\boldsymbol{x}_1 \cdots \mathrm{d}\boldsymbol{x}_N = 1$ has been used here. If we denote by $\delta_{\boldsymbol{x}_i}$ the Dirac measure at the point $\boldsymbol{x}_i$ we can write, in a somewhat formal but illuminating fashion,

$$\sum_{i=1}^{N} \int_{\mathbb{R}^3} D(Q, \delta_{\boldsymbol{x}_i}) Q_i(\boldsymbol{x}_i) \mathrm{d}\boldsymbol{x}_i = D(Q,Q). \tag{6.4.7}$$

Hence, by adding and subtracting we get from (6.4.6)

$$E_P \geq -F_1 - F_2 \tag{6.4.8}$$

where

$$F_1 = 2\sum_{i=1}^{N} \int_{\mathbb{R}^3} D(Q, \delta_{\boldsymbol{x}_i} - \mu_{\boldsymbol{x}_i}) Q_i(\boldsymbol{x}_i) \mathrm{d}\boldsymbol{x}_i \tag{6.4.9}$$

and

$$F_2 = \sum_{i=1}^{N} e_i \int_{\mathbb{R}^3} D(\mu_{\boldsymbol{x}_i}, \mu_{\boldsymbol{x}_i}) Q_i(\boldsymbol{x}_i) \mathrm{d}\boldsymbol{x}_i. \tag{6.4.10}$$

Clearly F_2 is positive. Observe that F_1 is also positive (by Newton's Theorem) for any choice of the $\mu_{\boldsymbol{x}_i}$.

6.5 Proof of Theorem 6.1, a First Bound

In this section, we shall prove Theorem 6.1, but with a worse bound on the constant C than the one stated in the theorem. In the next section we will explain how to improve the method in order to obtain the better bound.

We start with the bound (6.4.8) proved in the previous section. In (6.4.9) and (6.4.10) we are still free to choose the functions $\mu_{\boldsymbol{x}_i}$. It is clear that $\mu_{\boldsymbol{x}_i}$ has to depend on Q_i for otherwise F_1 would be quadratic in Q_i and would not be proportional to the $4/3$ power of Q_i.

Let $\mu : \mathbb{R}^3 \to \mathbb{R}$ be a non-negative bounded function satisfying

(a) μ is spherically symmetric about the origin,
(b) $\int_{\mathbb{R}^3} \mu(\boldsymbol{y})\mathrm{d}\boldsymbol{y} = 1$,
(c) $\mu(\boldsymbol{y}) = 0$ if $|\boldsymbol{y}| > 1$.

As already said, we assume, without loss of generality, that all e_i are positive. Given μ and some positive number λ, we choose the $\mu_{\boldsymbol{x}_i}$ to be

$$\mu_{\boldsymbol{x}_i}(\boldsymbol{y}) = \lambda^3 Q(\boldsymbol{x}_i)\mu(\lambda Q(\boldsymbol{x}_i)^{1/3}(\boldsymbol{x}_i - \boldsymbol{y})). \tag{6.5.1}$$

Note that since Q is an integrable function, this is well defined for almost every $\boldsymbol{x}_i$. Moreover, it is easily seen that the functions $\mu_{\boldsymbol{x}_i}$ obtained in this way satisfy the assumptions of Lemma 6.1.

The $\mu_{\boldsymbol{x}_i}$ constructed in this way describes a charge distributed over a length scale determined by the value of Q at $\boldsymbol{x}_i$. The factor λ could be set equal to one, but we introduce it in order to optimize the resulting constant at the end.

If we denote

$$R_\lambda(a, r) = ar^{-1} - \lambda a^{4/3}\phi(\lambda a^{1/3} r), \tag{6.5.2}$$

where ϕ is the potential associated with μ, i.e.,

$$\phi(|\boldsymbol{x}|) = \int_{\mathbb{R}^3} \frac{\mu(\boldsymbol{y})}{|\boldsymbol{x} - \boldsymbol{y}|}\mathrm{d}\boldsymbol{y} = \int_{\mathbb{R}^3} \min\left\{\frac{1}{|\boldsymbol{x}|}, \frac{1}{|\boldsymbol{y}|}\right\} \mu(\boldsymbol{y})\mathrm{d}\boldsymbol{y} \tag{6.5.3}$$

(see Theorem 5.2), then a simple computation shows that

$$F_1 = \iint_{\mathbb{R}^3\times\mathbb{R}^3} Q(\boldsymbol{y})R_\lambda(Q(\boldsymbol{x}), |\boldsymbol{x} - \boldsymbol{y}|)\mathrm{d}\boldsymbol{x}\mathrm{d}\boldsymbol{y}. \tag{6.5.4}$$

Moreover, since

$$\begin{aligned} D(\mu_{\boldsymbol{x}_i}, \mu_{\boldsymbol{x}_i}) &= \lambda^6 Q(\boldsymbol{x}_i)^2 \iint_{\mathbb{R}^3\times\mathbb{R}^3} \frac{\mu(\lambda Q(\boldsymbol{x}_i)^{1/3}(\boldsymbol{x}_i - \boldsymbol{y}))\mu(\lambda Q(\boldsymbol{x}_i)^{1/3}(\boldsymbol{x}_i - \boldsymbol{z}))}{|\boldsymbol{y} - \boldsymbol{z}|}\mathrm{d}\boldsymbol{y}\mathrm{d}\boldsymbol{z} \\ &= \lambda Q(\boldsymbol{x}_i)^{1/3} D(\mu, \mu) \end{aligned} \tag{6.5.5}$$

we find that

$$F_2 = \lambda D(\mu, \mu) \int_{\mathbb{R}^3} Q(\boldsymbol{x})^{1/3} \left(\sum_{i=1}^{N} e_i Q_i(\boldsymbol{x})\right) \mathrm{d}\boldsymbol{x}. \tag{6.5.6}$$

In case $e_i = e$ for all $i = 1, \dots, N$, F_2 is already of the desired form. Otherwise, we use Hölder's inequality to conclude that

$$F_2 \leq \lambda D(\mu, \mu) \left(\int_{\mathbb{R}^3} Q(\boldsymbol{x})^{4/3} \mathrm{d}\boldsymbol{x} \right)^{1/4} \left(\int_{\mathbb{R}^3} \left(\sum_{i=1}^{N} e_i Q_i(\boldsymbol{x}) \right)^{4/3} \mathrm{d}\boldsymbol{x} \right)^{3/4} . \qquad (6.5.7)$$

We concentrate on F_1 and prove first a crude estimate which hopefully clarifies what the choice of $\mu_{\boldsymbol{x}_i}$ in (6.5.1) accomplishes. Since μ is non-negative and $\int \mu(\boldsymbol{y})\mathrm{d}\boldsymbol{y} = 1$, (6.5.3) shows that $\phi(r) \leq r^{-1}$ which implies that $R_\lambda(a, r) \geq 0$. Further, since $\mu(\boldsymbol{y}) = 0$ for $|\boldsymbol{y}| > 1$, we find again by (6.5.3) that $R_\lambda(a, r) = 0$ if $\lambda a^{1/3} r > 1$. Hence we have the simple bound

$$R_\lambda(a, r) \leq \begin{cases} ar^{-1} & \text{if } \lambda a^{1/3} r \leq 1 \\ 0 & \text{otherwise,} \end{cases} \qquad (6.5.8)$$

which implies that

$$F_1 \leq \iint_{\lambda Q(\boldsymbol{x})^{1/3}|\boldsymbol{x}-\boldsymbol{y}| \leq 1} \frac{Q(\boldsymbol{x})Q(\boldsymbol{y})}{|\boldsymbol{x} - \boldsymbol{y}|} \mathrm{d}\boldsymbol{x}\mathrm{d}\boldsymbol{y}. \qquad (6.5.9)$$

The restriction upon the integration in (6.5.9) obviously plays an important role. To make use of it we resort to the following device. Write $Q(\boldsymbol{x}) = \int_0^\infty \chi_\alpha(\boldsymbol{x})\mathrm{d}\alpha$ where χ_α is the characteristic function of the set $\{\boldsymbol{x} : Q(\boldsymbol{x}) \geq \alpha\}$, i.e.

$$\chi_\alpha(\boldsymbol{x}) = \begin{cases} 1 & \text{if } Q(\boldsymbol{x}) \geq \alpha, \\ 0 & \text{otherwise.} \end{cases} \qquad (6.5.10)$$

Using this and Fubini's theorem to change the order of integration, the right side of (6.5.9) becomes

$$\int_0^\infty \mathrm{d}\alpha \int_0^\infty \mathrm{d}\beta \iint_{\lambda Q(\boldsymbol{x})^{1/3}|\boldsymbol{x}-\boldsymbol{y}| \leq 1} \frac{\chi_\alpha(\boldsymbol{x})\chi_\beta(\boldsymbol{y})}{|\boldsymbol{x} - \boldsymbol{y}|} \mathrm{d}\boldsymbol{x}\mathrm{d}\boldsymbol{y}$$

$$\leq \int_0^\infty \mathrm{d}\alpha \int_0^\infty \mathrm{d}\beta \iint_{\lambda \alpha^{1/3}|\boldsymbol{x}-\boldsymbol{y}| \leq 1} \frac{\chi_\alpha(\boldsymbol{x})\chi_\beta(\boldsymbol{y})}{|\boldsymbol{x} - \boldsymbol{y}|} \mathrm{d}\boldsymbol{x}\mathrm{d}\boldsymbol{y}. \qquad (6.5.11)$$

Where $\alpha \le \beta$ we bound the integrand from above by

$$\iint\limits_{\lambda\alpha^{1/3}|x-y|\le 1} \frac{\chi_\beta(y)}{|x-y|}\mathrm{d}x\mathrm{d}y = \int\limits_{\mathbb{R}^3} \chi_\beta(y)\mathrm{d}y \int\limits_{|x|\le(\lambda\alpha^{1/3})^{-1}} |x|^{-1}\mathrm{d}x$$

$$= \frac{2\pi}{\lambda^2\alpha^{2/3}} \int\limits_{\mathbb{R}^3} \chi_\beta(y)\mathrm{d}y \tag{6.5.12}$$

and where $\alpha \ge \beta$ we bound it by

$$\iint_{\lambda\alpha^{1/3}|x-y|\le 1} \frac{\chi_\alpha(y)}{|x-y|}\mathrm{d}x\mathrm{d}y = \frac{2\pi}{\lambda^2\alpha^{2/3}} \int\limits_{\mathbb{R}^3} \chi_\alpha(y)\mathrm{d}y. \tag{6.5.13}$$

Therefore (6.5.11) is bounded above by

$$\frac{2\pi}{\lambda^2}\left\{\int\limits_0^\infty \mathrm{d}\beta \int\limits_0^\beta \mathrm{d}\alpha \int\limits_{\mathbb{R}^3} \chi_\beta(y)\alpha^{-2/3}\mathrm{d}y + \int\limits_0^\infty \mathrm{d}\alpha \int\limits_0^\alpha \mathrm{d}\beta \int\limits_{\mathbb{R}^3} \chi_\alpha(x)\alpha^{-2/3}\mathrm{d}x\right\}$$

$$= \frac{8\pi}{\lambda^2} \int\limits_0^\infty \alpha^{1/3}\mathrm{d}\alpha \int\limits_{\mathbb{R}^3} \chi_\alpha(x)\mathrm{d}x$$

$$= \frac{6\pi}{\lambda^2} \int\limits_{\mathbb{R}^3} Q(x)^{4/3}\mathrm{d}x. \tag{6.5.14}$$

Returning to (6.4.8) and using (6.5.7) we have that

$$E_P \ge -\frac{6\pi}{\lambda^2} \int\limits_{\mathbb{R}^3} Q(x)^{4/3}\mathrm{d}x$$

$$- \lambda D(\mu,\mu)\left(\int\limits_{\mathbb{R}^3} Q(x)^{4/3}\mathrm{d}x\right)^{1/4}\left(\int\limits_{\mathbb{R}^3}\left(\sum_{i=1}^N e_i Q_i(x)\right)^{4/3}\mathrm{d}x\right)^{3/4}. \tag{6.5.15}$$

Maximizing the right side over λ yields the bound

$$E_P \ge -\frac{3^{4/3}}{2^{1/3}}\pi^{1/3}(D(\mu,\mu))^{2/3}\left(\int\limits_{\mathbb{R}^3} Q(x)^{4/3}\mathrm{d}x\right)^{1/2}\left(\int\limits_{\mathbb{R}^3}\left(\sum_{i=1}^N e_i Q_i(x)\right)^{4/3}\mathrm{d}x\right)^{1/2}. \tag{6.5.16}$$

In order to minimize $D(\mu, \mu)$, the optimal choice for μ is

$$\mu(\boldsymbol{y}) = \frac{1}{4\pi}\delta(|\boldsymbol{y}| - 1), \tag{6.5.17}$$

in which case $D(\mu, \mu) = 1/2$. This yields

$$E_P \geq -\frac{3^{4/3}}{2}\pi^{1/3}\left(\int_{\mathbb{R}^3} Q(\boldsymbol{x})^{4/3}\mathrm{d}\boldsymbol{x}\right)^{1/2}\left(\int_{\mathbb{R}^3}\left(\sum_{i=1}^{N} e_i Q_i(\boldsymbol{x})\right)^{4/3}\mathrm{d}\boldsymbol{x}\right)^{1/2}. \tag{6.5.18}$$

Note that $3^{4/3}\pi^{1/3}/2 \approx 3.17$.

6.6 An Improved Bound

One can improve the constant 3.17 in the bound (6.5.18) by replacing inequality (6.5.8) by a more sophisticated treatment. Since $R_\lambda(a, r)$ in (6.5.2) is continuously differentiable we have, using the fundamental theorem of calculus, that

$$\int_{\mathbb{R}^3} R_\lambda(Q(\boldsymbol{x}), |\boldsymbol{x}-\boldsymbol{y}|)\mathrm{d}\boldsymbol{x} = \int_{\mathbb{R}^3}\mathrm{d}\boldsymbol{x}\int_0^{Q(\boldsymbol{x})}\left(\frac{\partial}{\partial\alpha}R_\lambda\right)(\alpha, |\boldsymbol{x}-\boldsymbol{y}|)\mathrm{d}\alpha \tag{6.6.1}$$

and, again using the definition of χ_α in (6.5.10), this can be written as

$$\int_{\mathbb{R}^3}\mathrm{d}\boldsymbol{x}\int_0^{\infty}\chi_\alpha(\boldsymbol{x})\left(\frac{\partial}{\partial\alpha}R_\lambda\right)(\alpha, |\boldsymbol{x}-\boldsymbol{y}|)\mathrm{d}\alpha. \tag{6.6.2}$$

By inspection of (6.5.2) and (6.5.3), one easily sees that $\frac{\partial R_\lambda}{\partial\alpha}(\alpha, r)$ is bounded and hence we can write

$$\begin{aligned} F_1 &= \iint_{\mathbb{R}^3\times\mathbb{R}^3} Q(\boldsymbol{y})R_\lambda(Q(\boldsymbol{x}), |\boldsymbol{x}-\boldsymbol{y}|)\mathrm{d}\boldsymbol{x}\mathrm{d}\boldsymbol{y} \\ &= \int_0^\infty\mathrm{d}\alpha\int_0^\infty\mathrm{d}\beta\iint_{\mathbb{R}^3\times\mathbb{R}^3}\chi_\alpha(\boldsymbol{x})\chi_\beta(\boldsymbol{y})\frac{\partial R_\lambda}{\partial\alpha}(\alpha, |\boldsymbol{x}-\boldsymbol{y}|)\mathrm{d}\boldsymbol{x}\mathrm{d}\boldsymbol{y}. \end{aligned} \tag{6.6.3}$$

An upper bound is obtained by replacing $\frac{\partial R_\lambda}{\partial\alpha}$ by its positive part, which we denote as $\left[\frac{\partial R_\lambda}{\partial\alpha}\right]_+ = \max\{\frac{\partial R_\lambda}{\partial\alpha}, 0\}$.

The right side of (6.5.11) can be recovered from (6.6.3) by replacing $\left[\frac{\partial R_\lambda}{\partial \alpha}\right]_+(\alpha, r)$ by the function $r^{-1}\Theta(1-\lambda^{1/3}\alpha r)$, where Θ is the Heaviside step function

$$\Theta(t) = \begin{cases} 0 & \text{if } t < 0 \\ 1 & \text{if } t \geq 0. \end{cases} \tag{6.6.4}$$

The integral $\int_{\mathbb{R}^3} \Theta(1-\lambda^{1/3}\alpha|\boldsymbol{x}|)|\boldsymbol{x}|^{-1}\mathrm{d}\boldsymbol{x}$ in (6.5.12) is now replaced by

$$\int_{\mathbb{R}^3} \left[\frac{\partial R_\lambda}{\partial \alpha}\right]_+ (\alpha, |\boldsymbol{x}|)\mathrm{d}\boldsymbol{x}. \tag{6.6.5}$$

To compute this integral observe that

$$R_\lambda(\alpha, r) = \lambda\alpha^{4/3}G(\lambda\alpha^{1/3}r), \tag{6.6.6}$$

where $G(t) = \frac{1}{t} - \phi(t)$ for $t > 0$, with $\phi(t)$ given in (6.5.3). Hence

$$\frac{\partial R_\lambda}{\partial \alpha}(\alpha, r) = \lambda\alpha^{1/3}\left(\frac{\partial R_1}{\partial \alpha}\right)(1, \lambda\alpha^{1/3}r) \tag{6.6.7}$$

and, therefore, (6.6.5) becomes

$$(\lambda\alpha^{1/3})^{-2}\int_{\mathbb{R}^3}\left[\frac{\partial R_1}{\partial \alpha}\right]_+ (1, |\boldsymbol{x}|)\mathrm{d}\boldsymbol{x} = (\lambda\alpha^{1/3})^{-2}K(\mu), \tag{6.6.8}$$

where $K(\mu)$ depends only on μ. Following through all the steps in the previous section with the corresponding replacements we end up with the following estimate (compare with (6.5.16))

$$E_P \geq -\frac{3}{2}[6K(\mu)D(\mu,\mu)^2]^{1/3}\left(\int_{\mathbb{R}^3}\left(\sum_{i=1}^N e_i Q_i(\boldsymbol{x})\right)^{4/3}\mathrm{d}\boldsymbol{x}\right)^{1/2}\left(\int_{\mathbb{R}^3} Q(\boldsymbol{x})^{4/3}\mathrm{d}\boldsymbol{x}\right)^{1/2}. \tag{6.6.9}$$

In the previous section, we chose μ to be a constant surface charge distribution on the unit sphere which, by making a crude estimate on $K(\mu)$, yielded the constant 3.17. Computing directly from formula (6.6.8) yields the constant 1.81.

By choosing μ more cleverly this constant can be improved. We choose μ to be the uniform distribution of a unit charge smeared out in a ball instead of a sphere, i.e., $\mu(\boldsymbol{y}) = \mu_0(\boldsymbol{y}) := (3/4\pi)\Theta(1-|\boldsymbol{y}|)$. A computation of $\phi(r)$ in

(6.5.3) yields

$$\phi(r) = \begin{cases} \frac{3}{2}\left(1 - \frac{1}{3}r^2\right) & \text{if } 0 < r \le 1 \\ \frac{1}{2} & \text{if } r \ge 1, \end{cases} \tag{6.6.10}$$

and hence

$$R_\lambda(\alpha, r) = \begin{cases} \alpha r^{-1} - \frac{3}{2}\lambda\alpha^{4/3}\left(1 - \frac{1}{3}(\lambda\alpha^{1/3}r)^2\right) & \text{if } 0 \le r \le (\lambda\alpha^{1/3})^{-1}, \\ 0 & \text{if } r \ge (\lambda\alpha^{1/3})^{-1}. \end{cases} \tag{6.6.11}$$

Therefore

$$\frac{\partial R_1}{\partial \alpha}(1, r) = \begin{cases} \frac{1}{r}(1-r)(1-r-r^2) & 0 \le r \le 1, \\ 0 & 1 \le r. \end{cases} \tag{6.6.12}$$

This function is non-negative for $0 \le r \le \frac{\sqrt{5}-1}{2}$ and so

$$K(\mu_0) = 4\pi\left(\frac{59}{60} - \frac{5}{12}\sqrt{5}\right) = 0.6489. \tag{6.6.13}$$

An elementary calculation shows that

$$D(\mu_0, \mu_0) = \frac{3}{5} \tag{6.6.14}$$

and hence

$$C \le 1.68, \tag{6.6.15}$$

which proves Theorem 6.1. ■

CHAPTER 7

Stability of Non-Relativistic Matter

With the necessary preliminaries in place we can now turn to proofs of stability of matter. The simplest model of matter for which one wants to prove stability of the second kind is the conventional non-relativistic Hamiltonian dating back to the earliest days of quantum mechanics. It is also the first example for which stability was proved; this was done by Dyson and Lenard in 1967 [44]. With the aid of the inequalities proved in the previous chapters, the proof turns out to be rather short.

One consequence of stability of the second kind is that matter is *extensive*, and this connection will be explained in Section 7.5. Alternative proofs of stability will also be briefly discussed. This chapter will conclude with a demonstration that stability of the second kind does not hold for bosons.

In units described in detail in Section 2.1.7, the Hamiltonian under consideration is

$$H = \frac{1}{2}\sum_{j=1}^{N}\left(-i\nabla_j + \sqrt{\alpha}\boldsymbol{A}(\boldsymbol{x}_j)\right)^2 + \alpha V_C(\underline{\boldsymbol{X}}, \underline{\boldsymbol{R}}), \tag{7.0.1}$$

where $\alpha > 0$ is the fine structure constant, and where $V_C(\underline{\boldsymbol{X}}, \underline{\boldsymbol{R}})$ is the total Coulomb potential energy, defined in (2.1.21),

$$V_C(\underline{\boldsymbol{X}}, \underline{\boldsymbol{R}}) = \sum_{1\le i<j\le N}\frac{1}{|\boldsymbol{x}_i - \boldsymbol{x}_j|} - \sum_{i=1}^{N}\sum_{j=1}^{M}\frac{Z_j}{|\boldsymbol{x}_i - \boldsymbol{R}_j|} + \sum_{1\le k<l\le M}\frac{Z_k Z_l}{|\boldsymbol{R}_k - \boldsymbol{R}_l|}. \tag{7.0.2}$$

Here, N is the number of electrons, and there are M nuclei with charges $Z_j > 0$ located at distinct points $\boldsymbol{R}_j$. The magnetic vector potential $\boldsymbol{A}$ is an arbitrary function in $L^2_{\mathrm{loc}}(\mathbb{R}^3)$ with values in $\mathbb{R}^3$.

If the interaction of the spin of the particles with the magnetic field is taken into account, there is a contribution $\sqrt{\alpha}\boldsymbol{B}\cdot\boldsymbol{\sigma}$ to the energy, as explained in

Section 2.1.6. In this case, stability still holds, but only if one takes the energy of the magnetic field into account. Even then, it will be necessary to have a bound on $Z\alpha^2$ and α. The proof of stability is quite a bit harder in this case, and we defer its discussion until Chapter 9.

7.1 Proof of Stability of Matter

Theorem 7.1 (Stability of Non-Relativistic Matter). *Let* $Z = \max_j\{Z_j\}$. *For all normalized, antisymmetric wave functions* ψ *with* q *spin states,*

$$\boxed{(\psi, H\psi) \geq -0.231\,\alpha^2 N q^{2/3}\left(1 + 2.16\,Z(M/N)^{1/3}\right)^2.} \tag{7.1.1}$$

We note that since $N^{1/3}M^{2/3} \leq N + M$, this yields the desired linear lower bound. The importance of the linear dependence on the total number of particles was explained in the Prologue and in Section 3.2.

The bound (7.1.1) is not optimal, in the sense that for $N \gg M$ the lower bound should depend only on M, whereas for $N \ll M$ it should only depend on N. We note that a bound depending only on N can be deduced from relativistic stability discussed in Chapter 8, as explained in Remark 8.6. Moreover, a bound that depends only on M can be deduced from Theorem 12.1 in Chapter 12, where it is shown that the number of electrons that can be bound is at most $\sum_{j=1}^{M}(2Z_j + 1) \leq (2Z + 1)M$, that is, the ground state energy of H for $N \geq (2Z + 1)M$ is equal to the ground state energy with N equal to the largest integer $\leq (2Z + 1)M$.

In the case of neutral hydrogen, $Z = 1$, $M = N$ and $q = 2$, the bound (7.1.1) yields $\boxed{\text{7.29 Rydbergs}}$ per nucleus as a lower bound. (Recall from Section 2.1.7 that in our units an energy of α^2 is 2 Rydbergs.) In case the optimal constant in the kinetic energy inequality (4.2.10) equals the conjectured semiclassical value $(3/5)(6\pi^2)^{2/3}$, the bound could be improved to $\boxed{\text{4.90 Rydbergs}}$ for hydrogen.

Proof. We start by noting that without loss of generality, we may assume that all the Z_j equal a common value Z. This follows from the monotonicity of the ground state energy in the nuclear charges, which was discussed in Section 3.2.3. Thus we can assume that $Z_j = Z$ henceforth.

According to Corollary 4.1 in Chapter 4, we have

$$\frac{1}{2}\left(\psi, \sum_j (-i\nabla_j + \sqrt{\alpha}\boldsymbol{A}(\boldsymbol{x}_j))^2\,\psi\right) \geq \frac{K}{2}q^{-2/3}\int_{\mathbb{R}^3} \varrho_\psi(\boldsymbol{x})^{5/3}\mathrm{d}\boldsymbol{x}, \qquad (7.1.2)$$

with $K \geq (9/5)(4\pi^2)^{1/3} \approx 3.065$. Here, ϱ_ψ denotes the particle density, defined in (3.1.4). We point out that for this bound to hold, the antisymmetric nature of the wave functions is essential. It allows us to bound the spin-summed one-particle density matrix of ψ from above by q times the identity.

To obtain a lower bound on the Coulomb energy, we first employ the exchange estimate of Theorem 6.1, which states that

$$\left(\psi, \sum_{1\leq i<j\leq N} \frac{1}{|\boldsymbol{x}_i - \boldsymbol{x}_j|}\,\psi\right) \geq D(\varrho_\psi, \varrho_\psi) - 1.68\int_{\mathbb{R}^3} \varrho_\psi(\boldsymbol{x})^{4/3}\mathrm{d}\boldsymbol{x}, \qquad (7.1.3)$$

with D denoting the direct electrostatic energy (5.1.3). The electron–nuclear interaction is

$$\left(\psi, \sum_{i=1}^{N}\sum_{k=1}^{M} \frac{Z}{|\boldsymbol{x}_i - \boldsymbol{R}_k|}\,\psi\right) = \sum_{k=1}^{M}\int_{\mathbb{R}^3} \frac{Z}{|\boldsymbol{x} - \boldsymbol{R}_k|}\varrho_\psi(\boldsymbol{x})\mathrm{d}\boldsymbol{x}, \qquad (7.1.4)$$

so the total Coulomb energy is bounded below as

$$\begin{aligned}(\psi, V_C\,\psi) \geq D(\varrho_\psi, \varrho_\psi) &- \sum_{k=1}^{M}\int_{\mathbb{R}^3} \frac{Z}{|\boldsymbol{x} - \boldsymbol{R}_k|}\varrho_\psi(\boldsymbol{x})\mathrm{d}\boldsymbol{x} \\ &+ U(\underline{\boldsymbol{R}}) - 1.68\int_{\mathbb{R}^3} \varrho_\psi(\boldsymbol{x})^{4/3}\mathrm{d}\boldsymbol{x}. \qquad (7.1.5)\end{aligned}$$

(Recall the definition of the nuclear repulsion energy $U(\underline{\boldsymbol{R}})$ in (2.1.24).)

A lower bound to the ground state energy of H can be obtained by combining (7.1.2) and (7.1.5) and minimizing

$$\begin{aligned}\mathcal{E}^{\mathrm{TFD}}(\varrho) := \frac{K}{2}q^{-2/3}\int_{\mathbb{R}^3} \varrho(\boldsymbol{x})^{5/3}\mathrm{d}\boldsymbol{x} + \alpha D(\varrho, \varrho) + \alpha U(\underline{\boldsymbol{R}}) \\ - \sum_{k=1}^{M}\int_{\mathbb{R}^3} \frac{Z\alpha}{|\boldsymbol{x} - \boldsymbol{R}_k|}\varrho(\boldsymbol{x})\mathrm{d}\boldsymbol{x} - 1.68\alpha\int_{\mathbb{R}^3} \varrho(\boldsymbol{x})^{4/3}\mathrm{d}\boldsymbol{x} \qquad (7.1.6)\end{aligned}$$

under the condition that $\int_{\mathbb{R}^3} \varrho(\boldsymbol{x}) \mathrm{d}\boldsymbol{x} = N$. This is the Thomas–Fermi–Dirac problem, which is extensively studied in [113]. The simpler Thomas–Fermi functional (without the last term in (7.1.6)) will appear again in Section 7.3.

Instead of pursuing that route, it is simpler (for the purpose of our theorem) to use the electrostatic inequality of Theorem 5.3, which implies that

$$D(\varrho_\psi, \varrho_\psi) - \sum_{k=1}^{M} \int_{\mathbb{R}^3} \frac{Z}{|\boldsymbol{x} - \boldsymbol{R}_k|} \varrho_\psi(\boldsymbol{x}) \mathrm{d}\boldsymbol{x} + U(\underline{\boldsymbol{R}}) \geq - \int_{\mathbb{R}^3} \frac{Z}{\mathfrak{D}(\boldsymbol{x})} \varrho_\psi(\boldsymbol{x}) \mathrm{d}\boldsymbol{x}, \tag{7.1.7}$$

where $\mathfrak{D}(\boldsymbol{x}) = \min_k |\boldsymbol{x} - \boldsymbol{R}_k|$ denotes the distance of $\boldsymbol{x}$ to the nearest nucleus. The last positive term in (5.2.6) has been dropped for a lower bound, since it is not needed for the following argument. In brief we are using the bound

$$(\psi, V_C\, \psi) \geq -1.68 \int_{\mathbb{R}^3} \varrho_\psi(\boldsymbol{x})^{4/3} \mathrm{d}\boldsymbol{x} - \int_{\mathbb{R}^3} \frac{Z}{\mathfrak{D}(\boldsymbol{x})} \varrho_\psi(\boldsymbol{x}) \mathrm{d}\boldsymbol{x}. \tag{7.1.8}$$

The $\int \varrho_\psi^{4/3}$ term can be bounded via the Cauchy-Schwarz inequality as

$$\begin{aligned} \int_{\mathbb{R}^3} \varrho_\psi(\boldsymbol{x})^{4/3} \mathrm{d}\boldsymbol{x} &\leq \left(\int_{\mathbb{R}^3} \varrho_\psi(\boldsymbol{x})^{5/3} \mathrm{d}\boldsymbol{x} \right)^{1/2} \left(\int_{\mathbb{R}^3} \varrho_\psi(\boldsymbol{x}) \mathrm{d}\boldsymbol{x} \right)^{1/2} \\ &\leq \frac{a}{2} \int_{\mathbb{R}^3} \varrho_\psi(\boldsymbol{x})^{5/3} \mathrm{d}\boldsymbol{x} + \frac{N}{2a} \end{aligned} \tag{7.1.9}$$

for arbitrary $a > 0$. With the choice $a = Kq^{-2/3}\varepsilon/(1.68\alpha)$ for some $0 < \varepsilon < 1$, this implies the lower bound

$$\begin{aligned} (\psi, H\psi) \geq -(0.84)^2 \frac{2\alpha^2 q^{2/3} N}{K\varepsilon} + \frac{K}{2} q^{-2/3}(1-\varepsilon) \int_{\mathbb{R}^3} \varrho_\psi(\boldsymbol{x})^{5/3} \mathrm{d}\boldsymbol{x} \\ - Z\alpha \int_{\mathbb{R}^3} \frac{1}{\mathfrak{D}(\boldsymbol{x})} \varrho_\psi(\boldsymbol{x}) \mathrm{d}\boldsymbol{x}. \end{aligned} \tag{7.1.10}$$

We pick $b > 0$ and write $\mathfrak{D}(\boldsymbol{x})^{-1}$ as $(\mathfrak{D}(\boldsymbol{x})^{-1} - b) + b$. The contribution of the last term to the integral in (7.1.10) is simply bN. Moreover, by minimizing over all non-negative functions $\varrho_\psi(\boldsymbol{x})$, we find that

$$\frac{K}{2}q^{-2/3}(1-\varepsilon)\int_{\mathbb{R}^3}\varrho_\psi(\boldsymbol{x})^{5/3}\mathrm{d}\boldsymbol{x} - Z\alpha\int_{\mathbb{R}^3}\left(\frac{1}{\mathfrak{D}(\boldsymbol{x})}-b\right)\varrho_\psi(\boldsymbol{x})\mathrm{d}\boldsymbol{x}$$
$$\geq -\frac{2q}{5}\left(\frac{6}{5K(1-\varepsilon)}\right)^{3/2}(Z\alpha)^{5/2}\int_{\mathbb{R}^3}\left[\frac{1}{\mathfrak{D}(\boldsymbol{x})}-b\right]_+^{5/2}\mathrm{d}\boldsymbol{x}. \quad (7.1.11)$$

Recall that $[\,\cdot\,]_+$ denotes the positive part. Since

$$\left[\frac{1}{\mathfrak{D}(\boldsymbol{x})}-b\right]_+^{5/2} = \max_{1\leq k\leq M}\left[\frac{1}{|\boldsymbol{x}-\boldsymbol{R}_k|}-b\right]_+^{5/2} \leq \sum_{k=1}^{M}\left[\frac{1}{|\boldsymbol{x}-\boldsymbol{R}_k|}-b\right]_+^{5/2}$$

we can bound

$$\int_{\mathbb{R}^3}\left[\frac{1}{\mathfrak{D}(\boldsymbol{x})}-b\right]_+^{5/2}\mathrm{d}\boldsymbol{x} \leq M\int_{|\boldsymbol{x}|\leq 1/b}\left(\frac{1}{|\boldsymbol{x}|}-b\right)^{5/2}\mathrm{d}\boldsymbol{x} = \frac{5\pi^2}{4}Mb^{-1/2}. \quad (7.1.12)$$

After choosing the optimal value of b, we thus obtain the lower bound

$$(\psi, H\psi) \geq -(0.84)^2\frac{2\alpha^2q^{2/3}N}{K\varepsilon} - N^{1/3}M^{2/3}(Z\alpha)^2\frac{18}{5K(1-\varepsilon)}\left(\frac{\pi^2 q}{4}\right)^{2/3}. \quad (7.1.13)$$

Finally, after optimizing over ε, we have

$$(\psi, H\psi) \geq -\frac{2q^{2/3}\alpha^2}{K}N\left(0.84 + Z\sqrt{D}(M/N)^{1/3}\right)^2, \quad (7.1.14)$$

where $D = (9/10)\pi^{4/3}2^{-1/3} \approx 3.287$. As stated in Corollary 4.1, $K \geq (9/5)(4\pi^2)^{1/3} \approx 3.065$. This yields (7.1.1). ∎

7.2 An Alternative Proof of Stability

The proof of stability of non-relativistic matter given in the previous subsection relied on three essential ingredients. One is the LT inequality, which relates the kinetic energy of fermions to their density $\varrho_\psi(\boldsymbol{x})$. The second is the estimate on the indirect (or exchange) part of the Coulomb energy, which gives a lower bound on the electron interaction in terms of the classical electrostatic energy of the electron charge distribution. Finally, the electrostatic inequality of Section 5.2 relates the total classical electrostatic energy to the interaction energy of the electrons to the nearest nucleus only. In the end, one has to deal with the $\varrho^{5/3}$ semiclassical energy, the negative $\varrho^{4/3}$ exchange estimate, and the negative nearest neighbor electron–nucleus interaction.

One can eliminate the use of the exchange estimate in Section 7.1 by replacing Theorem 5.3 by its corollary, Theorem 5.4. This is the most direct route to the proof of stability, but it obscures the role of the non-classical exchange (or correlation) energy, and it leads to slightly worse constants. The idea for this route to stability is due to Solovej [173].

The electrostatic inequality of Theorem 5.4 implies that

$$V_C(\underline{\boldsymbol{X}}, \underline{\boldsymbol{R}}) \geq -(2Z+1) \sum_{i=1}^{N} \frac{1}{\mathfrak{D}(\boldsymbol{x}_i)}$$

where $\mathfrak{D}(\boldsymbol{x}) = \min_k |\boldsymbol{x} - \boldsymbol{R}_k|$, as before. We have dropped the positive last term in (5.3.2), since it will not be necessary in the following. We shall write again $\mathfrak{D}(\boldsymbol{x})^{-1} = (\mathfrak{D}(\boldsymbol{x})^{-1} - b) + b$ for $b > 0$. The LT inequality in Theorem 4.1 implies that for antisymmetric functions with q spin states,

$$\begin{aligned} &\sum_{i=1}^{N} \left[-\frac{1}{2}\Delta_i - (2Z+1)\alpha \left(\frac{1}{\mathfrak{D}(\boldsymbol{x}_i)} - b \right) \right] \\ &\geq -q L_{1,3} 2^{3/2} \alpha^{5/2} (2Z+1)^{5/2} \int_{\mathbb{R}^3} \left[\frac{1}{\mathfrak{D}(\boldsymbol{x})} - b \right]_+^{5/2} \mathrm{d}\boldsymbol{x}. \end{aligned} \tag{7.2.1}$$

The additional factor $2^{3/2}$ results from the factor $1/2$ in front of the Laplacian, which is absent in (4.1.11). As explained in Section 4.1.2, $L_{1,3} \leq \frac{\pi}{\sqrt{3}} L_{1,3}^{\mathrm{cl}} \approx 0.0123$. After using the bound (7.1.12) on the last integral we obtain the lower bound

$$H \geq -q L_{1,3} 2^{3/2} \frac{5\pi^2}{4} (2Z+1)^{5/2} \alpha^{5/2} M b^{-1/2} - (2Z+1)\alpha N b.$$

After inserting the optimal value of b, this yields

$$\begin{aligned} H &\geq -\frac{3}{2} (5\pi^2 L_{1,3})^{2/3} q^{2/3} \, [(2Z+1)\alpha]^2 \, M^{2/3} N^{1/3} \\ &\geq -1.073 \, q^{2/3} \, [(2Z+1)\alpha]^2 \, M^{2/3} N^{1/3}. \end{aligned} \tag{7.2.2}$$

In the case of neutral hydrogen, $Z = 1$, $M = N$ and $q = 2$, this bound yields 30.52 Rydbergs per nucleus as a lower bound, which is about a factor 4 worse than our previous bound.

We remark that the method for proving stability of matter given above can be extended to yield a lower bound on H that depends only on N and not on M. In order to do this, one must not drop the positive nearest-nucleus repulsion in Theorem 5.4, which we could afford to do above. This was shown in [83, Thm. 3 of Part II]. Alternatively, one can use the stability of relativistic matter to obtain such a bound, as explained in Remark 8.6 in the next chapter.

7.3 Stability of Matter via Thomas–Fermi Theory

The electrostatic inequality of Theorem 5.3, together with the exchange estimate of Theorem 6.1, allowed us to bound the total Coulomb potential energy $V_C(\underline{\boldsymbol{X}}, \underline{\boldsymbol{R}})$ from below by another kind of potential energy, which is just the sum of one-particle terms corresponding to the attraction to the nearest nucleus. For this latter potential energy it is then clear that binding between atoms does not occur, and hence stability of the second kind reduces to stability of the first kind. That is to say, in such a model it is clear that the lowest energy occurs when the Voronoi cells are as large as possible, which occurs when the nuclei are infinitely far apart. This reasoning is one of the key ingredients in the proofs of stability of matter given in the previous two sections, the other being the kinetic energy inequality in Corollary 4.1. In real life, atoms do bind together, of course, but there is nothing wrong with a lower bound on the total energy obtained by estimating the energy from below by a model that does not lead to binding.

There is another way of bounding the total energy from below by an expression that prohibits binding. This is the original route taken by Lieb and Thirring in [134] and uses the fact that binding between atoms does not occur in *Thomas–Fermi theory*. Since we have already given a complete proof of stability of non-relativistic matter in the previous section, we shall only give a brief account of the proof via Thomas–Fermi theory here. Our exposition will not be complete, since we refrain from proving the *no-binding theorem*. For details, we refer the interested reader to [130, 108, 113].

The Thomas–Fermi functional has, as its motivation, a semiclassical approximation to the energy of quantum-mechanical system with particle density $\varrho(\boldsymbol{x})$.

It is given by

$$\mathcal{E}^{\mathrm{TF}}(\varrho) = \frac{3}{10}\gamma q^{-2/3} \int_{\mathbb{R}^3} \varrho(\boldsymbol{x})^{5/3} \mathrm{d}\boldsymbol{x} - \sum_{j=1}^{M} Z_j \alpha \int_{\mathbb{R}^3} \frac{\varrho(\boldsymbol{x})}{|\boldsymbol{x} - \boldsymbol{R}_j|} \mathrm{d}\boldsymbol{x}$$

$$+ \frac{\alpha}{2} \iint_{\mathbb{R}^3 \times \mathbb{R}^3} \frac{\varrho(\boldsymbol{x})\varrho(\boldsymbol{y})}{|\boldsymbol{x} - \boldsymbol{y}|} \mathrm{d}\boldsymbol{x}\mathrm{d}\boldsymbol{y} + \sum_{i<j}^{M} \frac{Z_i Z_j \alpha}{|\boldsymbol{R}_i - \boldsymbol{R}_j|}. \tag{7.3.1}$$

The first term, with $\gamma = (6\pi^2)^{2/3}$, is the semiclassical approximation to the kinetic energy, as explained in Chapter 4. It will be convenient not to fix γ but rather keep it as an adjustable parameter, as will become clear below. We will always assume that $\gamma > 0$, however.

If N denotes the number of electrons, the ground state energy in Thomas–Fermi theory is given by

$$E^{\mathrm{TF}}(N, \underline{Z}, \underline{\boldsymbol{R}}, \gamma) = \inf\left\{ \mathcal{E}^{\mathrm{TF}}(\varrho) : \varrho \in L^{5/3}(\mathbb{R}^3),\ \varrho(\boldsymbol{x}) \geq 0, \int_{\mathbb{R}^3} \varrho(\boldsymbol{x})\mathrm{d}\boldsymbol{x} = N \right\}. \tag{7.3.2}$$

It has several remarkable properties.

- $E^{\mathrm{TF}}(N, \underline{Z}, \underline{\boldsymbol{R}}, \gamma) = E^{\mathrm{TF}}(\sum_i Z_i, \underline{Z}, \underline{\boldsymbol{R}}, \gamma)$ for all $N \geq \sum_i Z_i$. That is, the Thomas–Fermi energy can not be decreased by adding more electrons than the total nuclear charge of the system. In fact, one can show that a minimizer in (7.3.2) only exists for $N \leq \sum_i Z_i$, not for larger N. The excess charge simply moves to infinity, and the system stays neutral.
- $E^{\mathrm{TF}}(N, \underline{Z}, \underline{\boldsymbol{R}}, \gamma) \geq \min\left\{ \sum_{i=1}^{M} E^{\mathrm{TF}}(N_i, Z_i, \boldsymbol{R}_i, \gamma) : N_i \geq 0, \sum_{i=1}^{M} N_i = N \right\}$. This is the *no-binding theorem*, since it says that moving the nuclei infinitely far apart will lower the energy. Note that the Thomas–Fermi energy for a single atom, $E^{\mathrm{TF}}(N, Z, \boldsymbol{R}, \gamma)$, is independent of the nuclear position R.
- $E^{\mathrm{TF}}(N, Z, 0, \gamma) = Z^{7/3}\gamma^{-1} E^{\mathrm{TF}}(N/Z, 1, 0, 1)$. This simple scaling property follows easily by scaling (7.3.1) appropriately.

In combination, these properties imply the lower bound

$$E^{\mathrm{TF}}(N, \underline{Z}, \underline{\boldsymbol{R}}, \gamma) \geq \gamma^{-1} E^{\mathrm{TF}}(1, 1, 0, 1) \sum_{i=1}^{M} Z_i^{7/3} \geq \gamma^{-1} E^{\mathrm{TF}}(1, 1, 0, 1) Z^{7/3} M, \tag{7.3.3}$$

which is stability of the second kind. A numerical computation shows that $E^{\mathrm{TF}}(1, 1, 0, 1) = -7.356\,\alpha^2$.

The importance of the Thomas–Fermi functional for the problem of stability of matter comes from the observation in [134] that the Thomas–Fermi energy (for an appropriate value of γ) is a lower bound to the quantum-mechanical ground state energy. In fact, we already have all the necessary inequalities at our disposal to show this.

The kinetic energy inequality (4.2.10) implies that the first term in the Thomas–Fermi functional, with $\gamma = 6(\pi^2/2)^{1/3}$, is a lower bound to the kinetic energy of N electrons. The Lieb–Oxford inequality[1] (6.2.14), in combination with a simple Cauchy–Schwarz estimate as in (7.1.9), then implies that

$$(\psi, H\psi) \geq E^{\mathrm{TF}}(N, \underline{Z}, \underline{\boldsymbol{R}}, 6(\pi^2/2)^{1/3} - 5q^{2/3}1.68a/3) - 1.68\frac{N\alpha^2}{2a} \quad (7.3.4)$$

for any $a > 0$. By using (7.3.3) and optimizing over a one obtains the bound

$$(\psi, H\psi) \geq -0.231\, q^{2/3}\alpha^2 N \left(1 + 1.77\sqrt{\frac{1}{N}\sum_{j=1}^{M} Z_j^{7/3}}\right)^2 . \quad (7.3.5)$$

For the hydrogen gas, (7.3.5) implies a bound of 5.60 Rydbergs per atom, which is slightly better than the one obtained in Theorem 7.1.

7.4 Other Routes to a Proof of Stability

In the preceding three subsections, we have given several proofs of stability of non-relativistic matter with electrostatic forces. Historically, the first proof of this fact was given by Dyson and Lenard in 1967. Subsequently, many other proofs were given. None of these proofs, however, yields a bound on the optimal constant (in the neutral case) that is as good as the one obtained in the previous subsection using Thomas–Fermi theory and the Lieb–Oxford inequality. We give a brief summary of some of the early work here.

[1] In the original proof of stability of matter by Lieb and Thirring, the Lieb–Oxford bound was not available. Instead, a weaker bound on the exchange energy of the form $(\int \varrho^{5/3})^{1/2}(\int \varrho)^{1/2}$ in place of $\int \varrho^{4/3}$ was used.

7.4.1 Dyson–Lenard, 1967

The first proof of stability of non-relativistic matter was given by Dyson and Lenard in 1967 [44]. This proof was one of the most complicated proofs ever to appear in mathematical physics. For many years, it was regarded with awe, like Onsager's solution of the two-dimensional Ising model in 1944 [146]. The key question that had to be addressed was how to bring in the Pauli exclusion principle. In the proof by Lieb and Thirring, it is this principle that lies behind the $\int \varrho^{5/3}$ bound on the kinetic energy. This is a global result. In the Dyson–Lenard proof, in contrast, the Pauli principle is used locally to mandate that for $N \geq 2$ (spinless) fermions in a ball of radius ℓ, the kinetic energy (with Neumann boundary conditions) is at least $(N-1)/\ell^2$, since at most one such fermion can be in the zero energy state.

The lower bound obtained in [44] yields about 10^{14} Ry per particle for a hydrogen gas. This is, of course, an unrealistically large number and resulted from the many different inequalities that had to be used.

7.4.2 Federbush, 1975

Federbush [55] realized that the methods of constructive quantum field theory could be utilized to solve the stability problem. Although his paper is not very transparent, it is much shorter than the original proof by Dyson and Lenard. An estimate on the constant appearing in the inequality is not given, however.

7.4.3 Some Later Work

Other approaches for proving stability of non-relativistic matter were developed by Fefferman [56] and Graf [79], although without estimates of the constant in the inequality. In [56] results on extensivity of matter and properties of the ground state density are also obtained by dividing space suitably into boxes. In [79] the necessary electrostatic inequality is obtained by decomposing space into a lattice of disjoint simplices and, by averaging over translations and rotations of this lattice, a lower bound on the Coulomb energy is obtained by restricting the Coulomb interaction to pairs of particles belonging to the same simplex. The use of averaging to get a lower bound (but with cubes and only with the translational and not the rotational averaging) goes back to [34]. Finally, the Pauli principle is used locally in every simplex in the form of a lower bound on the kinetic energy in terms of the $5/3$ power of the particle number in the simplex.

7.5 Extensivity of Matter

Suppose we have a non-relativistic Coulomb system composed of several species of particles, of which the negative ones are fermions. The nuclei can be static or dynamic. By Theorem 7.1 and Corollary 4.1 there are constants $A \geq 0$ and $K > 0$ such that

$$\mathcal{E}(\psi) \geq -AN \qquad \text{and} \qquad T_\psi \geq K \int_{\mathbb{R}^3} \varrho_\psi^{5/3}(\boldsymbol{x}) \mathrm{d}\boldsymbol{x} \tag{7.5.1}$$

for any normalized ψ. In fact, $K \geq (9/10)(2\pi/q)^{2/3}$. Here $\mathcal{E}(\psi)$ is the energy functional $(\psi, H\psi)$, T_ψ is the kinetic energy of the negative particles, and ϱ_ψ is the density of the negative particles (see Eq. (3.1.4) in Chapter 3). There are N electrons and M nuclei. The constant A depends on Z and the ratio M/N.

The following theorem shows that any wave function ψ with bounded energy per particle is necessarily spatially extensive, in the sense that the particles occupy a volume proportional to N.

Theorem 7.2 (Extensivity of Matter). *Let ψ be any normalized function with energy $E = \mathcal{E}(\psi)$, and let A and K be constants such that (7.5.1) holds.*

(i) For each $p > 0$ there is a universal constant γ_p such that

$$\left\{ \frac{1}{N} \int_{\mathbb{R}^3} \varrho_\psi(\boldsymbol{x}) |\boldsymbol{x}|^p \mathrm{d}\boldsymbol{x} \right\}^{1/p} \geq \gamma_p K^{1/2} \left(\sqrt{E/N + A} + \sqrt{A} \right)^{-1} N^{1/3}. \tag{7.5.2}$$

Thus, the average distance to the origin, $\mathbf{0} \in \mathbb{R}^3$, among all the particles is at least as great as $N^{1/3}$. Since the choice of the origin is arbitrary, we can make the same statement relative to any point $\boldsymbol{y} \in \mathbb{R}^3$. The 'volume of matter' is, therefore, proportional to the number of particles.

(ii) If Ω is any measurable set in $\mathbb{R}^3$

$$\frac{1}{N} \int_{\Omega} \varrho_\psi(\boldsymbol{x}) \mathrm{d}\boldsymbol{x} \leq K^{-3/5} \left(\sqrt{E/N + A} + \sqrt{A} \right)^{6/5} \left(\frac{\mathrm{vol}(\Omega)}{N} \right)^{2/5}. \tag{7.5.3}$$

Thus, if Ω contains a non-vanishing fraction of the particles then vol(Ω) *has to be bounded below by a constant times N.*

Remark 7.1. A special case of (7.5.2) was given in [135]. The general case appeared in [108]. Inequality (7.5.3) is due to Thirring [178, Sect. 4.3.5].

Proof. (i) First we derive an upper bound on the kinetic energy T_ψ. Write

$$E = \mathcal{E}(\psi) = \lambda T_\psi + (1-\lambda)T_\psi + V_\psi \tag{7.5.4}$$

with $0 \le \lambda \le 1$, and note that the last two terms are bounded below by $-AN/(1-\lambda)$. This follows from the stability bound and scaling, i.e., the fact that the kinetic energy has the dimension of the square of an inverse length and the potential energy has the dimension of an inverse length. Thus we have

$$T_\psi \le \frac{E}{\lambda} + \frac{AN}{\lambda(1-\lambda)} = \frac{E+AN}{\lambda} + \frac{AN}{1-\lambda}. \tag{7.5.5}$$

Note that the right side of (7.5.5) tends to $+\infty$ as λ approaches 0 or 1 since $E + AN > 0$. Minimization of (7.5.5) with respect to λ yields the bound

$$T_\psi \le \left(\sqrt{E+AN} + \sqrt{AN}\right)^2. \tag{7.5.6}$$

With the second inequality in (7.5.1) we reach the basic inequality

$$K \int_{\mathbb{R}^3} \varrho_\psi(\boldsymbol{x})^{5/3} \mathrm{d}\boldsymbol{x} \le T_\psi \le \left(\sqrt{E+AN} + \sqrt{AN}\right)^2. \tag{7.5.7}$$

Next, we assert the following general fact [108, p. 563] about functions ϱ from $\mathbb{R}^3$ to $\mathbb{R}^+$:

$$\left(\int_{\mathbb{R}^3} \varrho(\boldsymbol{x})^{5/3} \mathrm{d}\boldsymbol{x}\right)^{p/2} \int_{\mathbb{R}^3} |\boldsymbol{x}|^p \varrho(\boldsymbol{x}) \mathrm{d}\boldsymbol{x} \ge C_p \left(\int_{\mathbb{R}^3} \varrho(\boldsymbol{x}) \mathrm{d}\boldsymbol{x}\right)^{1+5p/6}, \tag{7.5.8}$$

for all $p > 0$. The sharp constant, C_p, is obtained by inserting $\overline{\varrho}(\boldsymbol{x}) = (1 - |\boldsymbol{x}|^p)^{3/2}$ for $|\boldsymbol{x}| \le 1$ and $\overline{\varrho}(\boldsymbol{x}) = 0$ for $|\boldsymbol{x}| \ge 1$ into (7.5.8). This is a simple exercise in the calculus of variations which we shall leave to the reader.

By combining (7.5.7) and (7.5.8), together with $\int \varrho_\psi = N$, we obtain (7.5.2) with $\gamma_p = C_p^{1/p}$.

(ii) By Hölder's inequality, $\int_\Omega \varrho_\psi \le (\int_\Omega 1)^{2/5} (\int_\Omega \varrho_\psi^{5/3})^{3/5} \le \mathrm{vol}(\Omega)^{2/5} (\int_{\mathbb{R}^3} \varrho_\psi^{5/3})^{3/5}$. Inequality (7.5.3) is now an immediate consequence of (7.5.7). ∎

7.6 Instability for Bosons

7.6.1 The $N^{5/3}$ Law

If one allows the particles to have $q = N$ different spin states, then the bound (7.1.1) holds regardless of the statistics of the wave functions. This was discussed in Chapter 3, Subsection 3.1.3.1. In particular, this shows that for *bosons* a lower bound of the form $-CN^{5/3}$ holds. This was already pointed out by Dyson and Lenard [44] and an alternative proof was later given by Brydges and Federbush [24]. The power $5/3$ can be proved to be optimal in the case of infinitely massive nuclei [110], and we shall show this here. That is, in the bosonic case also an upper bound of this form holds. In particular, *non-relativistic matter made out of bosons is stable of the first kind, but unstable of the second kind. This is the primary example of the distinction between the two kinds of stability.*

Proposition 7.1 ($N^{5/3}$ Instability for Bosons). *Let $Z_i = Z$ for $1 \leq i \leq M$. There exists a normalized symmetric wave function ψ of N boson coordinates, and positions $\underline{\boldsymbol{R}} = (\boldsymbol{R}_1, \ldots, \boldsymbol{R}_M)$ of M nuclei in $\mathbb{R}^3$, such that*

$$\mathcal{E}(\psi) = (\psi, H\psi) \leq -C\alpha^2 Z^{4/3} \min\{N, ZM\}^{5/3} \tag{7.6.1}$$

for some constant $C > 0$ independent of all the parameters.

Proof. For simplicity, we shall assume that $M = n^3$ for some integer n, and $N = ZM$. In order to show the desired upper bound, we can use a trial function of the form

$$\psi(\boldsymbol{x}_1, \ldots, \boldsymbol{x}_N) = \prod_{i=1}^{N} \phi_\lambda(\boldsymbol{x}_i).$$

Here $\lambda > 0$ is a scaling parameter which will be optimized at the end. That is, we fix some function g with $\int_{\mathbb{R}^3} |g(\boldsymbol{x})|^2 \mathrm{d}\boldsymbol{x} = 1$ and write

$$\phi_\lambda(\boldsymbol{x}) = \lambda^{3/2} g(\lambda \boldsymbol{x}).$$

The energy of our trial function ψ is easy to calculate. We obtain

$$(\psi, H\psi) = N\lambda^2 \int_{\mathbb{R}^3} |\nabla g(\boldsymbol{x})|^2 \mathrm{d}\boldsymbol{x} + \lambda\alpha W(N, \underline{\boldsymbol{R}}), \tag{7.6.2}$$

where

$$W(N, \underline{\boldsymbol{R}}) = \frac{1}{2}N(N-1)\iint\limits_{\mathbb{R}^3\times\mathbb{R}^3} \frac{|g(\boldsymbol{x})|^2|g(\boldsymbol{y})|^2}{|\boldsymbol{x}-\boldsymbol{y}|}\mathrm{d}\boldsymbol{x}\mathrm{d}\boldsymbol{y}$$

$$- ZN\sum_{k=1}^{M}\int\limits_{\mathbb{R}^3} \frac{|g(\boldsymbol{x})|^2}{|\boldsymbol{x}-\boldsymbol{R}_k|}\mathrm{d}x + U(\underline{\boldsymbol{R}}).$$

The nuclear repulsion $U(\underline{\boldsymbol{R}})$ is defined in (2.1.24). We want to show that we can find an $\underline{\boldsymbol{R}}$ such that $W(N, \underline{\boldsymbol{R}}) \leq -CZ^{2/3}N^{4/3}$ for some constant $C > 0$. The desired upper bound (7.6.1) then follows immediately from (7.6.2) by optimizing over λ.

Choose g to be compactly supported, and divide the support of g into M cells Γ_k, $1 \leq k \leq M$, such that

$$\int\limits_{\Gamma_k} |g(\boldsymbol{x})|^2\mathrm{d}\boldsymbol{x} = \frac{1}{M} \quad \text{for all } k. \tag{7.6.3}$$

We place one nucleus in each cell Γ_k, and average its position with respect to the weight $M|g(\boldsymbol{x})|^2$, restricted to Γ_k. The average value of $W(N, \underline{\boldsymbol{R}})$ is then equal to

$$\left(\frac{1}{2}N(N-1) - ZNM + \frac{1}{2}Z^2M^2\right)\iint\limits_{\mathbb{R}^3\times\mathbb{R}^3} \frac{|g(\boldsymbol{x})|^2|g(\boldsymbol{y})|^2}{|\boldsymbol{x}-\boldsymbol{y}|}\mathrm{d}\boldsymbol{x}\mathrm{d}\boldsymbol{y}$$

$$-\frac{1}{2}Z^2M^2\sum_k\iint\limits_{\Gamma_k\times\Gamma_k} \frac{|g(\boldsymbol{x})|^2|g(\boldsymbol{y})|^2}{|\boldsymbol{x}-\boldsymbol{y}|}\mathrm{d}\boldsymbol{x}\mathrm{d}\boldsymbol{y}, \tag{7.6.4}$$

which implies that $W(N, \underline{\boldsymbol{R}})$ is less or equal to this value for some choice of $\underline{\boldsymbol{R}}$ (because an average is never less than the minimum).

Under the assumption $N = ZM$, the first term in (7.6.4) is negative and can be dropped for an upper bound. We are left with the problem of finding a lower bound on the self-energy terms

$$\frac{1}{2}\iint\limits_{\Gamma_k\times\Gamma_k} \frac{|g(\boldsymbol{x})|^2|g(\boldsymbol{y})|^2}{|\boldsymbol{x}-\boldsymbol{y}|}\mathrm{d}\boldsymbol{x}\mathrm{d}\boldsymbol{y}. \tag{7.6.5}$$

Let r_k denote the radius of the smallest ball containing Γ_k. Then (7.6.5) is certainly greater than the smallest possible self-energy of a charge distribution of total charge $1/M$ confined to a ball of radius r_k. It is well known that the smallest self-energy occurs when the charge is uniformly distributed over the boundary of the ball, and the minimal value equals $(1/M)^2/(2r_k)$ [118,

Section 11.15]. In particular,

$$\frac{1}{2}\iint\limits_{\Gamma_k\times\Gamma_k}\frac{|g(\boldsymbol{x})|^2|g(\boldsymbol{y})|^2}{|\boldsymbol{x}-\boldsymbol{y}|}\mathrm{d}\boldsymbol{x}\mathrm{d}\boldsymbol{y}\geq\frac{1}{2M^2r_k}.$$

Using Jensen's inequality,[2] $M^{-1}\sum_k r_k^{-1}\geq M(\sum_k r_k)^{-1}$. We thus have

$$\frac{1}{2}Z^2M^2\sum_k\iint\limits_{\Gamma_k\times\Gamma_k}\frac{|g(\boldsymbol{x})|^2|g(\boldsymbol{y})|^2}{|\boldsymbol{x}-\boldsymbol{y}|}\mathrm{d}\boldsymbol{x}\mathrm{d}\boldsymbol{y}\geq\frac{1}{2}Z^2M\frac{1}{\frac{1}{M}\sum_k r_k}.$$

It remains to estimate $(\sum_k r_k)/M$, the mean value of the radius of the smallest ball containing Γ_k. Note that we are free to choose the decomposition of the support of g into the cells Γ_k subject to the normalization constraint (7.6.3). It is fairly obvious that for sufficiently regular g one can choose the decomposition such that $(\sum_k r_k)/M\leq CM^{-1/3}$ for some constant $C>0$, which proves the desired bound. In fact, in [110] the following explicit construction is given.

For $\boldsymbol{x}=(x^1,x^2,x^3)\in\mathbb{R}^3$, let $g(\boldsymbol{x})=f(x^1)f(x^2)f(x^3)$ for some function f supported in the interval $[-1,1]$. Under the assumption that $M=n^3$ for some integer n, we can first decompose $[-1,1]$ into n intervals I_j with the property that $\int_{I_j}|f(t)|^2\mathrm{d}t=1/n$ for all $1\leq j\leq n$, and take the Γ_k to be products of intervals. If Γ_k is a rectangle with side lengths s, t and u, the radius of the smallest ball containing Γ_k is $r_k=(s^2+t^2+u^2)^{1/2}/2\leq(s+t+u)/2$. Note that the average value of s, t and u for all the Γ_k equals $1/n=M^{-1/3}$. Hence

$$\frac{1}{M}\sum_{m=1}^{M}r_k\leq\frac{3}{2}M^{-1/3}$$

in this case. This is precisely of the desired form. ∎

7.6.2 The $N^{7/5}$ Law

The $N^{5/3}$ instability is mitigated somewhat if the finite nuclear masses are taken into account. The system is still unstable, but the energy goes as $-CN^{7/5}$ in this case. It was Dyson in 1967 [43] who proved that the energy was at least as negative $-CN^{7/5}$ for charged bosons, but it wasn't until 1988 that a rigorous *lower* bound of this form was demonstrated by Conlon, Lieb and Yau [34]. Finally, in 2004, the correct value of the constant C for large N, as conjectured by Dyson, was proved as a lower bound by Lieb and Solovej [132], and as an

[2] If f is a convex function and μ is a positive measure on $\mathbb{R}$ then $\int f(x)\mu(\mathrm{d}x)/\mu(\mathbb{R})\geq f(\int x\,\mu(\mathrm{d}x)/\mu(\mathbb{R}))$. See [118, Sect. 2.2].

upper bound by Solovej in 2006 [172]. All these works consider the special case of positive and negative charges of equal magnitude, and equal masses of the particles. In principle, the analysis applies to a more general choice of charges and masses, however. For simplicity, we will only discuss the simpler case of equal masses and charges here. The Hamiltonian of the system under consideration is given by

$$\boxed{H = -\frac{1}{2}\sum_{i=1}^{N}\Delta_i + \sum_{1\le i<j\le N}\frac{e_i e_j}{|\boldsymbol{x}_i - \boldsymbol{x}_j|}} \tag{7.6.6}$$

where e_j is either $+1$ or -1. Note that here N denotes the *total* number of particles, both positive and negative. It follows from Theorem 3.3 that the infimum of $(\psi, H\psi)$ over all wave functions ψ is the same as the infimum over wave functions that are symmetric in the coordinates of the positive and of the negative particles. Hence Bose symmetry is not really relevant here and one might as well forget all symmetry requirements. The ground state energy of H is defined by minimizing not only over all wave functions ψ but also over the values of $e_j = \pm 1$.

If one accepts the $N^{7/5}$ law then, by scaling, the system contracts to a diameter $N^{-1/5}$ because the kinetic energy has to be $N\times$ diameter^{-2}. What is not so obvious is that the potential energy can be made as negative as $-N^{7/5}$. The Coulomb energy of the average charge density, which is zero, vanishes and, therefore, the Coulomb energy is entirely exchange-correlation energy in this case. Thus, in order to achieve this very low energy, it is necessary to employ a sophisticated wave function, namely the one suggested by Bogoliubov in 1947 [21].[3] In the end, one would expect that the one-particle density of the particles is given by some mean field equation. In fact, the density will be asymptotically equal to the ϱ that minimizes

$$\boxed{\frac{1}{2}\int_{\mathbb{R}^3}\left|\nabla\sqrt{\varrho(\boldsymbol{x})}\right|^2 \mathrm{d}\boldsymbol{x} - I_0\,\alpha^{5/4}\int_{\mathbb{R}^3}\varrho(\boldsymbol{x})^{5/4}\mathrm{d}\boldsymbol{x},} \tag{7.6.7}$$

[3] A decade later Bardeen, Cooper and Schrieffer [7] used the fermionic analogue of this wave function to explain superconductivity.

under the condition that $\int_{\mathbb{R}^3} \varrho(\boldsymbol{x})\mathrm{d}\boldsymbol{x} = N$. Here, I_0 is Foldy's constant [68]

$$I_0 = \left(\frac{2}{\pi}\right)^{3/4} \int_0^\infty \left(1 + t^4 - t^2\sqrt{t^4+2}\right) \mathrm{d}t = \frac{2^{3/2}\Gamma(3/4)}{5\pi^{1/4}\Gamma(5/4)} \approx 0.5744, \quad (7.6.8)$$

which we will explain below. Note that ϱ is the total density of $N/2$ positively charged and $N/2$ negatively charged particles.

If one defines $\Phi(\boldsymbol{x}) = N^{-4/5}\varrho(N^{-1/5}\boldsymbol{x})^{1/2}$, then $\int_{\mathbb{R}^3} \Phi(\boldsymbol{x})^2\mathrm{d}\boldsymbol{x} = 1$, and Φ satisfies a Lane–Emden equation

$$-\Delta\Phi(\boldsymbol{x}) - \frac{5}{2} I_0\, \alpha^{5/4}\, \Phi(\boldsymbol{x})^{3/2} + \mu\Phi(\boldsymbol{x}) = 0$$

for some $\mu > 0$. It is possible to prove that the minimizer of (7.6.7) is unique up to translations [13, 100, 141, 189].

The proof that this is the correct large N asymptotics is very complicated. Even the imprecise bounds found in [43] and [34] are very complicated, and we will not attempt to reproduce them here. There are no simple proofs in this subject, comparable to the stability of matter proofs for fermions presented in the beginning of this chapter.

The constant I_0 in (7.6.8) was derived by Foldy in 1961 [68]. He considered a different problem in which there is only one species of bosons with negative charge, free to move in a background charge of uniform positive density ϱ. This is also known as the **jellium model**; it will be discussed further in Section 14.7. By a straightforward application of Bogoliubov's 1947 procedure (using creation and annihilation operators on Fock space), Foldy arrived at the formula

$$\frac{E_0}{N} \approx -I_0\, \alpha^{5/4}\, \varrho^{1/4}, \quad (7.6.9)$$

which he conjectured to be asymptotically correct for large ϱ. A lower bound of this form, but with a different constant, was also proved in [34], and the correct lower bound (7.6.9) for large ϱ was proved in [131]. The corresponding upper bound was derived in [172]. There is apparently no simple way to understand the exponent $1/4$ in (7.6.9). It is subtle quantum-mechanical correlation effect and cannot be understood in classical terms.

The connection between the two problems was recognized by Dyson in his paper in 1967 [43]. His considerations led to the energy minimization problem in (7.6.7). We can make this plausible in the following way. The kinetic energy consists of two parts. One is very local on a length scale of order $\varrho^{-1/3}$ and is responsible for the correlations among the particles. Second, there is a global kinetic energy, which is not accounted for in the Foldy energy, which is the energy of a homogeneous system. This extra envelope energy is simply $(1/2)\int |\nabla\sqrt{\varrho}|^2$. The potential energy is slightly more subtle. According to (7.6.9) the energy per unit volume is $-I_0\alpha^{5/4}\varrho^{5/4}$ and hence one might expect that the second term in the energy functional (7.6.7) should be $2I_0\alpha^{5/4}\int(\varrho(\boldsymbol{x})/2)^{5/4}\mathrm{d}\boldsymbol{x}$. This is the answer one would get in a mean-field picture in which the positively charged particles form a uniform background for the negatively charged particles, and *vice versa*. The truth is that at high density, the correlation energy can actually be understood perturbatively. In the jellium case, the particles stay out of each other's way, and this gives rise to the Coulomb correlation energy. In the two-component case, like charges will also stay out of each other's way but, in addition, they will prefer being close to the opposite charge. Both give a negative Coulomb correlation energy. The point about the perturbation theoretic argument is that the correlation energy for both these effects is the same, and both give a negative contribution to the energy. If one accepts this argument, the total correlation energy should contain the total ϱ, and not $\varrho/2$, as given by the second term in (7.6.7).

Poetry aside, it is rigorously established in [132, 172] that the energy for large N is given by minimizing (7.6.7).

One last remark. It would seem that the jellium problem should be easier than the two component case. This is indeed so in the rigorous proofs in [130] and [132], as well as in Dyson's original heuristic arguments [43]. In [34], however, where the correct exponents in the N dependence where first proved, it was found that it was easier to prove the two-component case first, and derive the jellium case from it.

CHAPTER 8

Stability of Relativistic Matter

8.1 Introduction

When electrons are treated relativistically the stability of matter question will obviously be more delicate than in the non-relativistic case. The reason is that $|\boldsymbol{p}|$ is much less than $|\boldsymbol{p}|^2$ when $|\boldsymbol{p}|$ is large. (Recall that $|\boldsymbol{p}| - m \leq \sqrt{\boldsymbol{p}^2 + m^2} - m \leq |\boldsymbol{p}|$ so, for our purposes, we may as well consider just $|\boldsymbol{p}|$ instead of $\sqrt{\boldsymbol{p}^2 + m^2} - m$ for the kinetic energy of a particle of mass m and momentum $\boldsymbol{p}$.) The premonition that this subject has significant dangers is borne out by the fact that the stability of a single hydrogenic atom, which always occurs in the non-relativistic case, is found to occur relativistically if and only if $Z\alpha \leq 2/\pi$.[1] The implication for the many-body problem is, of course, that even stability of the first kind requires that $Z_j\alpha$ must be less than or equal to $2/\pi$ for every $1 \leq j \leq M$. But is this enough? The surprising (or not so surprising, depending on your point of view) fact is that a bound on α *itself* is needed, no matter how small the maximal Z_j might be. Recall from Section 3.2.1 that stability of the first kind and stability of the second kind are equivalent for relativistic matter.

The fact that a separate bound on α is needed was unknown in the physics literature, and is largely unknown in the physics community even to this day. Given that α is below some critical value, is it true that stability will hold all the way up to $Z_j\alpha = 2/\pi$ for all j, independent of the number of nuclei (and independent of α)? The answer is yes, and this was proved (in the absence of magnetic fields) in 1988 [138]. This fact remains true in the presence of arbitrary magnetic fields (ignoring the $\boldsymbol{\sigma} \cdot \boldsymbol{B}$ term), but it took two decades more to prove it [71]. Section 8.7 contains a summary of known results and a short history of the problem.

[1] A bound $Z\alpha \leq 1$ is known to be needed for the Dirac theory of the hydrogenic atom, and the need for a bound in both cases has the same origin. Cf. Chapter 10.

The Hamiltonian under consideration, as described in Chapter 3, is

$$H = \sum_{i=1}^{N} T(\boldsymbol{p}_i) + \alpha V_C(\underline{\boldsymbol{X}}, \underline{\boldsymbol{R}}), \tag{8.1.1}$$

with V_C given in (2.1.21). The kinetic energy is $T(\boldsymbol{p}) = \sqrt{(\boldsymbol{p} + \sqrt{\alpha}\boldsymbol{A}(\boldsymbol{x}))^2 + m^2} - m$, including the possibilities that $m = 0$ or $\boldsymbol{A} = 0$.

Before investigating stability of the Hamiltonian (8.1.1), we shall first study the relativistic one-body problem in detail in Section 8.2. Sections 8.3 and 8.4 contain additional bounds which will be useful in the sequel. The proof of stability of relativistic matter is given in Section 8.5. The proof given there does not yield the best known results, but it has the virtue of being the most simple and direct, and is the closest to being a natural generalization of the proof of the stability of non-relativistic matter given in Section 7.1 in Chapter 7. In particular, we do not give the rather complicated proof of stability up to $Z\alpha = 2/\pi$ in the many-body case. An alternative proof, which will be important later in Chapter 11, will be given in Section 8.6. We summarize the best known results concerning stability of relativistic matter in Section 8.7.

Another question, which will be addressed in Section 8.8, is an upper bound on the allowed values of α for which stability of relativistic matter holds. We shall show that instability of the first kind really does occur if α is too large; the requirement of a bound on α in Section 8.5 is not just an artifact of our proof but actually reflects reality.

8.1.1 Heuristic Reason for a Bound on α Itself

As will be shown in the next section, stability requires a bound $Z\alpha \leq 2/\pi$ for each nucleus. This fact, alone, leads to the conclusion that α itself must be bounded, as we now discuss heuristically. Indeed, it was shown in [38] that one can create an arbitrarily negative energy using only one electron and many nuclei if α is greater than a certain value of the order of unity, no matter how small Z might be.

Imagine that we have M nuclei of charge Z and that $Z\alpha = \varepsilon \ll 1$. Let M be such that $M\varepsilon \approx 4/\pi$. Thus, if we can bring the M nuclei together we will create, effectively, a nucleus with $Z'\alpha \approx 4/\pi$ and then one electron can have an arbitrarily negative energy. (Recall that whatever energy bound we have is supposed to be independent of the location of the nuclei.) What might prevent this nuclear fusion from happening is the Coulomb repulsion of the nuclei. Suppose the nuclei are located in a small ball of radius δ, in which case the nuclear repulsion energy is about C/δ with $C = M^2Z^2\alpha = M^2(Z\alpha)^2/\alpha \approx (4/\pi)^2\,\alpha^{-1}$. This number will be small if α is large. In short, the effective repulsion, which supposedly prevents collapse, will be small if α is large. This argument suggests that if we regard $Z\alpha$ as fixed then an upper bound on α is needed because a tiny value of Z can be offset by a large number of nuclei. It turns out that this conclusion is correct, even if one does not believe the argument.

For two nuclei, with $Z\alpha = 2/\pi$ for each, this argument can be made rigorous, in fact, as was done in [38]. This will be further discussed in Section 8.8 and the requirement of an N-independent bound on α, first proved in [138], will be proved there. In particular, the dependence of this bound on q, the number of spin states of each electron, will be shown to be $\sim 1/q^{1/3}$. This, in turn, implies that there is *never* stability of the first kind for relativistic bosons because N bosons can be thought of as N fermions with $q = N$, as explained in Section 3.1.3.1. (Recall that *non*-relativistic bosons are always stable of the first kind, but not the second.)

8.2 The Relativistic One-Body Problem

Before discussing the problem of stability of matter with relativistic kinematics we consider the corresponding one-body problem. Since $|\boldsymbol{p}|$ and $1/|\boldsymbol{x}|$ both scale like an inverse length, there is a critical coupling constant above which even stability of the first kind fails.

In the non-relativistic case the relevant function space for one particle was $H^1(\mathbb{R}^3)$, namely functions f such that $f(\boldsymbol{x}) \in L^2(\mathbb{R}^3)$ and $|\boldsymbol{k}|\widehat{f}(\boldsymbol{k}) \in L^2(\mathbb{R}^3)$. In the relativistic case under consideration here, the relevant space is $H^{1/2}(\mathbb{R}^3)$, namely $f(\boldsymbol{x}) \in L^2(\mathbb{R}^3)$ and $|\boldsymbol{k}|^{1/2}\widehat{f}(\boldsymbol{k}) \in L^2(\mathbb{R}^3)$. This can be summarized as $(1 + |\boldsymbol{k}|^{1/2})\widehat{f}(\boldsymbol{k}) \in L^2(\mathbb{R}^3)$.

We start with the following simple observation.

Lemma 8.1 (Equivalent Form of Relativistic Kinetic Energy). *For $f \in H^{1/2}(\mathbb{R}^d)$, we have*

$$\begin{aligned}(f, |\boldsymbol{p}| f) &= \int_{\mathbb{R}^d} |2\pi \boldsymbol{k}||\widehat{f}(\boldsymbol{k})|^2 \mathrm{d}\boldsymbol{k} \\ &= \frac{1}{2}\Gamma\left(\frac{d+1}{2}\right)\pi^{-(d+1)/2} \iint_{\mathbb{R}^d\times\mathbb{R}^d} \frac{|f(\boldsymbol{x}) - f(\boldsymbol{y})|^2}{|\boldsymbol{x} - \boldsymbol{y}|^{d+1}} \mathrm{d}\boldsymbol{x}\mathrm{d}\boldsymbol{y}.\end{aligned} \tag{8.2.1}$$

Remark 8.1. Part of the lemma is the assertion that the integral on the right is well defined for an $H^{1/2}$ function, despite the apparent non-integrable singularity $|\boldsymbol{x}-\boldsymbol{y}|^{-(d+1)}$.

Proof. The following proof is the same as in [118, Sect. 7.12]. For an alternative proof, see [72, Lemma 3.1]. We first note that

$$(f, |\boldsymbol{p}| f) = \lim_{t\to 0} \frac{1}{t}(f, (1-e^{-t|\boldsymbol{p}|})f),$$

which follows by writing out the expectation values as integrals in Fourier space, and using dominated convergence to justify the limit $t \to 0$.

Recall that with our convention for the Fourier transform, the operator $|\boldsymbol{p}| = \sqrt{-\Delta}$ acts via multiplication by $2\pi|\boldsymbol{k}|$ in Fourier space. The inverse Fourier transform of $g_t(\boldsymbol{k}) = e^{-2\pi t|\boldsymbol{k}|}$ on $\mathbb{R}^d$ is given by [118, Sect. 7.11]

$$\widehat{g_t}(\boldsymbol{x}) = \Gamma\left(\frac{d+1}{2}\right)\pi^{-(d+1)/2}\frac{t}{(t^2+|\boldsymbol{x}|^2)^{(d+1)/2}}, \tag{8.2.2}$$

which can easily be obtained from the formula (4.4.10) in Chapter 4. It can also be computed directly by elementary means. Hence

$$(f, (1-e^{-t|\boldsymbol{p}|})f) = \frac{1}{2}\iint_{\mathbb{R}^d\times\mathbb{R}^d} |f(\boldsymbol{x})-f(\boldsymbol{y})|^2\,\widehat{g_t}(\boldsymbol{x}-\boldsymbol{y})\,\mathrm{d}\boldsymbol{x}\mathrm{d}\boldsymbol{y}.$$

The lemma follows by first dividing by t and then taking $t \to 0$. The monotone convergence theorem allows us to take the limit $t \to 0$ of the integrand. ■

Remark 8.2. We record, for the reader's possible interest, the analogous Fourier transform of $g_t^{(m)}(\boldsymbol{k}) = \exp(-t\sqrt{(2\pi\boldsymbol{k})^2+m^2})$ in $\mathbb{R}^3$ (see [118, Sect. 7.11] for the general $\mathbb{R}^d$ case):

$$\widehat{g_t^{(m)}}(\boldsymbol{x}) = \frac{m^2}{2\pi^2}\frac{t}{t^2+\boldsymbol{x}^2}K_2(m\sqrt{\boldsymbol{x}^2+t^2}), \tag{8.2.3}$$

where K_2 is the modified Bessel function of the third kind. Notice that (8.2.2) and (8.2.3) have a nice Euclidean form (after multiplication by t^{-1}), i.e., they are functions only of the $d+1$ dimensional distance $\sqrt{t^2+\boldsymbol{x}^2}$.

The representation (8.2.1) is useful for establishing the following inequality, which can be interpreted as a relativistic uncertainty principle.

Lemma 8.2 (Relativistic Hardy Inequality). *Let $d \geq 2$, and let f be a non-zero function in $H^{1/2}(\mathbb{R}^d)$. Then there is the strict inequality*

$$(f, |\boldsymbol{p}| f) > 2\frac{\Gamma(\frac{d+1}{4})^2}{\Gamma(\frac{d-1}{4})^2} \int_{\mathbb{R}^d} \frac{|f(\boldsymbol{x})|^2}{|\boldsymbol{x}|} \mathrm{d}\boldsymbol{x}. \tag{8.2.4}$$

Moreover, the constant is sharp, i.e., for any bigger constant the inequality fails for some function in $H^{1/2}(\mathbb{R}^d)$. For $d = 3$ the constant in (8.2.4) is $2\,\Gamma(1)^2/\Gamma(1/2)^2 = 2/\pi$.

Remark 8.3. This inequality goes back to Kato [99] and Herbst [89]. See also [184, 12, 188, 72]. The non-relativistic analogue is Hardy's inequality $\int_{\mathbb{R}^d} |\nabla f(\boldsymbol{x})|^2 \mathrm{d}\boldsymbol{x} \geq [(d-2)^2/4] \int_{\mathbb{R}^d} |f(\boldsymbol{x})|^2 |\boldsymbol{x}|^{-2} \mathrm{d}\boldsymbol{x}$ for $d \geq 3$.

Proof. For simplicity, we restrict our attention to the case $d = 3$, which will be the relevant case for our considerations on stability of relativistic matter. For general d, the proof works the same way, but the integrals involved are slightly more complicated. See [89, 188, 72].

According to Lemma 8.1,

$$(f, |\boldsymbol{p}| f) = \frac{1}{2\pi^2} \iint_{\mathbb{R}^3 \times \mathbb{R}^3} \frac{|f(\boldsymbol{x}) - f(\boldsymbol{y})|^2}{|\boldsymbol{x} - \boldsymbol{y}|^4} \mathrm{d}\boldsymbol{x}\, \mathrm{d}\boldsymbol{y}, \tag{8.2.5}$$

which we can write as

$$(f, |\boldsymbol{p}| f) = \lim_{\varepsilon \to 0} \frac{1}{2\pi^2} \iint_{\mathbb{R}^3 \times \mathbb{R}^3} \frac{|f(\boldsymbol{x}) - f(\boldsymbol{y})|^2}{(|\boldsymbol{x} - \boldsymbol{y}|^2 + \varepsilon)^2} \mathrm{d}\boldsymbol{x}\, \mathrm{d}\boldsymbol{y}. \tag{8.2.6}$$

The purpose of ε is to avoid the singularity at $\boldsymbol{x} = \boldsymbol{y}$. The limit $\varepsilon \to 0$ will be taken at the end of the calculation.

We can write

$$|f(\boldsymbol{x}) - f(\boldsymbol{y})|^2 = \frac{||\boldsymbol{x}| f(\boldsymbol{x}) - |\boldsymbol{y}| f(\boldsymbol{y})|^2}{|\boldsymbol{x}||\boldsymbol{y}|} + |f(\boldsymbol{x})|^2 \left(1 - \frac{|\boldsymbol{x}|}{|\boldsymbol{y}|}\right) + |f(\boldsymbol{y})|^2 \left(1 - \frac{|\boldsymbol{y}|}{|\boldsymbol{x}|}\right). \tag{8.2.7}$$

Hence the integral in (8.2.6) can be written as

$$\iint_{\mathbb{R}^3\times\mathbb{R}^3} \frac{||x|f(x)-|y|f(y)|^2}{(|x-y|^2+\varepsilon)^2}\frac{\mathrm{d}x\,\mathrm{d}y}{|x|\,|y|} + 2\iint_{\mathbb{R}^3\times\mathbb{R}^3} \frac{|f(x)|^2}{(|x-y|^2+\varepsilon)^2}\left(1-\frac{|x|}{|y|}\right)\mathrm{d}x\mathrm{d}y. \tag{8.2.8}$$

After performing the y angular integration, the second integral equals

$$8\pi\int_{\mathbb{R}^3}\mathrm{d}x\,|f(x)|^2\int_0^\infty \mathrm{d}t\,\frac{t(t-|x|)}{(|x|^2-t^2)^2+2\varepsilon(|x|^2+t^2)+\varepsilon^2}. \tag{8.2.9}$$

The t integral is non-negative and converges, as $\varepsilon\to 0$, to the principle-value integral

$$\text{p.v.}\int_0^\infty \frac{t(t-|x|)}{(|x|^2-t^2)^2}\mathrm{d}t = \frac{1}{2|x|}. \tag{8.2.10}$$

It is tedious but elementary to justify exchanging the x integration and the $\varepsilon\to 0$ limit.

We conclude that

$$(f,|p|f)-\frac{2}{\pi}\int_{\mathbb{R}^3}|f(x)|^2\frac{1}{|x|}\mathrm{d}x = \lim_{\varepsilon\to 0}\frac{1}{2\pi^2}\iint_{\mathbb{R}^3\times\mathbb{R}^3}\frac{||x|f(x)-|y|f(y)|^2}{(|x-y|^2+\varepsilon)^2}\frac{\mathrm{d}x\,\mathrm{d}y}{|x|\,|y|}. \tag{8.2.11}$$

Since the right side is positive, this proves (8.2.4). The inequality is strict since the right side of (8.2.11) is zero only if $f(x)=c/|x|$, which is not an $H^{1/2}$ function for $c\neq 0$.

Note that the right side of (8.2.11) is finite for a compactly supported function f that diverges like $1/|x|$ for small $|x|$, while $\int|f(x)|^2|x|^{-1}\mathrm{d}x$ would be infinite for such a function. A simple limiting argument thus shows that if $2/\pi$ is replaced by a larger number the inequality (8.2.4) cannot hold. The constant $2/\pi$ is, therefore, sharp. ■

From Lemma 8.2 we conclude that the energy functional

$$(f,|p|f)-Z\alpha\int_{\mathbb{R}^3}|f(x)|^2\frac{1}{|x|}\mathrm{d}x$$

for $f\in H^{1/2}(\mathbb{R}^3)$ with $\int_{\mathbb{R}^3}|f(x)|^2\mathrm{d}x=1$ is bounded from below (in fact, positive) for $Z\alpha\le 2/\pi$, while it is unbounded for $Z\alpha>2/\pi$. The same is true if the mass is taken into consideration, i.e., $|\boldsymbol{p}|$ is replaced by $\sqrt{\boldsymbol{p}^2 + m^2} - m$, since

$$|\boldsymbol{p}| - m \leq \sqrt{\boldsymbol{p}^2 + m^2} - m \leq |\boldsymbol{p}|. \tag{8.2.12}$$

Hence, stability of the first kind holds for a relativistic hydrogenic atom if and only if $Z\alpha \leq 2/\pi$.

8.3 A Localized Relativistic Kinetic Energy

Before we turn our attention to the problem of stability of matter in the relativistic case, we state and prove a local version of the inequality (8.2) which will be needed in the following. It is taken from [138, Theorem 7].

Lemma 8.3 (Localized Relativistic Kinetic Energy). *Let B be a ball of radius D centered at some $\boldsymbol{w} \in \mathbb{R}^3$. Then*

$$\boxed{\begin{aligned} &\frac{1}{2\pi^2} \iint\limits_{B\times B} \frac{|f(\boldsymbol{x}) - f(\boldsymbol{y})|^2}{|\boldsymbol{x} - \boldsymbol{y}|^4} \mathrm{d}\boldsymbol{x}\mathrm{d}\boldsymbol{y} \\ &\quad \geq \frac{2}{\pi} \int\limits_B \frac{|f(\boldsymbol{x})|^2}{|\boldsymbol{x} - \boldsymbol{w}|} \mathrm{d}\boldsymbol{x} - \frac{1}{D} \int\limits_B Y(|\boldsymbol{x} - \boldsymbol{w}|/D) |f(\boldsymbol{x})|^2 \mathrm{d}\boldsymbol{x}, \end{aligned}} \tag{8.3.1}$$

where

$$Y(t) = \frac{2}{\pi(1+t)} + \frac{1+3t^2}{\pi(1+t^2)t} \ln(1+t) - \frac{1-t^2}{\pi(1+t^2)t} \ln(1-t) - \frac{4t}{\pi(1+t^2)} \ln t. \tag{8.3.2}$$

Note that $Y(t)$ is non-negative, continuous and bounded for $0 \leq t \leq 1$. Numerical evaluation shows that $Y(t) \leq 1.57$.

Proof. Without loss of generality, we can set $\boldsymbol{w} = 0$ and $D = 1$. Let h be a strictly positive function on $[0, 1]$. Similarly to (8.2.7), we can write

$$\begin{aligned} |f(\boldsymbol{x}) - f(\boldsymbol{y})|^2 &= h(|\boldsymbol{x}|)h(|\boldsymbol{y}|) \left| \frac{f(\boldsymbol{x})}{h(|\boldsymbol{x}|)} - \frac{f(\boldsymbol{y})}{h(|\boldsymbol{y}|)} \right|^2 \\ &\quad + |f(\boldsymbol{x})|^2 \left(1 - \frac{h(|\boldsymbol{y}|)}{h(|\boldsymbol{x}|)}\right) + |f(\boldsymbol{y})|^2 \left(1 - \frac{h(|\boldsymbol{x}|)}{h(|\boldsymbol{y}|)}\right). \end{aligned}$$

Using positivity of the first term, and performing the angle integration as in the proof of Lemma 8.2, we obtain

$$\frac{1}{2\pi^2}\iint\limits_{B\times B} \frac{|f(\boldsymbol{x})-f(\boldsymbol{y})|^2}{|\boldsymbol{x}-\boldsymbol{y}|^4} \mathrm{d}\boldsymbol{x}\mathrm{d}\boldsymbol{y}$$

$$\geq \frac{2}{\pi}\int\limits_B \mathrm{d}\boldsymbol{x}\, |f(\boldsymbol{x})|^2 \text{ p.v.} \int\limits_0^1 \frac{2t^2}{(|\boldsymbol{x}|^2-t^2)^2}\left(1-\frac{h(t)}{h(|\boldsymbol{x}|)}\right)\mathrm{d}t. \quad (8.3.3)$$

(Compare with the proof of Lemma 8.2, where the choice $h(t) = 1/t$ was used.)

In the following, we will choose $h(t) = 1/t + t$. This particular choice is motivated by the fact that the derivative of h at $t = 1$ vanishes. It is easy to see that the potential on the right side of (8.3.3) is bounded at $|\boldsymbol{x}| = 1$ only if h has this vanishing derivative property. Upon evaluating the integrals, we find that

$$\text{p.v.} \int\limits_0^1 \frac{2t^2}{(|\boldsymbol{x}|^2-t^2)^2}\left(1-\frac{h(t)}{h(|\boldsymbol{x}|)}\right)\mathrm{d}t = \frac{1}{|\boldsymbol{x}|} - \frac{\pi}{2}Y(|\boldsymbol{x}|),$$

which finishes the proof. ■

8.4 A Simple Kinetic Energy Bound

The following simple inequality gives a lower bound on the many-body kinetic energy in terms of the kinetic energy of the square root of the one-particle density. This bound has proved to be useful in several contexts. For $\delta = 2$, it was first discovered by M. & T. Hoffman-Ostenhof [92]. It works both in the relativistic case and in the non-relativistic case, and uses the positivity of the Fourier transform of $\exp(-t|\boldsymbol{p}|^\delta)$ for $0 \leq \delta \leq 2$ [150, Example 2 in Section XIII.12].[2] Since $\exp(-t\sqrt{|\boldsymbol{p}|^2+m^2})$ is also positive definite, it also works in the relativistic case with non-zero mass. It applies to any many-particle wave function, irrespective of the symmetry type. It is partly also a diamagnetic

[2] In [150] this positivity of the Fourier transform of $\exp(-t|\boldsymbol{p}|^\delta)$ is stated for $0 \leq \delta \leq 2$ for $d \geq 2$ while it is only stated for $1 \leq \delta \leq 2$ in the case $d = 1$. With the help of Eq. (4.4.10), however, the result in [150] can be easily extended to $0 < \delta \leq 2$ for $d = 1$. For all d, a direct proof of the positivity for $\delta = 1$, the case of interest to us, can be obtained directly from the formula (4.4.10), starting with the known positivity for $\delta = 2$; further use of that formula will yield positivity for $\delta = 1/2, 1/4, \ldots$.

inequality, since the magnetic vector potential $\boldsymbol{A}$ appears only on the left side of the inequality, not on the right.

Lemma 8.4 (Kinetic Energy Bound in terms of the Density). *For* $0 \leq \delta \leq 2$, *the bound*

$$\boxed{\left(\psi, \sum_{i=1}^{N} \left|\boldsymbol{p}_i + \sqrt{\alpha}\boldsymbol{A}(\boldsymbol{x}_i)\right|^{\delta} \psi\right) \geq \left(\sqrt{\varrho_\psi}, |\boldsymbol{p}|^{\delta} \sqrt{\varrho_\psi}\right)} \tag{8.4.1}$$

holds for any N*-particle wave-function* $\psi \in H^{\delta/2}(\mathbb{R}^{Nd})$ *with density* ϱ_ψ *(defined in (3.1.4), and any magnetic vector potential* $\boldsymbol{A}$*. (The Pauli exclusion principle plays no role here.)*

Proof. We first consider the case $\boldsymbol{A} = 0$. In terms of the one-particle density matrix $\gamma^{(1)}$ of ψ (defined in Eq. (3.1.27)), Inequality (8.4.1) says in this case that

$$\int\limits_{\mathbb{R}^d} \mathrm{d}\boldsymbol{k}\, |2\pi \boldsymbol{k}|^{\delta} \iint\limits_{\mathbb{R}^d \times \mathbb{R}^d} \mathrm{d}\boldsymbol{x}\mathrm{d}\boldsymbol{y}\, e^{2\pi i \boldsymbol{k}\cdot(\boldsymbol{x}-\boldsymbol{y})} \left(\gamma^{(1)}(\boldsymbol{x}, \boldsymbol{y}) - \sqrt{\varrho_\psi(\boldsymbol{x})\varrho_\psi(\boldsymbol{y})}\right) \geq 0. \tag{8.4.2}$$

Note that the positive definiteness of $\gamma^{(1)}$ implies that $|\gamma^{(1)}(\boldsymbol{x}, \boldsymbol{y})|^2 \leq \gamma^{(1)}(\boldsymbol{x}, \boldsymbol{x})\,\gamma^{(1)}(\boldsymbol{y}, \boldsymbol{y}) = \varrho_\psi(\boldsymbol{x})\varrho_\psi(\boldsymbol{y})$, with equality for $\boldsymbol{x} = \boldsymbol{y}$, of course. Since $|\boldsymbol{p}|^{\delta} = \lim_{t\to 0}(1 - \exp(-t|\boldsymbol{p}|^{\delta})/t$, (8.4.2) follows if we can show that

$$\int\limits_{\mathbb{R}^d} \mathrm{d}\boldsymbol{k}\, e^{-t|2\pi \boldsymbol{k}|^{\delta}} \iint\limits_{\mathbb{R}^d \times \mathbb{R}^d} \mathrm{d}\boldsymbol{x}\mathrm{d}\boldsymbol{y}\, e^{2\pi i \boldsymbol{k}\cdot(\boldsymbol{x}-\boldsymbol{y})} \left(\gamma^{(1)}(\boldsymbol{x}, \boldsymbol{y}) - \sqrt{\varrho_\psi(\boldsymbol{x})\varrho_\psi(\boldsymbol{y})}\right) \leq 0$$

for all $t > 0$. But this inequality follows immediately from the fact that $\exp(-t|\boldsymbol{p}|^{\delta})$ has a non-negative Fourier transform for $0 \leq \delta \leq 2$.

In order to extend the inequality to the magnetic case $\boldsymbol{A} \neq 0$, it is sufficient to show that

$$\begin{aligned}\iint\limits_{\mathbb{R}^3 \times \mathbb{R}^3} g_t^{\boldsymbol{A}}(\boldsymbol{x}, \boldsymbol{y})\gamma^{(1)}(\boldsymbol{x}, \boldsymbol{y})\mathrm{d}\boldsymbol{x}\mathrm{d}\boldsymbol{y} &\leq \iint\limits_{\mathbb{R}^3 \times \mathbb{R}^3} g_t^{0}(\boldsymbol{x}, \boldsymbol{y})|\gamma^{(1)}(\boldsymbol{x}, \boldsymbol{y})|\mathrm{d}\boldsymbol{x}\mathrm{d}\boldsymbol{y} \\ &\leq \iint\limits_{\mathbb{R}^3 \times \mathbb{R}^3} g_t^{0}(\boldsymbol{x}, \boldsymbol{y})\sqrt{\varrho_\psi(\boldsymbol{x})}\sqrt{\varrho_\psi(\boldsymbol{y})}\mathrm{d}\boldsymbol{x}\mathrm{d}\boldsymbol{y},\end{aligned} \tag{8.4.3}$$

where $g_t^A(\boldsymbol{x}, \boldsymbol{y})$ denotes the integral kernel of the operator $e^{-t|\boldsymbol{p}+\sqrt{\alpha}\boldsymbol{A}(\boldsymbol{x})|}$. This inequality follows from the fact that $|g_t^A(\boldsymbol{x}, \boldsymbol{y})| \leq g_t^0(\boldsymbol{x}, \boldsymbol{y})$ for all $\boldsymbol{x}, \boldsymbol{y} \in \mathbb{R}^3$ and $t > 0$, which is stated in Theorem 4.4. ∎

8.5 Proof of Relativistic Stability

In the relativistic case, the proof of stability of matter is more complicated than in the non-relativistic case mainly for one reason: although the relativistic kinetic energy dominates the Coulomb potential for small enough coupling, as shown in Lemma 8.2, the semiclassical expression for the relativistic kinetic energy, which was shown in Chapter 4 to equal $\int \varrho_\psi(\boldsymbol{x})^{4/3}\mathrm{d}\boldsymbol{x}$, can not dominate the Coulomb energy, $-\int \varrho_\psi(\boldsymbol{x})|\boldsymbol{x}|^{-1}\mathrm{d}\boldsymbol{x}$. More precisely, among all non-negative densities ϱ with $\int \varrho(\boldsymbol{x})\mathrm{d}\boldsymbol{x} \leq 1$, say, the quantity $\mathcal{E}(\varrho) = \int \varrho(\boldsymbol{x})^{4/3}\mathrm{d}\boldsymbol{x} - \lambda \int \varrho(\boldsymbol{x})|\boldsymbol{x}|^{-1}\mathrm{d}\boldsymbol{x}$ is *not* bounded from below, no matter how small λ is. If it were bounded, it would have to be non-negative, for otherwise $\mathcal{E}(\varrho)$ can be made arbitrarily negative by rescaling $\varrho(\boldsymbol{x}) \mapsto \gamma^3\varrho(\gamma\boldsymbol{x})$, which preserves $\int \varrho(\boldsymbol{x})\mathrm{d}\boldsymbol{x}$ but drives $\mathcal{E}(\varrho)$ to $-\infty$ as $\gamma \to \infty$. But for any given ϱ, $\mathcal{E}(\varrho)$ can be made negative simply by replacing $\varrho(\boldsymbol{x})$ by $\varepsilon\varrho(\boldsymbol{x})$ and choosing ε sufficiently small.

Simply mimicking the proof of stability of matter in the non-relativistic case will not yield a proof in the relativistic case. The Coulomb singularity can not be controlled by the semiclassical kinetic energy. As shown by Lieb, Loss and Siedentop in [122] stability can, indeed, be proved in a somewhat similar way, however, if part of the kinetic energy is bounded from below as in Lemma 8.4 above, and the remaining part is estimated via the Lieb–Thirring type bound (4.2.11) in terms of the semiclassical expression. We will now explain this method in detail, following closely the strategy in [122]. We repeat what was already said in the introduction: This method does not yield the best results, but it is the simplest route to a proof of relativistic stability.

For simplicity, we consider here only the case $m = 0$, i.e., the case of massless particles. The necessary modifications for $m > 0$ will be discussed in Remark 8.5. We note that in the absence of mass, all terms in the Hamiltonian scale like an inverse length. Hence, under the scaling transformation $\psi(\underline{\boldsymbol{X}}) \mapsto \gamma^{3N/2}\psi(\gamma\underline{\boldsymbol{X}})$, $\underline{\boldsymbol{R}} \mapsto \gamma\underline{\boldsymbol{R}}$ and $\boldsymbol{A}(\boldsymbol{x}) \to \gamma^{-1}\boldsymbol{A}(\gamma\boldsymbol{x})$, H transforms to γH for $\gamma > 0$. In particular, the ground state energy E_0 of H satisfies (for all N, M, α and all choices of the Z_j)

$$E_0 = 0 \quad \text{or} \quad E_0 = -\infty. \tag{8.5.1}$$

(As defined in Chapter 3, we understand E_0 to be the infimum of $(\psi, H\,\psi)$ over all choices of ψ with the right symmetry, all choices of the nuclear coordinates $\underline{\boldsymbol{R}}$ and all vector potentials $\boldsymbol{A}$.) This also means that *stability of the first and the second kind coincide in the relativistic case*, as we said. This situation is not changed by the introduction of a mass, as explained in Remark 8.5 below.

We shall offer two proofs of relativistic stability. Neither one gives optimal results – which are recorded in Section 8.7. In this and the next section we shall prove the following theorem.

Theorem 8.1 (Conditions for Relativistic Stability). *Let $m = 0$. For any antisymmetric wave function ψ with q spin states, and any vector potential $\boldsymbol{A}$,*

$$\boxed{\left(\psi, \sum_{i=1}^{N} \left|\boldsymbol{p}_i + \sqrt{\alpha}\boldsymbol{A}(\boldsymbol{x}_i)\right| \psi\right) + \alpha\,(\psi, V_C\,\psi) \geq 0}$$

as long as three conditions are satisfied:

- $Z\alpha < 2/\pi$, *with* $Z = \max\{Z_j\}$
- $1.68\alpha < (1 - \alpha Z\pi/2)\,Kq^{-1/3}$
- $\dfrac{(\alpha Z)^2\alpha}{[(1 - \alpha Z\pi/2)\,Kq^{-1/3} - 1.68\alpha]^3}\left[7.6245\left(\dfrac{\pi}{2}\right)^4 + 4\pi\right] \leq \dfrac{1}{2}(4/3)^3.$

$$(8.5.2)$$

Here, K is the optimal constant in the relativistic kinetic energy inequality (4.2.11), which is known to satisfy $K \geq (6.08)^{-1/3}\frac{3}{4}(6\pi^2)^{1/3} \approx 1.60$. For this value of K, the three conditions (8.5.2) are satisfied if

$$\boxed{\frac{1}{\alpha} \geq \frac{Z\pi}{2} + 2.2566\,q^{1/3}Z^{2/3} + 1.0307\,q^{1/3}.} \tag{8.5.3}$$

Remarks 8.4. We note that condition (8.5.2) is fulfilled, for any fixed value of $Z\alpha < 2/\pi$, if α is sufficiently small (α has to go to zero as $Z\alpha \to 2/\pi$). On the other hand, for the physical values $q = 2$ and $\alpha = 1/137$, Ineq. (8.5.3) is satisfied for all $Z < 58.5$ and hence stability holds for these parameter values. This number should be compared with $2/(\pi\alpha) \approx 87.2$, which will be achieved later (see (8.7.1)).

Note that if $1.68\alpha \geq Kq^{-1/3}$ then the conditions of Theorem 8.1 cannot be satisfied for any choice of $Z \geq 0$.

We can summarize the conclusions of Theorem 8.1 by saying that for any fixed value of $\nu < 2/\pi$ there is an $\alpha_c(\nu)$ such that stability holds whenever $Z\alpha \leq \nu$ and $\alpha \leq \alpha_c(\nu)$. Moreover, the lower bound on $\alpha_c(\nu)$ derived in Theorem 8.1 depends on q according to $\alpha_c \sim q^{-1}$ for large q. Although our lower bound on $\alpha_c(\nu)$ unfortunately goes to zero as $\nu \to 2\pi$, we shall see in Section 8.7 that the actual $\alpha_c(\nu)$ does *not* go to zero as $\nu \to 2/\pi$, even with an arbitrary magnetic field. In Section 8.8 we shall see that the large q asymptotic behavior $\alpha_c \sim q^{-1}$ is, in fact, correct. This is not to be confused with the behavior of the critical α *for fixed* Z (as opposed to fixed $Z\alpha$), which is $q^{-1/3}$, not q^{-1}.

The heart of the proof of Theorem 8.1 is the following lemma, which will also be useful later in Chapter 13.

Lemma 8.5 (Energy Bound with the Nearest Nucleus Potential). *For fixed* $\underline{\boldsymbol{R}} = (\boldsymbol{R}_1, \ldots, \boldsymbol{R}_M)$, *let* $\mathfrak{D}(\boldsymbol{x}) = \min_j |\boldsymbol{x} - \boldsymbol{R}_j|$. *Let* ε *and* λ *satisfy* $\varepsilon > \pi\lambda/2 > 0$. *For any normalized, fermionic wave function* $\psi \in \bigwedge^N L^2(\mathbb{R}^3; \mathbb{C}^q)$, *and any vector potential* $\boldsymbol{A}$,

$$\boxed{\left(\psi, \sum_{i=1}^{N}\left(\varepsilon\left|\boldsymbol{p}_i + \sqrt{\alpha}\boldsymbol{A}(\boldsymbol{x}_i)\right| - \frac{\lambda}{\mathfrak{D}(\boldsymbol{x}_i)}\right)\psi\right) \geq -C\frac{\lambda^4 q}{(\varepsilon - \pi\lambda/2)^3}\sum_{j=1}^{M}\frac{1}{D_j},} \tag{8.5.4}$$

with

$$C = \left(\frac{3}{4}\right)^3\left[7.63\left(\frac{\pi}{2}\right)^4 + 4\pi\right]\frac{1}{4K^3} \tag{8.5.5}$$

and K *as in Theorem 8.1. As before,* $D_j = (1/2)\min_{k\neq j}|\boldsymbol{R}_k - \boldsymbol{R}_j|$ *denotes half the distance of* $\boldsymbol{R}_j$ *to its nearest neighbor nucleus. For* $K = (6.08)^{-1/3}\frac{3}{4}(6\pi^2)^{1/3}$ *we have* $C = 1.514$.

Proof. Without loss of generality we can set $\varepsilon = 1$. For $0 \leq \lambda \leq 2/\pi$, we split the kinetic energy into two parts, $|\boldsymbol{p}| = (1 - \pi\lambda/2)|\boldsymbol{p}| + (\pi\lambda/2)|\boldsymbol{p}|$. After applying the relativistic Lieb–Thirring inequality (4.2.6) to the first part and Lemma 8.4 to the second, we obtain

$$\left(\psi, \sum_{i=1}^{N}\left|\boldsymbol{p}_i + \sqrt{\alpha}\boldsymbol{A}(\boldsymbol{x}_i)\right|\psi\right)$$
$$\geq (1 - \pi\lambda/2)Kq^{-1/3}\int_{\mathbb{R}^3}\varrho_\psi(\boldsymbol{x})^{4/3}\mathrm{d}\boldsymbol{x} + \pi\lambda/2\left(\sqrt{\varrho_\psi}, |\boldsymbol{p}|, \sqrt{\varrho_\psi}\right). \tag{8.5.6}$$

In order to reduce the problem with many nuclei to the one with just one nucleus, and hence reduce the problem of stability of the second kind to stability of the first kind, we have to localize the last term in (8.5.6). Note that the M balls of radius D_j centered at $\boldsymbol{R}_j$ are non-overlapping. Hence Lemma 8.1 implies the lower bound

$$\left(\sqrt{\varrho_\psi}, |\boldsymbol{p}|, \sqrt{\varrho_\psi}\right) \geq \sum_{j=1}^{M} \frac{1}{2\pi^2} \iint_{\substack{|\boldsymbol{x}-\boldsymbol{R}_j|<D_j \\ |\boldsymbol{y}-\boldsymbol{R}_j|<D_j}} \frac{\left|\sqrt{\varrho_\psi(\boldsymbol{x})} - \sqrt{\varrho_\psi(\boldsymbol{y})}\right|^2}{|\boldsymbol{x}-\boldsymbol{y}|^4} \mathrm{d}\boldsymbol{x}\mathrm{d}\boldsymbol{y}. \quad (8.5.7)$$

With the aid of Lemma 8.3, we can further bound the right side from below as

$$\left(\sqrt{\varrho_\psi}, |\boldsymbol{p}|, \sqrt{\varrho_\psi}\right) \geq \sum_{j=1}^{M} \left(\frac{2}{\pi} \int_{|\boldsymbol{x}-\boldsymbol{R}_j|\leq D_j} \frac{\varrho_\psi(\boldsymbol{x})}{|\boldsymbol{x}-\boldsymbol{R}_j|} \mathrm{d}\boldsymbol{x} - \frac{1}{D_j} \int_{|\boldsymbol{x}-\boldsymbol{R}_j|\leq D_j} Y(|\boldsymbol{x}-\boldsymbol{R}_j|/D_j)\varrho_\psi(\boldsymbol{x})\mathrm{d}\boldsymbol{x} \right), \quad (8.5.8)$$

with Y defined in (8.3.2). After multiplying this inequality by $\pi\lambda/2$, we see that the singularity of the right side at the nuclei is exactly $\lambda/|\boldsymbol{x}-\boldsymbol{R}_j|$, which is what we want in (8.5.4). In fact, by combining the bounds (8.5.6)–(8.5.8) we obtain

$$\left(\psi, \sum_{i=1}^{N} \left|\boldsymbol{p}_i + \sqrt{\alpha}\boldsymbol{A}(\boldsymbol{x}_i)\right| \psi\right) \geq \lambda \int_{\mathbb{R}^3} \left(\frac{1}{\mathfrak{D}(\boldsymbol{x})} + \mathcal{U}(\boldsymbol{x})\right) \varrho_\psi(\boldsymbol{x})\mathrm{d}\boldsymbol{x} + (1-\pi\lambda/2)Kq^{-1/3} \int_{\mathbb{R}^3} \varrho_\psi(\boldsymbol{x})^{4/3}\mathrm{d}\boldsymbol{x}, \quad (8.5.9)$$

where $\mathcal{U}$ is a bounded function, which takes the value

$$\mathcal{U}(\boldsymbol{x}) = \begin{cases} -|\boldsymbol{x}-\boldsymbol{R}_j|^{-1} & \text{if } |\boldsymbol{x}-\boldsymbol{R}_j| > D_j \\ -(\pi/2D_j)Y(|\boldsymbol{x}-\boldsymbol{R}_j|/D_j) & \text{if } |\boldsymbol{x}-\boldsymbol{R}_j| \leq D_j \end{cases} \quad (8.5.10)$$

in the Voronoi cell $\Gamma_j = \{\boldsymbol{x} \in \mathbb{R}^3 : |\boldsymbol{x}-\boldsymbol{R}_j| < |\boldsymbol{x}-\boldsymbol{R}_k| \text{ for all } k \neq j\}$.

We are left with deriving a lower bound on the two terms containing $\mathcal{U}$ and $\varrho_\psi^{4/3}$ in (8.5.9). For convenience, let $A = 4Kq^{-1/3}(1-\pi\lambda/2)/3$. Using Hölder's inequality, it is easy to see that

$$\frac{3}{4}A \int_{\mathbb{R}^3} \varrho_\psi(\boldsymbol{x})^{4/3}\mathrm{d}\boldsymbol{x} + \lambda \int_{\mathbb{R}^3} \mathcal{U}(\boldsymbol{x})\varrho_\psi(\boldsymbol{x})\mathrm{d}\boldsymbol{x} \geq -\frac{\lambda^4}{4A^3} \int_{\mathbb{R}^3} |\mathcal{U}(\boldsymbol{x})|^4\mathrm{d}\boldsymbol{x}. \quad (8.5.11)$$

From the definition (8.5.10) of $\mathcal{U}$ we can bound

$$\int_{\mathbb{R}^3} |\mathcal{U}(\boldsymbol{x})|^4 \mathrm{d}\boldsymbol{x}$$

$$\leq \sum_{j=1}^{M} \left[\left(\frac{\pi}{2D_j} \right)^4 \int_{|\boldsymbol{x}-\boldsymbol{R}_j| \leq D_j} Y(|\boldsymbol{x}-\boldsymbol{R}_j|/D_j)^4 \mathrm{d}\boldsymbol{x} + \int_{|\boldsymbol{x}-\boldsymbol{R}_j| \geq D_j} \frac{1}{|\boldsymbol{x}-\boldsymbol{R}_j|^4} \mathrm{d}\boldsymbol{x} \right]. \tag{8.5.12}$$

The last integral equals $4\pi/D_j$. The first integral is

$$\int_{|\boldsymbol{x}-\boldsymbol{R}_j| \leq D_j} Y(|\boldsymbol{x}-\boldsymbol{R}_j|/D_j)^4 \mathrm{d}\boldsymbol{x} = D_j^3 \int_{|\boldsymbol{x}| \leq 1} Y(|\boldsymbol{x}|)^4 \mathrm{d}\boldsymbol{x} \approx 7.63\, D_j^3. \tag{8.5.13}$$

Hence we arrive at the lower bound

$$(8.5.11) \geq -\frac{\lambda^4}{4A^3} \left[7.63 \left(\frac{\pi}{2} \right)^4 + 4\pi \right] \sum_{j=1}^{M} \frac{1}{D_j},$$

which, in combination with (8.5.9), proves the lemma. ■

The estimate above can be improved slightly. In fact, the 4π in (8.5.5) can be replaced by 3π, as argued in [122]. This can be achieved by an improved bound on the integral in (8.5.12) in the region $|\boldsymbol{x}-\boldsymbol{R}_j| > D_j$, by taking into account the restriction of $\boldsymbol{x}$ to the Voronoi cell Γ_j. The same applies to Eq. (8.5.2) in Theorem 8.1.

With the aid of Lemma 8.5, the proof of Theorem 8.1 is now quite short.

Proof of Theorem 8.1. As discussed in Section 3.2.3, we can assume that $Z_j = Z$ for all $1 \leq j \leq M$ without loss of generality. To bound the potential energy from below, we use the same strategy as in the non-relativistic case, Eq. (7.1.3)–(7.1.8). In the relativistic case considered here it is important, however, to *retain* the positive last term in (5.2.6). More precisely, by combining the electrostatic inequality of Theorem 5.3 and the exchange estimate of Theorem 6.1, we obtain

$$(\psi, V_C \psi) \geq -1.68 \int_{\mathbb{R}^3} \varrho_\psi(\boldsymbol{x})^{4/3} \mathrm{d}\boldsymbol{x} - \int_{\mathbb{R}^3} \frac{Z}{\mathfrak{D}(\boldsymbol{x})} \varrho_\psi(\boldsymbol{x}) \mathrm{d}\boldsymbol{x} + \frac{1}{8} Z^2 \sum_{j=1}^{M} \frac{1}{D_j}. \tag{8.5.14}$$

Here, $D_j = (1/2)\min_{k\neq j}|\boldsymbol{R}_k - \boldsymbol{R}_j|$ denotes half the distance of $\boldsymbol{R}_j$ to its nearest neighbor nucleus and hence equals the distance of $\boldsymbol{R}_j$ to the boundary of Γ_j, while $\mathfrak{D}(\boldsymbol{x})$ is the distance from $\boldsymbol{x}$ to the closest nucleus, i.e., $\mathfrak{D}(\boldsymbol{x}) = \min_j |\boldsymbol{x} - \boldsymbol{R}_j|$.

In order to control the negative exchange term, we can apply the relativistic kinetic energy inequality of Corollary 4.1, which implies that

$$\frac{1.68\,\alpha q^{1/3}}{K}\left(\psi, \sum_{i=1}^{N}\left|\boldsymbol{p}_i + \sqrt{\alpha}\boldsymbol{A}(\boldsymbol{x}_i)\right|\,\psi\right) - 1.68\,\alpha\int_{\mathbb{R}^3}\varrho_\psi(\boldsymbol{x})^{4/3}\mathrm{d}\boldsymbol{x} \geq 0. \tag{8.5.15}$$

We are thus left with the problem of deriving a lower bound on

$$\left(\psi, \sum_{i=1}^{N}\left((1 - 1.68\,\alpha q^{1/3}K^{-1})\left|\boldsymbol{p}_i + \sqrt{\alpha}\boldsymbol{A}(\boldsymbol{x}_i)\right| - \frac{Z\alpha}{\mathfrak{D}(\boldsymbol{x}_i)}\right)\psi\right). \tag{8.5.16}$$

Under the assumption that $1 - 1.68\,\alpha q^{1/3}/K > \pi Z\alpha/2$ we can apply Lemma 8.5 to conclude that

$$(8.5.16) \geq -C\frac{(Z\alpha)^4 q}{(1 - \pi Z\alpha/2 - 1.68\alpha q^{1/3}/K)^3}\sum_{j=1}^{M}\frac{1}{D_j}, \tag{8.5.17}$$

with C given in (8.5.5). In combination with (8.5.14) this proves positivity of $(\psi, H\psi)$ and hence stability as long as

$$C\frac{(\alpha Z)^4 q}{\left(1 - \alpha Z\pi/2 - 1.68\alpha q^{1/3}/K\right)^3} \leq \frac{Z^2\alpha}{8},$$

which is equivalent to (8.5.2). ■

Remark 8.5. In the previous theorem, we proved stability of relativistic matter in the massless case, in which the kinetic energy of particle of momentum $\boldsymbol{p}$ is equal to $|\boldsymbol{p} + \sqrt{\alpha}\boldsymbol{A}(\boldsymbol{x})|$. As already pointed out in Eq. (8.2.12), the simple inequality $\sqrt{\boldsymbol{p}^2 + m^2} - m \geq |\boldsymbol{p}| - m$ shows that stability also holds in the case of non-zero mass whenever it holds with zero mass. The bound on the energy obtained in this way is not very good, however, at least in the case of small $Z\alpha$. It is simply $-mN$, which equals $-N$ in our units. This is the rest mass energy mc^2 of all the electrons, which may be huge compared to the expected energy which is of the order $(Z\alpha)^2$. A better bound can be obtained by including the mass in the appropriate Lieb–Thirring inequality, which was done in [37] (see also [122].) See the discussion at the end of Section 4.1.

Remark 8.6 (Stability of Relativistic Matter Implies Stability of Non-Relativistic Matter). The elementary inequality

$$\frac{1}{2}\boldsymbol{p}^2 \geq a|\boldsymbol{p}| - \frac{a^2}{2}$$

which holds for any $a > 0$, allows one to deduce stability of non-relativistic matter from stability of relativistic matter. If we choose a big enough such that both α/a and $Z\alpha/a$ are small enough to satisfy the stability criteria of Theorem 8.1, we conclude that the non-relativistic Hamiltonian (7.0.1) is bounded from below by $-(a^2/2)N$. Compared with Theorem 7.1, this has the advantage of yielding *a bound that depends only on N and not on M*. We shall discuss this question further in Chapter 12.

8.6 Alternative Proof of Relativistic Stability

In this section, we shall demonstrate an alternative proof of relativistic stability of matter. It is closer to the proof in [138] and uses the refined electrostatic inequality in Theorem 5.5 instead of using the simpler Theorem 5.3 and the Lieb–Oxford inequality (6.2.14), which were employed in the proof in the previous section. Although not yielding better stability bounds, this alternative method of proof will turn out to be useful later in Chapter 11 when we discuss the stability of non-relativistic quantum electrodynamics.

We proceed as in the proof in the previous section, but instead of using (8.5.14) we use Theorem 5.5, which states that

$$(\psi, V_C\,\psi) \geq -\int\limits_{\mathbb{R}^3} \left(\frac{Z}{\mathfrak{D}(\boldsymbol{x})} + F^{\lambda}(\boldsymbol{x})\right) \varrho_{\psi}(\boldsymbol{x})\mathrm{d}\boldsymbol{x} + \frac{Z^2}{8}\sum_{j=1}^{M}\frac{1}{D_j}$$

for any $0 < \lambda < 1$, where $F^{\lambda}(x) = W^{\lambda}(\boldsymbol{x}) - Z/\mathfrak{D}(\boldsymbol{x})$ which, in the Voronoi cell Γ_j, takes the value

$$F^{\lambda}(\boldsymbol{x}) = \begin{cases} \frac{1}{2}D_j^{-1}(1 - D_j^{-2}|\boldsymbol{x} - \boldsymbol{R}_j|^2)^{-1} & \text{for } |\boldsymbol{x} - \boldsymbol{R}_j| \leq \lambda D_j \\ \left(\sqrt{2Z} + \frac{1}{2}\right)|\boldsymbol{x} - \boldsymbol{R}_j|^{-1} & \text{for } |\boldsymbol{x} - \boldsymbol{R}_j| > \lambda D_j. \end{cases} \tag{8.6.1}$$

After following the same steps as in the proof of Lemma 8.5 in the previous section, we conclude that

$$\left(\psi, \left(\sum_{i=1}^{N} \left|\boldsymbol{p}_i + \sqrt{\alpha}\boldsymbol{A}(\boldsymbol{x}_i)\right| + \alpha V_C\right)\psi\right)$$

$$\geq -\frac{\alpha^4}{4\Lambda^3} \int\limits_{\mathbb{R}^3} |\mathcal{U}(\boldsymbol{x}) + F^\lambda(\boldsymbol{x})|^4 \mathrm{d}\boldsymbol{x} + \frac{Z^2\alpha}{8} \sum_{j=1}^{M} \frac{1}{D_j},$$

where $\Lambda = (4/3)(1 - \alpha Z\pi/2)Kq^{-1/3}$, and $\mathcal{U}$ is defined in (8.5.10). Using the triangle inequality for the L^4 norm, we have

$$\int\limits_{\mathbb{R}^3} |\mathcal{U}(\boldsymbol{x}) + F^\lambda(\boldsymbol{x})|^4 \mathrm{d}\boldsymbol{x} \leq \left[\left(\int\limits_{\mathbb{R}^3} |\mathcal{U}(\boldsymbol{x})|^4 \mathrm{d}\boldsymbol{x}\right)^{1/4} + \left(\int\limits_{\mathbb{R}^3} |F^\lambda(\boldsymbol{x})|^4 \mathrm{d}\boldsymbol{x}\right)^{1/4}\right]^4.$$

The L^4 norm of $\mathcal{U}$ was already estimated (8.5.12)–(8.5.13), where it was found that

$$\int\limits_{\mathbb{R}^3} |\mathcal{U}(\boldsymbol{x})|^4 \mathrm{d}\boldsymbol{x} \leq \left[7.63\left(\frac{\pi}{2}\right)^4 + 4\pi\right] \sum_{j=1}^{M} \frac{1}{D_j}.$$

To bound the L^4 norm of F^λ, we split the integral over each Γ_j into a part where $|\boldsymbol{x} - \boldsymbol{R}_j| \leq \lambda D_j$ and a part where $|\boldsymbol{x} - \boldsymbol{R}_j| > \lambda D_j$. For the former, we obtain

$$\int\limits_{\Gamma_j \cap \{|\boldsymbol{x}-\boldsymbol{R}_j| \leq \lambda D_j\}} |F^\lambda(\boldsymbol{x})|^4 \mathrm{d}\boldsymbol{x} = \frac{\pi}{4D_j} \int\limits_0^\lambda \frac{t^2}{(1-t^2)} \mathrm{d}t$$

$$\leq \frac{\pi\lambda}{4D_j} \int\limits_0^\lambda \frac{t}{(1-t^2)} \mathrm{d}t = \frac{\pi\lambda}{24\, D_j} \left(\frac{1}{(1-\lambda^2)^3} - 1\right).$$

For the latter, we bound

$$\int\limits_{\Gamma_j \cap \{|\boldsymbol{x}-\boldsymbol{R}_j| \geq \lambda D_j\}} |F^\lambda(\boldsymbol{x})|^4 \mathrm{d}\boldsymbol{x} \leq \int\limits_{\{|\boldsymbol{x}-\boldsymbol{R}_j| \geq \lambda D_j\}} |F^\lambda(\boldsymbol{x})|^4 \mathrm{d}\boldsymbol{x} = \frac{4\pi}{\lambda D_j} \left(\sqrt{2Z} + 1/2\right)^4.$$

Summing up the contributions of all the Voronoi cells, we conclude that

$$\int_{\mathbb{R}^3} |U(\boldsymbol{x}) + F^\lambda(\boldsymbol{x})|^4 \mathrm{d}\boldsymbol{x} \le A_\lambda(Z) \sum_{j=1}^{M} \frac{1}{D_j},$$

where $A_\lambda(Z)$ is given by

$$A_\lambda(Z) := \pi \left(\left[\frac{\lambda}{24} \left(\frac{1}{(1-\lambda)^3} - 1 \right) + \frac{4}{\lambda} \left(\sqrt{2Z} + 1/2 \right)^4 \right]^{1/4} + \left[7.63 \frac{\pi^3 Z^4}{16} + 4Z^4 \right]^{1/4} \right)^4 .$$

Stability thus holds if

$$\min_{0<\lambda<1} A_\lambda(Z) \le \frac{Z^2}{8\alpha^3} \left(\frac{4}{3} \right)^3 \frac{K^3}{q} \left(1 - \frac{Z\alpha\pi}{2} \right)^3 . \tag{8.6.2}$$

■

If we set $q = 2$ and $\alpha = 1/137$, use the known bound $K \ge 1.60$ and compute the minimum over λ numerically, we have stability as long as $0.0014 < Z < 43.3$. If we invoke, in addition, the monotonicity argument explained in Section 3.2.3 in Chapter 3, this method proves stability if $0 \le Z_j < 43.3$ for all $1 \le j \le M$.

For later use in Chapter 11, we display explicitly the main inequality proved in this section.

Lemma 8.6 (Improved Bound with the Nearest Nucleus Potential). *Let α, Z and q satisfy the stability condition (8.6.2). For all $0 < \lambda < 1$, all vector potentials $\boldsymbol{A}$ and all nuclear coordinates $\underline{\boldsymbol{R}} = (\boldsymbol{R}_1, \ldots, \boldsymbol{R}_M)$,*

$$\boxed{\left(\psi, \sum_{i=1}^{N} \left(\frac{1}{\alpha} |\boldsymbol{p}_i + \sqrt{\alpha}\boldsymbol{A}(\boldsymbol{x}_i)| - W^\lambda(\boldsymbol{x}_i) \right) \psi \right) \ge -\frac{Z^2}{8} \sum_{j=1}^{M} \frac{1}{D_j}} \tag{8.6.3}$$

for all normalized, fermionic wave functions $\psi \in \bigwedge^N L^2(\mathbb{R}^3; \mathbb{C}^q)$. The function W^λ is defined in (5.4.1).

8.7 Further Results on Relativistic Stability

The proof we have given of the stability of relativistic matter was chosen for its simplicity and for its direct connection with the various fundamental inequalities proved in earlier chapters. The emphasis was on trying to present, to the extent possible, a unified picture of non-relativistic and relativistic stability.

The actual historical path was quite different. As time went on better results were obtained, and some of them will be discussed here. Indeed, it was only in 2007 that stability for values up to $Z\alpha = 2/\pi$ with arbitrary magnetic fields was finally obtained by Frank, Lieb and Seiringer [71].

Perhaps the earliest result was that of Daubechies and Lieb [38] in 1983 that relativistic stability holds for one electron and $M > 1$ nuclei provided $Z_j\alpha \leq 2/\pi$ for all j and $\alpha \leq 1/3\pi$. It was also shown that stability of the first kind fails if α is large and $M = 2$ (and hence for N electrons and $2N$ nuclei). This appears to be the first mention of the need for a bound on α, and we discuss this $M = 2$ case further in the next section.

The first proof of stability for all N and M is Conlon's [32] which gave stability if $Z = 1, q = 1$ and $\alpha < 10^{-200}$. Despite this very poor constant, Conlon's 1984 paper was a major achievement and showed that stability had an affirmative solution – which was *not* beyond doubt at the time. His paper was followed in 1986 by Fefferman and de la Llave's proof [60] of stability for $Z = 1, q = 1$ and $\alpha \leq 1/(2.06\pi)$.

In 1988 Lieb and Yau [138] proved two stability theorems. The simpler one showed stability for all $Z\alpha < \nu < 2/\pi$ and $\alpha < \alpha_c(\nu)$, but with $\alpha_c(\nu) \to 0$ as $\nu \to 2/\pi$. This simpler proof can easily be extended (although this was not stated explicitly) to include a magnetic field. The more complicated theorem was valid all the way up to $Z\alpha = 2/\pi$ for $\alpha < 1/(47q)$, which includes the physical case $q = 2$, $\alpha = 1/137$. Unfortunately, the proof of this second theorem does *not* generalize easily to include a magnetic field. The problem of modifying the proof to include a magnetic field without weakening the $Z\alpha \leq 2/\pi$ condition was finally solved in [71], but at a price of reducing α: For the Hamiltonian

$$H = \sum_{i=1}^{N} \left| \boldsymbol{p}_i + \sqrt{\alpha}\boldsymbol{A}(\boldsymbol{x}_i) \right| + \alpha V_C(\underline{\boldsymbol{X}}, \underline{\boldsymbol{R}}), \tag{8.7.1}$$

stability holds for all $\boldsymbol{A}(\boldsymbol{x})$ if $Z\alpha \leq 2/\pi$ and $q\alpha \leq \dfrac{1}{66.5}$.

This just barely covers the physical case $q = 2$ and $\alpha = 1/137$.

The 1996 paper by Lieb, Loss and Siedentop [122] is the one that takes the simplifying track that we followed in Section 8.5.

The necessity of α being small in all these results is not an artifact of the method. This was already clear from the early result in [38], as was discussed in Section 8.1.1. One can show that for α large enough, there is instability for large enough M, irrespective of how small $Z\alpha$ might be. Since $Z \geq 1$ in nature, this result may not seem to be of practical importance (because the condition $Z\alpha \leq 2/\pi$ already implies the need for a bound on α, namely $\alpha \leq 2/\pi$), but it is important conceptually. It says that relativistic stability of matter necessarily implies a bound on the value of the fine structure constant α. The *instability* for large α was proved in two theorems in the Lieb–Yau paper [138], and we outline these two results in Theorem 8.2 below, along with a summary of the known stability theorems.

8.8 Instability for Large α, Large q or Bosons

The question addressed in this section is whether instability occurs if α is large enough, regardless of what $Z\alpha > 0$ might be. For one electron and one nucleus, i.e., a hydrogenic atom, only $Z\alpha$ appears in the Hamiltonian and hence no separate bound on α is needed for stability. The simplest non-trivial case to consider, therefore, is one electron and two nuclei. As stated above, this was the first case in which instability was noted [38]. Suppose the two nuclear charges are separated by a distance $R = |\boldsymbol{R}_1 - \boldsymbol{R}_2|$. Let $E(R)$ be the ground state energy of

$$|\boldsymbol{p}| - \frac{Z_1\alpha}{|\boldsymbol{x} - \boldsymbol{R}_1|} - \frac{Z_2\alpha}{|\boldsymbol{x} - \boldsymbol{R}_2|}.$$

By scaling, $E(R) = C/R$ for some constant $C \leq 0$, depending on $Z_1\alpha$ and $Z_2\alpha$. Now, if $(Z_1 + Z_2)\alpha \leq 2/\pi$, then $C = 0$. If, however, $(Z_1 + Z_2)\alpha > 2/\pi$, then $C < 0$, since we can find a wave function ψ with $(\psi, |\boldsymbol{p}|\psi) < (Z_1 + Z_2)\alpha(\psi, |\boldsymbol{x}|^{-1}\psi)$ and put the two nuclei arbitrarily close to the origin to make the energy of ψ negative. As long as $Z_1\alpha \leq 2/\pi$ and $Z_2\alpha \leq 2/\pi$, C is finite, as can be easily deduced from Lemma 8.3.

In opposition to this negative energy C/R, which attracts the nuclei, there is the nuclear repulsion, which equals

$$\frac{Z_1 Z_2\alpha}{R} = \frac{(Z_1\alpha)(Z_2\alpha)}{\alpha}\frac{1}{R}.$$

Hence, for fixed $Z_1\alpha$ and $Z_2\alpha$, we can make α big enough so that this nuclear repulsion is smaller than $|C|/R$. For this value of α, the ground state energy E_0

of one electron and two nuclei is thus negative. Hence $E_0 = -\infty$ by scaling, as remarked earlier.

The same argument works for any number of nuclei. *One electron will cause instability (of the first kind) if α is too large.* The number of nuclei needed for instability will increase as $Z\alpha \to 0$ (because we require the *total* nuclear charge $\sum_{k=1}^{M} Z_k$ to exceed $2/(\pi\alpha)$) but the important point is that there is a *fixed* number α_c such that instability always occurs (with sufficiently many nuclei) when $\alpha > \alpha_c$, regardless of how small the values of $Z_i\alpha$ may be.

In combination, Theorem 8.1 in Section 8.5 together with its improvements discussed in Section 8.7 state that the relativistic Hamiltonian (8.7.1) is stable, i.e., non-negative, for all M, N and *all magnetic fields* if the condition

$$\boxed{\frac{1}{\alpha} \geq \frac{Z\pi}{2} + 2.2566\, q^{1/3} Z^{2/3} + 1.0307\, q^{1/3}}$$

or the condition

$$\boxed{q\alpha \leq \frac{1}{66.5} \quad \text{and} \quad Z\alpha \leq 2/\pi}$$

is satisfied. For comparison with the following instability theorem, 8.2, it is convenient to reformulate the stability conditions as follows. There is stability if *all* of the following three **stability conditions** hold: For appropriate $C > 0$ and $C' > 0$,

$$\boxed{\begin{array}{l} (1)\ Z_j\alpha \leq 2/\pi \text{ for all } j = 1, \ldots, M, \\ (2)\ (\frac{\pi}{2} Z_j\alpha)^2\alpha < C/q \text{ for all } j = 1, \ldots, M, \\ (3)\ \alpha < C' q^{-1/3}. \end{array}}$$

That this simple set of stability conditions (almost) correctly describes the true state of affairs is the content of the next theorem, proved in [138, Thms. 3 and 4].

Theorem 8.2 (Conditions for Instability). *There are constants $D < 115120$ and $D' < 128/(15\pi)$ such that the Hamiltonian (8.1.1) is unstable of the first kind for large enough N and M if one of the following three conditions* *holds:*

$$
\begin{array}{l}
(i)\ Z_j\alpha > 2/\pi \quad \textit{for some } j, \\
(ii)\ (\tfrac{\pi}{2}Z_j\alpha)^2\alpha > D/q \quad \textit{for all } j, \\
(iii)\ \alpha > D'.
\end{array}
$$

As will be demonstrated in the proof of Theorem 8.2, under condition (ii) it suffices to have only q electrons to cause the collapse. Under conditions (i) and (iii) only one electron is needed for instability!

The three instability conditions (i), (ii) and (iii) in Theorem 8.2 are (almost) the negation of the three stability conditions (1), (2) and (3) displayed above, except for the numerical values of the constants, and the factor $q^{-1/3}$ in the third condition. *For fixed* Z both the stability and the instability bounds show that the critical α is of the order $q^{-1/3}$ for large q. But if we want a bound that is uniform in Z (valid for arbitrarily small Z) is the critical α of order 1 or of order $q^{-1/3}$? We conjecture that the former is correct.

We believe that the factor $q^{-1/3}$ should actually not be part of the stability condition (3); that is, $\alpha < C'$ for some constant C', together with stability conditions (1) and (2), should be sufficient for stability. *This is an open problem.* We note that if the Lieb–Oxford inequality used in the proof of Theorem 8.1, Eq. (8.5.14), held with a factor $q^{-1/3}$ multiplying the $\int \varrho^{4/3}$ term, our proof of Theorem 8.1 would actually yield the condition $\alpha < C'$ instead of $\alpha < C'q^{-1/3}$. This factor would naturally occur from considerations of the homogeneous electron gas; see the discussion in Chapter 6, Eq. (6.2.11). As noted there, a general exchange inequality cannot hold with this factor $q^{-1/3}$, however, but our conjecture is that it holds with $q^{-1/3}$ for the wave functions actually relevant for causing instability.

Note that if we define $\alpha_c(\nu)$ as the critical value of α for stability to hold *for all* $Z\alpha < \nu$, then the stability and instability results imply that $\alpha_c(\nu) \sim 1/q$ for large q, as mentioned in the Remark 8.4.

Proof of Theorem 8.2. We choose $A = 0$ and, without loss of generality, set $m = 0$. Instability under condition (i) has already been shown in Section 8.2.

We shall now prove instability under condition (iii). We take only one species of nuclei, i.e., we set $Z_j = Z > 0$ for all $j = 1, 2, \ldots, M$. Recall that it suffices merely to show that the energy can be made negative, after which it can be driven to $-\infty$ by length scaling. For simplicity of the notation, we shall ignore the spin degrees of freedom here. Alternatively, think of either $q = 1$ or all spin components but one be zero. Pick a $\phi \in L^2(\mathbb{R}^3)$ with $\|\phi\|_2 = 1$ and with finite kinetic energy, that is, $\tau = (\phi, |\boldsymbol{p}|\phi) < \infty$. Let $N = 1$. Then

$$(\phi, H\phi) = \tau - Z\alpha \sum_{k=1}^{M} \int_{\mathbb{R}^3} \frac{|\phi(\boldsymbol{x})|^2}{|\boldsymbol{x} - \boldsymbol{R}_k|} \mathrm{d}\boldsymbol{x} + Z^2\alpha \sum_{1 \le k < l \le M} \frac{1}{|\boldsymbol{R}_k - \boldsymbol{R}_l|}. \quad (8.8.1)$$

For a given ϕ, we try to choose the nuclear positions $\boldsymbol{R}_k$ so as to make this expression as small as possible. For an upper bound on the smallest possible value, we can average the expression (8.8.1) over the nuclear positions, with weight given by $\prod_{k=1}^{M} |\phi(\boldsymbol{R}_k)|^2$. In this way, instability is established if it can be shown that

$$\tau - \mathcal{I}\left(Z\alpha M - \frac{1}{2} Z^2 \alpha M(M-1) \right) < 0,$$

where $\mathcal{I}$ denotes

$$\mathcal{I} = \iint_{\mathbb{R}^3 \times \mathbb{R}^3} \frac{|\phi(\boldsymbol{x})|^2 |\phi(\boldsymbol{y})|^2}{|\boldsymbol{x} - \boldsymbol{y}|} \mathrm{d}\boldsymbol{x}\mathrm{d}\boldsymbol{y}.$$

For a given value of Z we can choose M such that $|M - 1/2 - 1/Z| \le 1/2$. Then

$$-Z\alpha M + \frac{1}{2} Z^2 \alpha M(M-1) = \frac{1}{2}\left(Z^2\alpha \left[M - \frac{1}{2} - \frac{1}{Z} \right]^2 - \frac{1}{4} Z^2 \alpha - \alpha(1+Z) \right)$$
$$\le -\frac{1}{2}\alpha.$$

Instability thus occurs for $\alpha > 2\tau/\mathcal{I}$. To obtain a numerical value for τ and $\mathcal{I}$, we choose $\phi(\boldsymbol{x}) = \pi^{-1/2} \exp(-|\boldsymbol{x}|)$. Then $\tau = 8/(3\pi)$ and $\mathcal{I} = 5/8$, and hence $2\tau/\mathcal{I} = 128/(15\pi)$.

Finally, we show that condition (ii) suffices for instability. We take $N = q$ electrons, and choose the wave function to be $\prod_{i=1}^{N} \phi(\boldsymbol{x}_i)$ times the totally antisymmetric spin wave function given by $\chi(\underline{\sigma}) = (N!)^{-1/2} \det g_i(\sigma_j)|_{i,j=1}^{N}$, where $g_i(\sigma) = \delta_{\sigma,i}$. With the same notation as above, the energy of this state is given by

$$N\tau + \alpha W(N, \underline{\boldsymbol{R}})$$

where

$$W(N, \underline{\boldsymbol{R}}) = -ZN\alpha \sum_{k=1}^{M} \int_{\mathbb{R}^3} \frac{|\phi(\boldsymbol{x})|^2}{|\boldsymbol{x} - \boldsymbol{R}_k|} \mathrm{d}\boldsymbol{x}$$

$$+ Z^2\alpha \sum_{1 \le k < l \le M} \frac{1}{|\boldsymbol{R}_k - \boldsymbol{R}_l|} + \frac{\alpha}{2} N(N-1)\mathcal{I}.$$

Here, the last term is the repulsive interaction among the electrons.

When discussing the instability of non-relativistic bosons in Chapter 7, Section 7.6, we already encountered the expression $W(N, \underline{\boldsymbol{R}})$ and showed that it is possible to choose the nuclear positions $\underline{\boldsymbol{R}}$ in such a way that

$$W(N, \underline{\boldsymbol{R}}) \le -CZ^{2/3}N^{4/3}$$

for some constant $C > 0$ depending only on ϕ. Therefore, instability occurs if

$$q\tau - C\alpha Z^{2/3} q^{4/3} < 0,$$

or

$$(Z\alpha)^2\alpha > \left(\frac{\tau}{C}\right)^3 \frac{1}{q}.$$

The numerical value for τ/C depends on the choice of ϕ. We refer to [138] for details of an appropriate choice of ϕ and a careful calculation of the constant C (which leads to the stated value of D). ■

We now turn our attention to relativistic stability for bosons. Recall that in the non-relativistic case discussed in the previous chapter, bosonic particles are unstable of the second (but not the first) kind, that is, the ground state energy goes like $-N^{5/3}$ in the case of static nuclei. In the relativistic case, the situation is even worse, in the sense that even stability of the first kind fails for large enough particle number.

Theorem 8.3 (Instability for Bosons). *For every $\alpha > 0$ and $Z > 0$ and for sufficiently large M and N (depending on α and Z), the fundamental Hamiltonian (8.1.1) for N bosons is unstable of the first kind as long as $Z_j \ge Z$ for all $1 \le j \le M$.*

Proof. As pointed out in Chapter 3, Section 3.1.3.1 and Corollary 3.1, the ground state energy of N bosons is the same as the one for N fermions with $q = N$ spin states for each particle. Hence by Theorem 8.2, condition (ii), there is instability for each fixed Z and α if $q = N$ is large enough. ∎

Remark 8.7. The instability result in Theorem 8.3 refers to static nuclei. If the nuclei have a finite mass and their kinetic energy is taken into account, the stability situation is a little more subtle. The nuclei can be either bosons or fermions. If they are fermions we can apply the stability theorem, 8.1, to the positive nuclei instead of the negative bosons and see that stability is restored in an appropriate parameter regime. On the other hand, if some of the nuclei are bosons, then instability always occurs. This follows from the consideration of the non-relativistic case in the previous chapter, where it was explained that the energy of non-relativistic bosons goes like $-N^{7/5}$ for large N. To apply this to the relativistic case one simply notes that $|\boldsymbol{p}| \leq (\boldsymbol{p}^2 + 1)/2$, and hence the ground state energy in the relativistic case is bounded from above by $+N - CN^{7/5}$ for some $C > 0$. It is thus negative for large enough N and, therefore, equals $-\infty$ by scaling, as explained in the beginning of Section 8.5, Eq. (8.5.1).

CHAPTER 9

Magnetic Fields and the Pauli Operator

9.1 Introduction

In the preceding chapters we proved the stability of matter for electrons (i.e., fermions) interacting with fixed nuclei, both relativistically and non-relativistically. We showed that the proofs go through even in the presence of magnetic fields, meaning that one can replace $\boldsymbol{p}$ by $\boldsymbol{p} + \sqrt{\alpha}\boldsymbol{A}(\boldsymbol{x})$ for any choice of the vector field $\boldsymbol{A}$. Indeed, the diamagnetic inequalities discussed in Section 4.4 show that the *bounds* obtained for the ground state energy can only improve when $\boldsymbol{A}$ is included (but this does not imply that the *true* ground state energy increases).

In real life the energy of matter actually decreases significantly when a field is added and it can decrease to $-\infty$ if the field is allowed to become arbitrarily large. The reason is that the electron *spins* interact with the magnetic field, and hence an additional term must be added to the kinetic energy to account for this interaction. What prevents the field from becoming large, and the energy becoming arbitrarily negative, is the positive self-energy of the field (i.e., the energy needed to create the field). The goal is to show that the additional negative interaction energy is balanced by the field energy and an 'equilibrium' is reached in which the total energy is finite. We shall see that this is achieved if and only if α and $Z\alpha^2$ are not too large – *even in the non-relativistic case*. Thus, the *non-relativistic* Hamiltonian we shall treat in this chapter mimics the *relativistic* one we treated in the previous chapter in the sense that a bound on α is needed.

Only the non-relativistic model of particles with spin in the presence of magnetic fields will be treated in this chapter. The discussion of relativistic electrons together with magnetic fields will be deferred to Chapter 10 where the Pauli operator, treated here, will be seen to be the non-relativistic limit of the Dirac operator.

9.2 The Pauli Operator and the Magnetic Field Energy

We recall the definition of the Pauli one-body kinetic energy operator given in Chapter 2, Eq. (2.1.45),

$$\boxed{T_A(\boldsymbol{p}) = \frac{1}{2}\left|\boldsymbol{\sigma}\cdot\left(\boldsymbol{p}+\sqrt{\alpha}\boldsymbol{A}(\boldsymbol{x})\right)\right|^2 = \frac{1}{2}\left(\boldsymbol{p}+\sqrt{\alpha}\boldsymbol{A}(\boldsymbol{x})\right)^2 + \frac{\sqrt{\alpha}}{2}\boldsymbol{\sigma}\cdot\boldsymbol{B}(\boldsymbol{x}).} \tag{9.2.1}$$

The wave functions on which this operator acts lie in $L^2(\mathbb{R}^3; \mathbb{C}^2 \otimes \mathbb{C}^q)$. We allow for q internal states (often called 'flavors') in addition to the two coming from the spin 1/2 nature of the particles. In order words, a wave function of one particle is a $\mathbb{C}^2$ valued function (instead of the usual complex-valued function) of $z = (\boldsymbol{x}, \sigma)$, with $\boldsymbol{x} \in \mathbb{R}^3$ and $1 \leq \sigma \leq q$, and the number of internal states, altogether, is $2q$. With this convention, we can write the quadratic form corresponding to (9.2.1) as

$$(\psi, T_A(\boldsymbol{p})\,\psi) = \frac{1}{2}\int \left|\left(\boldsymbol{p}+\sqrt{\alpha}\boldsymbol{A}(\boldsymbol{x})\right)\psi\right|^2 \mathrm{d}z + \frac{\sqrt{\alpha}}{2}\int \langle\psi, \boldsymbol{\sigma}\cdot\boldsymbol{B}(\boldsymbol{x})\psi\rangle \mathrm{d}z. \tag{9.2.2}$$

As explained in Chapter 2, $|\cdot|$ stands for the norm of a vector in $\mathbb{C}^2$, and $\langle\cdot,\cdot\rangle$ stands for the inner product on $\mathbb{C}^2$. Recall that the sum over σ is contained in the dz integral, as in (3.1.15).

We assume that $\boldsymbol{A} \in L^2_{\mathrm{loc}}(\mathbb{R}^3; \mathbb{R}^3)$. The magnetic field $\boldsymbol{B}(\boldsymbol{x})$ is given as $\boldsymbol{B}(\boldsymbol{x}) = \operatorname{curl} \boldsymbol{A}(\boldsymbol{x})$, which has to interpreted in the sense of distributions if $\boldsymbol{A}$ is not differentiable. We also assume that $\mathcal{E}_{\mathrm{mag}}(\boldsymbol{B}) < \infty$, i.e., $\boldsymbol{B} \in L^2(\mathbb{R}^3; \mathbb{R}^3)$. Recall that we choose units such that the electron mass, $\hbar$ and the speed of light all equal 1.

When discussing the question of stability of matter for arbitrary magnetic fields it is important to consider the energy cost of creating such a magnetic field, namely the self-energy of $\boldsymbol{B}$. As discussed in Chapter 2, Section 2.1.2, in our units this self-energy is given by

$$\boxed{\mathcal{E}_{\mathrm{mag}}(\boldsymbol{B}) = \frac{1}{8\pi}\int_{\mathbb{R}^3} |\boldsymbol{B}(\boldsymbol{x})|^2 \mathrm{d}\boldsymbol{x}.} \tag{9.2.3}$$

That is, when minimizing the energy over all possible wave functions ψ and magnetic fields $\boldsymbol{B}(\boldsymbol{x})$, it is important to add the self-energy in order to obtain a finite result, as we shall see later. This was not necessary in Chapter 7, where the $\boldsymbol{\sigma} \cdot \boldsymbol{B}$ term in the Hamiltonian was absent. The minimal energy defined in this way is called the **ground state energy** of the system. Stability of the first and second kind always refer to this energy of particles and field.

The last term in (9.2.1) can alternatively be written as

$$\frac{\sqrt{\alpha}}{2}\boldsymbol{\sigma} \cdot \boldsymbol{B}(\boldsymbol{x}) = \sqrt{\alpha}\,\frac{g}{2}\,\boldsymbol{S} \cdot \boldsymbol{B}(\boldsymbol{x})$$

where $\boldsymbol{S} = \boldsymbol{\sigma}/2$ is the **electron spin**, and $g = 2$ is the gyromagnetic ratio (or **Landé g factor**) of the electron. It is possible to study the Pauli operator (9.2.1) with a factor $g/2$ in front of the last term for general values of g, not just $g = 2$. One observes the following, however

- The sign of g is important. If $g < 0$, it is not clear whether

$$\frac{1}{2}\left(\boldsymbol{p} + \sqrt{\alpha}\boldsymbol{A}(\boldsymbol{x})\right)^2 + \frac{\sqrt{\alpha}g}{4}\boldsymbol{\sigma} \cdot \boldsymbol{B}(\boldsymbol{x}) \tag{9.2.4}$$

  is positive for all magnetic fields $\boldsymbol{B}(\boldsymbol{x})$.
- For $0 \leq g \leq 2$, the operator (9.2.4) can be written as a convex combination of (9.2.1) and $(\boldsymbol{p} + \sqrt{\alpha}\boldsymbol{A}(\boldsymbol{x}))^2$ and is thus positive. Moreover, by employing this fact the analysis of stability of matter can be reduced to the $g = 2$ case. Recall that in Chapter 7 we have shown the stability of matter in case $g = 0$, for any values of Z and α and any magnetic field.
- For $g > 2$, (9.2.4) is not positive, in general. Even worse, the energy of one electron and one nucleus can be driven to $-\infty$ by increasing $\boldsymbol{B}$, even if the self-energy of the magnetic field is taken into account! There exist particles in nature with g factors bigger than 2, or even negative. These are composite particles, however, like protons ($g = 5.6$) which are made out of three quarks. When analyzing the stability of matter with such particles, one can not ignore their finite size and internal structure.

9.3 Zero-Modes of the Pauli Operator

An important role in our analysis will be played by **zero-modes** of the Pauli operator (9.2.1), that is, a pair $(\psi(\boldsymbol{x}), \boldsymbol{A}(\boldsymbol{x}))$ with $\psi \in L^2(\mathbb{R}^3; \mathbb{C}^2)$ and $\operatorname{curl} \boldsymbol{A}(\boldsymbol{x}) \in L^2(\mathbb{R}^3; \mathbb{R}^3)$ such that

$$\boldsymbol{\sigma} \cdot (\boldsymbol{p} + \sqrt{\alpha}\boldsymbol{A}(\boldsymbol{x}))\psi(\boldsymbol{x}) = 0.$$

Such a ψ is an eigenfunction of the Pauli operator (9.2.1) with eigenvalue 0. In addition, we demand that the magnetic field energy (9.2.3) of $\boldsymbol{B} = \operatorname{curl} \boldsymbol{A}$ be finite, and we stress that ψ must be square integrable. In other words, for special magnetic fields there is a genuine normalizable (and, therefore, time-independent) state with zero kinetic energy. Such a state does not exist when $\boldsymbol{A} = 0$, i.e., for the Laplacian alone. There is no non-zero $f \in L^2(\mathbb{R}^d)$ such that $\Delta f = 0$.

The importance of zero-modes for stability was recognized in [74], but their existence was uncertain. The first examples were given by Loss and Yau in [139] and the first one cited in that paper was

$$\sqrt{\alpha}\, \boldsymbol{A}(\boldsymbol{x}) = -\frac{3}{(1+\boldsymbol{x}^2)^2}\left[(1-\boldsymbol{x}^2)\boldsymbol{w} + 2(\boldsymbol{w}\cdot\boldsymbol{x})\boldsymbol{x} + 2\boldsymbol{w}\wedge\boldsymbol{x}\right] \tag{9.3.1}$$

where $\boldsymbol{w} = (0, 0, 1)$. The magnetic field in this case equals

$$\boldsymbol{B}(\boldsymbol{x}) = \operatorname{curl} \boldsymbol{A}(\boldsymbol{x}) = \frac{4}{1+\boldsymbol{x}^2}\boldsymbol{A}(\boldsymbol{x}). \tag{9.3.2}$$

The corresponding spinor-valued function $\psi \in L^2(\mathbb{R}^3; \mathbb{C}^2)$ with zero kinetic energy is

$$\psi(\boldsymbol{x}) = (1+\boldsymbol{x}^2)^{-3/2}\,[1 + i\boldsymbol{\sigma}\cdot\boldsymbol{x}]\begin{pmatrix}1\\0\end{pmatrix}. \tag{9.3.3}$$

Many more zero-modes were found later [1, 48, 45, 6], but their complete classification is still unknown.

Notice that, by scaling, any zero-mode gives rise to a one-parameter family of zero-modes. This observation will be useful in the sequel. That is, if $(\psi, \boldsymbol{A})$ is a zero-mode then so is $(\psi_\lambda, \boldsymbol{A}_\lambda)$, for all $\lambda > 0$, where

$$\psi_\lambda(\boldsymbol{x}) = \lambda^{3/2}\psi(\lambda\boldsymbol{x}), \qquad \boldsymbol{A}_\lambda(\boldsymbol{x}) = \lambda \boldsymbol{A}(\lambda\boldsymbol{x}), \tag{9.3.4}$$

and

$$\boldsymbol{B}_\lambda(\boldsymbol{x}) = \lambda^2 \boldsymbol{B}(\lambda\boldsymbol{x}), \qquad \mathcal{E}_{\mathrm{mag}}(\boldsymbol{B}_\lambda) = \lambda\mathcal{E}_{\mathrm{mag}}(\boldsymbol{B}). \tag{9.3.5}$$

The scaling of ψ_λ is chosen such that $\|\psi_\lambda\|_2 = \|\psi\|_2$.

As in the example above, the dependence on α is rather simple. If $(\psi, \widetilde{\boldsymbol{A}})$ is a zero-mode for $\alpha = 1$, then $(\psi, \alpha^{-1/2}\widetilde{\boldsymbol{A}})$ is, by definition, a zero-mode for any $\alpha > 0$. Under this transformation $\widetilde{\boldsymbol{B}}$ also gets multiplied by $\alpha^{-1/2}$, and hence the field energy $\mathcal{E}_{\mathrm{mag}}(\widetilde{\boldsymbol{B}})$ gets multiplied by α^{-1}.

9.4 A Hydrogenic Atom in a Magnetic Field

Before studying the question of stability of matter with the Pauli operator in full generality let us consider the simplest case of one electron in the field of one nucleus of charge Z. In this case we shall already see that stability can only hold if the field energy is taken into account. The Hamiltonian, including the magnetic field energy, is given by

$$H_A = T_A(p) - \frac{Z\alpha}{|x|} + \mathcal{E}_{\mathrm{mag}}(B). \tag{9.4.1}$$

The fact that the field energy is necessary to stabilize the system is an immediate consequence of the existence of zero-modes discussed in the previous section. In fact, with $(\psi, \alpha^{-1/2}\widetilde{A})$ being a zero-mode of $T_A(p)$ and with $(\psi_\lambda, \alpha^{-1/2}\widetilde{A}_\lambda)$ scaled as in (9.3.4), we have

$$\begin{aligned}(\psi_\lambda, H_{\alpha^{-1/2}\widetilde{A}_\lambda}\psi_\lambda) &= -Z\alpha(\psi_\lambda, |x|^{-1}\psi_\lambda) + \mathcal{E}_{\mathrm{mag}}(\alpha^{-1/2}\widetilde{B}_\lambda) \\ &= -\lambda\, Z\alpha(\psi, |x|^{-1}\psi) + \frac{\lambda}{\alpha}\mathcal{E}_{\mathrm{mag}}(\widetilde{B}).\end{aligned} \tag{9.4.2}$$

Two things can be seen from this expression. First, if we omit the field energy, we can drive the energy to $-\infty$ by increasing λ to infinity. Second, Eq. (9.4.2) shows that there is a critical value of $Z\alpha^2$ beyond which the infimum (over ψ and A) of the total energy is $-\infty$.

Our goal in this section is to show that the critical $Z\alpha^2$ is not zero. We follow closely the discussion in [74]. In fact, we shall show the following.

Theorem 9.1 (Stability of Hydrogen for Small $Z\alpha^2$). *Let H_A be given in (9.4.1). Under the condition that*

$$Z\alpha^2 \le \frac{\pi}{2}\left(\frac{3}{4}\right)^{3/2} \tag{9.4.3}$$

we have

$$(\psi, H_A\psi) \ge -(Z\alpha)^2(\psi, \psi)$$

for any $\psi \in H^1(\mathbb{R}^3; \mathbb{C}^2)$ and for any magnetic vector potential $A(x)$.

Recall from (2.2.17) that if the Zeeman term $\sigma \cdot B$ is omitted the infimum of $(\psi, H_A\psi)/(\psi, \psi)$ (over all ψ and A) equals $-(1/2)(Z\alpha)^2$.

The theorem states that a hydrogenic atom is stable if $Z\alpha^2$ is not too large. For the physical value $\alpha = 1/137$, the critical Z is seen to be at least as big as $\alpha^{-2}\frac{\pi}{2}\left(\frac{3}{4}\right)^{3/2} \approx 19\,160$.

The following proof is somewhat convoluted, but at present there does not seem to be any straightforward, simple path to the conclusion.

Proof. We can assume that $(\psi, \psi) = 1$. Since $T_A(p) \geq 0$, we have the lower bound $T_A(p) \geq \varepsilon\, T_A(p)$ for any $0 \leq \varepsilon \leq 1$. Moreover, $\sigma \cdot B(x) \geq -|B(x)|$. Hence, for any $\psi \in H^1(\mathbb{R}^3; \mathbb{C}^2)$,

$$(\psi, H_A\psi) \geq \left(\psi, \left[\frac{\varepsilon}{2}(p + \sqrt{\alpha}A(x))^2 - \frac{\varepsilon}{2}\sqrt{\alpha}|B(x)| - Z\alpha|x|^{-1}\right]\psi\right) + \mathcal{E}_{\mathrm{mag}}(B).$$

For given $|\psi|^2$, we can minimize the second and fourth terms with respect to $|B(x)|$, which yields

$$\mathcal{E}_{\mathrm{mag}}(B) - \frac{\varepsilon}{2}\sqrt{\alpha}\int\limits_{\mathbb{R}^3} |\psi(x)|^2|B(x)|\mathrm{d}x \geq -\frac{\alpha\varepsilon^2\pi}{2}\int\limits_{\mathbb{R}^3} |\psi(x)|^4\mathrm{d}x.$$

On the other hand, the diamagnetic inequality (4.4.4) implies that

$$(\psi, (p + \sqrt{\alpha}A(x))^2\psi) \geq \int |\nabla|\psi(x)||^2\mathrm{d}x.$$

Thus, we are led to investigate a lower bound for the quantity

$$\frac{\varepsilon}{2}\int |\nabla|\psi(x)||^2\mathrm{d}x - \int \frac{Z\alpha}{|x|}|\psi(x)|^2\mathrm{d}x - \frac{\alpha\varepsilon^2\pi}{2}\int |\psi(x)|^4\mathrm{d}x. \tag{9.4.4}$$

We are free to choose any ε between 0 and 1 to make this expression as large as possible. For given ψ, the optimal choice of ε is

$$\varepsilon = \min\left\{1, \int |\nabla|\psi(x)||^2\mathrm{d}x\left[2\alpha\pi\int |\psi(x)|^4\mathrm{d}x\right]^{-1}\right\}. \tag{9.4.5}$$

Two cases have to be distinguished. If $\int |\nabla|\psi(x)||^2\mathrm{d}x \geq 2\alpha\pi\int |\psi(x)|^4\mathrm{d}x$, then $\varepsilon = 1$ and the expression (9.4.4) is bounded from below by

$$\frac{1}{4}\int |\nabla|\psi(x)||^2\mathrm{d}x - \int \frac{Z\alpha}{|x|}|\psi(x)|^2\mathrm{d}x \geq -(Z\alpha)^2\int |\psi(x)|^2\mathrm{d}x.$$

The last inequality is just the ground state energy of the hydrogen atom, discussed in Section 2.2.2. In the contrary case, where $\int |\nabla|\psi(x)||^2\mathrm{d}x <$

$2\alpha\pi \int |\psi(\boldsymbol{x})|^4 \mathrm{d}\boldsymbol{x}$, the expression (9.4.4) becomes

$$\frac{1}{8\pi\alpha} \frac{\left(\int |\nabla|\psi(\boldsymbol{x})||^2 \mathrm{d}\boldsymbol{x}\right)^2}{\int |\psi(\boldsymbol{x})|^4 \mathrm{d}\boldsymbol{x}} - \int \frac{Z\alpha}{|\boldsymbol{x}|} |\psi(\boldsymbol{x})|^2 \mathrm{d}\boldsymbol{x} \tag{9.4.6}$$

for our choice of ε in (9.4.5). The Sobolev inequality discussed in Chapter 2, Eq. (2.2.4), states that

$$\frac{3}{4}(4\pi^2)^{2/3} \left(\int |\psi(\boldsymbol{x})|^6 \mathrm{d}\boldsymbol{x}\right)^{1/3} \leq \int |\nabla|\psi(\boldsymbol{x})||^2 \mathrm{d}\boldsymbol{x}.$$

Using this and Schwarz's inequality we obtain

$$\begin{aligned} \int |\psi(\boldsymbol{x})|^4 \mathrm{d}\boldsymbol{x} &\leq \left(\int |\psi(\boldsymbol{x})|^2 \mathrm{d}\boldsymbol{x}\right)^{1/2} \left(\int |\psi(\boldsymbol{x})|^6 \mathrm{d}\boldsymbol{x}\right)^{1/2} \\ &\leq \left(\frac{4}{3}\right)^{3/2} \frac{1}{4\pi^2} \left(\int |\psi(\boldsymbol{x})|^2 \mathrm{d}\boldsymbol{x}\right)^{1/2} \left(\int |\nabla|\psi(\boldsymbol{x})||^2 \mathrm{d}\boldsymbol{x}\right)^{3/2}. \end{aligned}$$

In addition, from the ground state energy of the hydrogen atom we get

$$\left(\int |\nabla|\psi(\boldsymbol{x})||^2 \mathrm{d}\boldsymbol{x}\right)^{1/2} \left(\int |\psi(\boldsymbol{x})|^2 \mathrm{d}\boldsymbol{x}\right)^{1/2} \geq \int \frac{|\psi(\boldsymbol{x})|^2}{|\boldsymbol{x}|} \mathrm{d}\boldsymbol{x}$$

(see Eq. (2.2.18) in Chapter 2). Hence (9.4.6) is bounded from below by

$$\left(\frac{\pi}{2\alpha}\left(\frac{3}{4}\right)^{3/2} - Z\alpha\right) \int \frac{|\psi(\boldsymbol{x})|^2}{|\boldsymbol{x}|} \mathrm{d}\boldsymbol{x}.$$

The condition $\int |\psi(\boldsymbol{x})|^2 \mathrm{d}\boldsymbol{x} = 1$ has been used. Under condition (9.4.3) the last term is positive for any ψ, and hence (9.4.6) is positive. ■

The proof of Theorem 9.1 may not be the slickest possible, but it was historically the first and has the virtue of using only well-known inequalities used elsewhere in this book. The reader might wonder why we threw away the non-negative term $(1 - \varepsilon)T_A(\boldsymbol{p})$ in the lower bound. What has been done, in fact, is to control the dangerous $\boldsymbol{\sigma} \cdot \boldsymbol{B}$ term by an optimal combination of two methods; one way is to use $-\varepsilon|\boldsymbol{B}|$ as a partial lower bound, and the other way is to use $-(1 - \varepsilon)(\boldsymbol{p} + \sqrt{\alpha}\boldsymbol{A})^2$ as a lower bound on the remainder. The zero-modes of $T_A(\boldsymbol{p})$ play an important role for the stability question. If $(\psi, \boldsymbol{A})$ is a zero-mode, then $(\psi, T_A(\boldsymbol{p})\psi) = 0$ and hence nothing has been thrown away, in fact.

It was shown in [74] that the critical $Z\alpha^2$ is entirely determined by zero-modes; that is, the minimum of the right side of (9.4.2) is zero when taken over *all* zero-modes $(\psi, \boldsymbol{A})$ when $Z\alpha^2$ is less than the critical value; it is $-\infty$ beyond that value. The complete set of zero-modes is unknown, however, and therefore the exact value of the critical $Z\alpha^2$ remains unknown.

9.5 The Many-Body Problem with a Magnetic Field

The stability of the hydrogen atom with Pauli kinetic energy and magnetic fields was proved in [74] by Fröhlich, Lieb and Loss. The relation to zero-modes was also observed there. For one nucleus and arbitrarily many electrons (the large atom), as well as for one electron but arbitrary nuclei (and with a required bound on α as well as on $Z\alpha^2$), stability was proved by Lieb and Loss in [117]. The obvious next goal was to show stability for all N and M.

A solution to this problem was given by Lieb, Loss and Solovej in 1995 [123] for values of $Z\alpha^2$ and α that, while not quite as generous as the bound in the previous section, were more than sufficient to cover all physical situations. Earlier in 1995 Fefferman announced stability [57] for $Z = 1$ and sufficiently small α; the proof in 1996 [58] does not contain quantitative values of the constants.

The results in [123] were subsequently applied to the stability of matter with quantized magnetic fields (Quantum Electrodynamics) in [27], a subject to which we shall return in Chapter 11. The additional difficulty there is that the field energy is a bit more complicated than $(1/8\pi)\int |\boldsymbol{B}|^2$, but this problem can be overcome, as we shall see later.

Stability requires that $Z\alpha^2$ be sufficiently small (as is apparent from the one-body problem), but it also requires that α itself be small (independent of $Z\alpha^2$). Conditions for stability are given in Theorem 9.2, while the fact that α must not be too large in order to have stability of the first kind is the content of Theorem 9.3.

Paper [123] contains three proofs of stability of the Pauli Hamiltonian, and they are all remarkably short given what seemed earlier to be the difficult nature of the problem. We shall give the first of these here because it most closely utilizes the machinery already developed in this book. Another proof, which was developed in [127], will also be given in this section (after Corollary 9.1), and this is even shorter provided one uses a useful theorem of Birman and Solomyak [20], which will be proved in the appendix to this chapter. Interestingly, the numerical constants in the four proofs in [123, 127] (i.e., the required bounds on $Z\alpha^2$ and α) are not very far from each other.

The Hamiltonian under consideration is

$$H = \sum_{i=1}^{N} T_A(\boldsymbol{p}_i) + \alpha V_C(\underline{\boldsymbol{X}}, \underline{\boldsymbol{R}}) + \mathcal{E}_{\mathrm{mag}}(\boldsymbol{B}), \tag{9.5.1}$$

where $T_A(\boldsymbol{p})$ is the Pauli operator in (9.2.1) and the Coulomb potential $V_C(\underline{\boldsymbol{X}}, \underline{\boldsymbol{R}})$ is given in Eq. (2.1.21). As explained in Section 9.2, the wave functions are antisymmetric functions having $2q$ spin states.

Theorem 9.2 (Conditions for Stability). *Let $Z = \max_j\{Z_j\}$, and let*

$$\frac{1}{\kappa} = \frac{Z\pi}{2} + 2.26\,(2q)^{1/3}\, Z^{2/3} + 1.03\,(2q)^{1/3}. \tag{9.5.2}$$

The ground state energy of H in (9.5.1) satisfies

$$\boxed{E_0 \geq -4N\frac{\alpha^2}{\kappa^2}} \tag{9.5.3}$$

provided

$$\boxed{\frac{2q\,\alpha^2}{\kappa} \leq 8.62\frac{\sqrt{3}}{8\pi^2}.} \tag{9.5.4}$$

For the physical values $\alpha = 1/137$ and $2q = 2$, this theorem proves stability of matter for $Z \leq 953$.

First proof of Theorem 9.2. The first main ingredient in the proof of Theorem 9.2 is the relativistic stability of matter proved in Theorem 8.1 in the previous chapter. That is, for κ given in (9.5.2),

$$\sum_{i=1}^{N} |\boldsymbol{p}_i + \sqrt{\alpha}\boldsymbol{A}(\boldsymbol{x}_i)| + \kappa V_C(\underline{\boldsymbol{X}}, \underline{\boldsymbol{R}}) \geq 0. \tag{9.5.5}$$

The Hamiltonian (9.5.1) is, therefore, bounded from below by

$$H \geq \sum_{i=1}^{N} \Big(T_A(\boldsymbol{p}_i) - \frac{\alpha}{\kappa}|\boldsymbol{p}_i + \sqrt{\alpha}\boldsymbol{A}(\boldsymbol{x}_i)| \Big) + \mathcal{E}_{\text{mag}}(\boldsymbol{B}).$$

This is just a sum of one-body operators, plus the magnetic field energy. Let $h \otimes \mathbb{I}_{\mathbb{C}^q}$ be this one-body operator, namely

$$h = T_A(\boldsymbol{p}) - \frac{\alpha}{\kappa}|\boldsymbol{p} + \sqrt{\alpha}\boldsymbol{A}(\boldsymbol{x})|,$$

which acts on $L^2(\mathbb{R}^3; \mathbb{C}^2)$. On the space of totally antisymmetric wave functions, the lowest eigenvalue of $\sum_{i=1}^{N}(h \otimes \mathbb{I}_{\mathbb{C}^q})_i$ is equal to the sum of the lowest N eigenvalues of $h \otimes \mathbb{I}_{\mathbb{C}^q}$.

Let $\mu > 0$, and write $h = -\mu + (h + \mu)$. The sum of the lowest N eigenvalues of $(h + \mu) \otimes \mathbb{I}_{\mathbb{C}^q}$ is certainly bounded from below by the sum of all the negative eigenvalues of $(h + \mu) \otimes \mathbb{I}_{\mathbb{C}^q}$. The latter can be written as

$$-q \int_{\mu}^{\infty} N_e(h) \mathrm{d}e,$$

where $N_e(\mathcal{O})$ denotes the number of eigenvalues below $-e$ of an operator $\mathcal{O}$.

For any $\mu > 0$, the Hamiltonian H is thus bounded from below as

$$- N\mu - q \int_{\mu}^{\infty} N_e(h)\, \mathrm{d}e + \mathcal{E}_{\text{mag}}(\boldsymbol{B}). \tag{9.5.6}$$

Using the positivity of $T_A(\boldsymbol{p})$, we further estimate $T_A(\boldsymbol{p}) \geq (\mu/e) T_A(\boldsymbol{p})$ for $e \geq \mu$. Moreover,

$$\frac{\alpha}{\kappa} |\boldsymbol{p} - \sqrt{\alpha} \boldsymbol{A}(\boldsymbol{x})| \leq \frac{\alpha^2}{2e\kappa^2} (\boldsymbol{p} + \sqrt{\alpha} \boldsymbol{A}(\boldsymbol{x}))^2 + \frac{e}{2}$$

by Schwarz's inequality. The parameter μ is at our disposal and we choose it to be $\mu = 4(\alpha/\kappa)^2$. Using, in addition, that $\boldsymbol{\sigma} \cdot \boldsymbol{B}(\boldsymbol{x}) \geq -|\boldsymbol{B}(\boldsymbol{x})|$, the operator h is then bounded from below by

$$h \geq h_e \equiv \frac{3\alpha^2}{2e\kappa^2} (\boldsymbol{p} + \sqrt{\alpha} \boldsymbol{A}(\boldsymbol{x}))^2 - \frac{2\alpha^{5/2}}{e\kappa^2} |\boldsymbol{B}(\boldsymbol{x})| - \frac{e}{2}.$$

To get an upper bound on the number of eigenvalues of h_e below $-e$, we can use the CLR bound (the $\gamma = 0$ case of Theorem 4.1) to obtain

$$N_e(h) \leq N_e(h_e) \leq 2\, L_{0,3} \int_{\mathbb{R}^3} \left[\frac{4\sqrt{\alpha}}{3} |\boldsymbol{B}(\boldsymbol{x})| - \frac{e^2 \kappa^2}{3\alpha^2} \right]_+^{3/2} \mathrm{d}\boldsymbol{x}. \tag{9.5.7}$$

As usual, $[t]_+ = \max\{0, t\}$ denotes the positive part. The factor 2 on the right side results from the 2 spin states. Recall that the best known upper bound on $L_{0,3}$, given in Eq. (4.1.17), is $L_{0,3} \leq 0.116$.

After integration over e, (9.5.7) becomes

$$\int_{4(\alpha/\kappa)^2}^{\infty} \mathrm{d}e \int_{\mathbb{R}^3} \left[\frac{4\sqrt{\alpha}}{3} |\boldsymbol{B}(\boldsymbol{x})| - \frac{e^2\kappa^2}{3\alpha^2} \right]_+^{3/2} \mathrm{d}\boldsymbol{x}$$

$$\leq \int_{0}^{\infty} \mathrm{d}e \int_{\mathbb{R}^3} \left[\frac{4\sqrt{\alpha}}{3} |\boldsymbol{B}(\boldsymbol{x})| - \frac{e^2\kappa^2}{3\alpha^2} \right]_+^{3/2} \mathrm{d}\boldsymbol{x} = \frac{\pi}{\sqrt{3}} \frac{\alpha^2}{\kappa} \int_{\mathbb{R}^3} |\boldsymbol{B}(\boldsymbol{x})|^2 \mathrm{d}\boldsymbol{x}.$$

Consequently,

$$H \geq -4\frac{\alpha^2}{\kappa^2}N + \left(\frac{1}{8\pi} - 2q\, L_{0,3}\frac{\pi}{\sqrt{3}}\frac{\alpha^2}{\kappa}\right)\int\limits_{\mathbb{R}^3} |\boldsymbol{B}(\boldsymbol{x})|^2 \mathrm{d}\boldsymbol{x}.$$

Stability thus holds as long as (9.5.4) is satisfied. ■

For not too small Z our bound (9.5.3) is quite reasonable since the energy should be of the order of $(Z\alpha)^2$. For small Z, however, the result is far from optimal. The reason for this defect is that even if Z is very small (9.5.5) does not hold for large κ. This is a consequence of the instability results for relativistic matter in Theorem 8.2.

The previous proof uses the CLR bound, which was stated in Chapter 4, Theorem 4.1, without proof. If one wanted to use only results proved in this book, one could avoid the use of the CLR bound and use Corollary 4.2 instead. The divergence of the bound in Corollary 4.2 for small e does not cause any trouble here, since the integration in (9.5.6) is only over $e \geq \mu$ anyway. The bounds on $Z\alpha^2$ and α to ensure stability, obtained in this way, may turn out to be worse, however.

For use later we shall explicitly display the following inequality, which is at the heart of the proof of Theorem 9.2. We call it a corollary, although it is really a corollary of the proof and not the theorem *per se*.

Corollary 9.1 (Stability with Restricted Field Energy). *Let* $\psi \in \bigwedge^N L^2(\mathbb{R}^3; \mathbb{C}^{2q})$ *be a normalized, antisymmetric N electron wave function, whose density* ϱ_ψ *is supported in the set* $\Xi \subset \mathbb{R}^3$ *(which could be all of* $\mathbb{R}^3$*), and let* $\kappa > 0$*. If*

$$\frac{2q\,\alpha^2}{\kappa} \leq 8.62\frac{\sqrt{3}}{8\pi^2}$$

then

$$\left(\psi, \sum_{i=1}^N \left[T_A(\boldsymbol{p}_i) - \frac{\alpha}{\kappa}|\boldsymbol{p}_i + \sqrt{\alpha}\boldsymbol{A}(\boldsymbol{x}_i)|\right]\psi\right) + \frac{1}{8\pi}\int\limits_{\Xi} |\boldsymbol{B}(\boldsymbol{x})|^2 \mathrm{d}\boldsymbol{x} \geq -4N\frac{\alpha^2}{\kappa^2}. \tag{9.5.8}$$

The point of this corollary is that all the bounds hold even if *only* the magnetic field energy in the support of ϱ_ψ is utilized. An inspection of the proof of Theorem 9.2 shows that only the magnetic field on this support is relevant.

Inequality (9.5.8) will be important in Chapter 11, where the Pauli operator with quantized electromagnetic field will be studied.

As promised above, we shall now give an alternative proof of Theorem 9.2. It is taken from [127].

Second proof of Theorem 9.2. We start with the same observation as in the beginning of the first proof. Namely, from stability of relativistic matter we have

$$H \geq \sum_{i=1}^{N} \left(T_A(\boldsymbol{p}_i) - \frac{\alpha}{\kappa} |\boldsymbol{p}_i + \sqrt{\alpha} \boldsymbol{A}(\boldsymbol{x}_i)| \right) + \mathcal{E}_{\text{mag}}(\boldsymbol{B})$$

for κ given in (9.5.2). Since $T_A(\boldsymbol{p}) \geq 0$, we can bound $T_A(\boldsymbol{p})$ from below as

$$T_A(\boldsymbol{p}) \geq \lambda \sqrt{T_A(\boldsymbol{p})} - \frac{\lambda^2}{4}$$

for $\lambda \geq 0$. Thus

$$H \geq -\text{Tr} \left[\lambda \sqrt{T_A(\boldsymbol{p})} - \frac{\alpha}{\kappa} |\boldsymbol{p} + \sqrt{\alpha} \boldsymbol{A}(\boldsymbol{x})| \right]_- - \frac{\lambda^2}{4} N + \mathcal{E}_{\text{mag}}(\boldsymbol{B}).$$

The notation $\text{Tr}\,[\mathcal{O}]_-$ simply stands for the sum of the absolute value of the negative eigenvalues of a self-adjoint operator $\mathcal{O}$. Likewise, $\text{Tr}\,[\mathcal{O}]_-^p$ is the sum of the pth powers of the absolute value of the negative eigenvalues.

There is a useful inequality for traces of differences of fractional powers of operators, which was proved by Birman and Solomyak in [20]. It says that

$$\text{Tr}\,[B - A]_- \leq \text{Tr}\,[B^2 - A^2]_-^{1/2}$$

for positive operators A and B. For the convenience of the reader, we present a proof of this inequality in the appendix to this chapter. As a result

$$\text{Tr} \left[\lambda \sqrt{T_A(\boldsymbol{p})} - \frac{\alpha}{\kappa} |\boldsymbol{p} + \sqrt{\alpha} \boldsymbol{A}(\boldsymbol{x})| \right]_- \leq \text{Tr} \left[\lambda^2 T_A(\boldsymbol{p}) - \frac{\alpha^2}{\kappa^2} (\boldsymbol{p} + \sqrt{\alpha} \boldsymbol{A}(\boldsymbol{x}))^2 \right]_-^{1/2}$$

Since $\boldsymbol{\sigma} \cdot \boldsymbol{B}(\boldsymbol{x}) \geq -|\boldsymbol{B}(\boldsymbol{x})|$,

$$\lambda^2 T_A(\boldsymbol{p}) - \frac{\alpha^2}{\kappa^2} (\boldsymbol{p} + \sqrt{\alpha} \boldsymbol{A}(\boldsymbol{x}))^2 \geq \left(\frac{\lambda^2}{2} - \frac{\alpha^2}{\kappa^2} \right) (\boldsymbol{p} + \sqrt{\alpha} \boldsymbol{A}(\boldsymbol{x}))^2 - \lambda^2 \frac{\sqrt{\alpha}}{2} |\boldsymbol{B}(\boldsymbol{x})|.$$

We shall choose $\lambda > \sqrt{2}\alpha/\kappa$. The trace of the square root of the negative part of this operator can be bounded with the aid of the LT inequality (Theorem 4.1

for $\gamma = 1/2$). The result is

$$\operatorname{Tr}\left[\left(\frac{\lambda^2}{2} - \frac{\alpha^2}{\kappa^2}\right)(\boldsymbol{p} + \sqrt{\alpha}\boldsymbol{A}(\boldsymbol{x}))^2 - \lambda^2 \frac{\sqrt{\alpha}}{2}|\boldsymbol{B}(\boldsymbol{x})|\right]_-^{1/2}$$
$$\leq L_{1/2,3} \frac{2q\,\lambda^4\alpha}{4\left(\frac{\lambda^2}{2} - \frac{\alpha^2}{\kappa^2}\right)^{3/2}} \int_{\mathbb{R}^3} |\boldsymbol{B}(\boldsymbol{x})|^2 \mathrm{d}\boldsymbol{x}. \tag{9.5.9}$$

(Recall that there are $2q$ spin states.) Hence

$$H \geq -\frac{\lambda^2}{4}N + \left(\frac{1}{8\pi} - L_{1/2,3}\frac{2q\,\lambda^4\alpha}{4\left(\frac{\lambda^2}{2} - \frac{\alpha^2}{\kappa^2}\right)^{3/2}}\right)\int_{\mathbb{R}^3} |\boldsymbol{B}(\boldsymbol{x})|^2 \mathrm{d}\boldsymbol{x}.$$

With the choice $\lambda = \sqrt{8}\alpha/\kappa$, stability is thus shown if

$$\frac{2q\,\alpha^2}{\kappa} \leq \frac{1}{8\pi}\frac{3^{3/2}}{16\,L_{1/2,3}}. \tag{9.5.10}$$

As explained in Chapter 4, Eq. (4.1.18), $L_{1/2,3} \leq 6.87\,L^{\mathrm{cl}}_{1/2,3} = 0.0683$. With this estimate, the right side of (9.5.10) turns out to be the same as the right side of (9.5.4) up to four figure accuracy. ∎

The two proofs give remarkably similar results! The second one is infinitesimally better and shows stability in a slightly bigger parameter regime. For the physical value of α and q, the first proof shows stability for $Z \leq 953$, while the second proof works for $Z \leq 954$. This should be regarded as no more than a coincidence, since there are ways to improve the results in both methods.

Finally, we show that small α is actually needed for stability. The following theorem shows this and also shows that bosons are always unstable for any value of $Z\alpha^2 > 0$. This is analogous to the relativistic case discussed in Section 8.8.

Theorem 9.3 (Conditions for Instability). *There is a constant $D > 0$ such that the Hamiltonian (9.4.1) is unstable of the first kind for large enough N and M if one of the following two conditions hold:*

(1) $Z_j\alpha^2 > (\pi/2)(3/4)^{3/2}$ *for some* j,
(2) $\alpha > D/q^{1/2}$.

Proof. The proof follows closely the corresponding proof of Theorem 8.2 in the relativistic case, and is even simpler.

Instability under condition (1) has already been shown in Section 9.4. We shall now prove instability under condition (2). Given the instability under condition (1), it clearly suffices to consider the case $Z \le q$. Pick a zero-mode $(\psi, \alpha^{-1/2}\widetilde{\boldsymbol{A}})$ of the Pauli operator, with $\psi \in L^2(\mathbb{R}^3; \mathbb{C}^2)$. We take $N = q$, in which case all the electrons can be put in the state ψ (cf. the discussion in Section 3.1.3.1). That is, we can take the wave function to be $\prod_{i=1}^q \psi(\boldsymbol{x}_i)$ times a totally antisymmetric function in spin space. If $\tau = \mathcal{E}_{\text{mag}}(\widetilde{\boldsymbol{B}})$ denotes the magnetic field energy and $\mathcal{I}$ denotes twice the self-energy of a charge distribution $|\psi|^2$, i.e.,

$$\mathcal{I} = \iint_{\mathbb{R}^3 \times \mathbb{R}^3} \frac{|\psi(\boldsymbol{x})|^2 |\psi(\boldsymbol{y})|^2}{|\boldsymbol{x} - \boldsymbol{y}|} \mathrm{d}\boldsymbol{x}\,\mathrm{d}\boldsymbol{y},$$

the energy of the system is given by

$$\frac{\tau}{\alpha} + \frac{\alpha \mathcal{I}}{2} q(q-1) - Z\alpha \sum_{j=1}^{M} \int_{\mathbb{R}^3} \frac{|\psi(\boldsymbol{x})|^2}{|\boldsymbol{x} - \boldsymbol{R}_j|} \mathrm{d}\boldsymbol{x} + Z^2 \alpha \sum_{1 \le k < l \le M} \frac{1}{|\boldsymbol{R}_k - \boldsymbol{R}_l|}. \tag{9.5.11}$$

We can choose the nuclear positions $\boldsymbol{R}_k$ so as to make this expression as small as possible. For an upper bound on the smallest possible value, we can average the expression (9.5.11) over the nuclear positions (as we have done several times before), with weight given by $\prod_{k=1}^M |\psi(\boldsymbol{R}_k)|^2$. In this way, the energy is negative if

$$\frac{\tau}{\alpha} + \frac{\mathcal{I}\alpha}{2} \big(-2Z\alpha M q + Z^2 M(M-1) + q(q-1)\big) < 0.$$

Since $0 < Z \le q$, we can pick an M such that $|ZM - q| \le 1/2$. Then

$$-2Z\alpha M q + Z^2 M(M-1) + q(q-1) = (ZM - q)^2 - Z^2 M - q \le 1/4 - q.$$

Hence the energy is negative if

$$\alpha^2 > \frac{2\tau}{\mathcal{I}(q - 1/4)}.$$

Once the energy is negative, it can be driven to $-\infty$ by length scaling. (See the discussion at the end of Section 9.3.) ■

9.6 Appendix: BKS Inequalities

As a convenience to the reader we give a proof of some cases of the inequalities due to Birman, Koplienko, and Solomyak [19]. It is taken, with minor modifications, from [127, Appendix A]. The case needed in Section 9.5 corresponds to $p = 2$ below, which was proved earlier by Birman and Solomyak in [20]. In Section 9.5 we were interested in $(B - A)_-$, but here we treat $(B - A)_+$ to simplify keeping track of signs. The proof is the same. Recall that $X_+ := (|X| + X)/2$.

Theorem 9.4 (BKS Inequalities). *Let $p \geq 1$ and suppose that A and B are two non-negative, self-adjoint linear operators on a separable Hilbert space such that $(B^p - A^p)_+^{1/p}$ is trace class. Then $(B - A)_+$ is also trace class and*

$$\operatorname{Tr}(B - A)_+ \leq \operatorname{Tr}(B^p - A^p)_+^{1/p}.$$

Proof. Our proof will use essentially only two facts: $X \mapsto X^{-1}$ is operator monotone decreasing on the set of non-negative self-adjoint operators (i.e., $X \geq Y \geq 0 \Longrightarrow Y^{-1} \geq X^{-1}$) and $X \mapsto X^r$ is operator monotone increasing on the set of non-negative self-adjoint operators for all $0 < r \leq 1$. (See the footnote on page 87.) Consequently, $X \mapsto X^{-r}$ is operator monotone decreasing for $0 < r \leq 1$. We refer to [17, Chapter V] for details on operator monotone functions.

As a preliminary remark, we can suppose that $B \geq A$. To see this, write $B^p = A^p + D$. If we replace B by $[A^p + D_+]^{1/p}$ then $(B^p - A^p)_+ = D_+$ is unchanged, while $X := B - A$ becomes $[A^p + D_+]^{1/p} - A$ and this is bigger than $B - A$ because of operator monotonicity of the map $X \mapsto X^{1/p}$ on positive operators. Since the trace is also operator monotone, we can therefore suppose that $D = D_+$, i.e., $B^p = A^p + C^p$ with $A, B, C \geq 0$. Our goal, then, is to prove that

$$\operatorname{Tr}\left[(A^p + C^p)^{1/p} - A\right] \leq \operatorname{Tr} C, \tag{9.6.1}$$

under the assumption that C is trace class.

To prove (9.6.1) we consider the operator $X := [A^p + C^p]^{1/p} - A$, which is well defined on the domain of A. We assume, at first, that $A^p \geq \varepsilon^p$ for some positive number ε. Then, by the functional calculus, and with

$$E := [A^p + C^p]^{(1-p)/p} \quad \text{and} \quad P := A^{1-p} - E$$

we have

$$X = E[A^p + C^p] - A^{1-p}A^p = -PA^p + EC^p. \tag{9.6.2}$$

Clearly, $P \geq 0$ and $0 \leq P \leq \varepsilon^{1-p}$.

Let $Y := EC^p$. We claim that Y is trace class. This follows from $Y^\dagger Y = C^p E^2 C^p \leq C^p C^{2-2p} C^p = C^2$. Thus, $|Y| \leq C$, and hence $\operatorname{Tr} Y = \operatorname{Tr} C^{p/2} E C^{p/2} \leq \operatorname{Tr} C$.

It is also true that P is trace class. To see this, use the integral representation $A^{1-p} = c \int_0^\infty (t + A^p)^{-1} t^{(1-p)/p} \, \mathrm{d}t$ for suitable $c > 0$. Use this twice and then use the well known resolvent formula, which is

$$\frac{1}{A^p + t} - \frac{1}{A^p + C^p + 1} = \frac{1}{A^p + t} C^p \frac{1}{A^p + C^p + t}.$$

In this way we find that

$$P = c \int_0^\infty (A^p + t)^{-1} C^p (A^p + C^p + t)^{-1} t^{(1-p)/p} \, \mathrm{d}t.$$

Since C is trace class, so is C^p, and the integral converges because of our assumed lower bound on A. Thus, P is trace class and hence there is a complete, orthonormal family of vectors $v_1, v_2, \ldots$, each of which is an eigenvector of P.

Since $X \geq 0$, the trace of X is well defined by $\sum_{j=1}^\infty (v_j, Xv_j)$ for *any* complete, orthonormal family. The same remark applies to EC^p since it is trace class. To complete the proof of (9.6.1), therefore, it suffices to prove that $(v_j, PA^p \, v_j) \geq 0$ for each j. But this number is $\lambda_j(v_j, A^p \, v_j) \geq 0$, where λ_j is the (non-negative) eigenvalue of P, and the positivity follows from the positivity of A.

We now turn to the case of general $A \geq 0$. We can apply the above proof to the operator $A + \varepsilon$ for some positive number ε. Thus we have

$$\operatorname{Tr} \left[[(A + \varepsilon)^p + C^p]^{1/p} - (A + \varepsilon) \right] \leq \operatorname{Tr} C. \tag{9.6.3}$$

Let $\varphi_1, \varphi_2, \ldots$ be an orthonormal basis chosen from the domain of A^p. This basis then also belongs to the domain of A and the domain of $[(A + \varepsilon)^p + C^p]^{1/p}$ for all $\varepsilon \geq 0$. We then have

$$\operatorname{Tr} X = \sum_j (\varphi_j, X\varphi_j).$$

Note that *a priori* we do not know that the trace is finite, but since the operator is non-negative this definition of the trace is meaningful. Operator monotonicity of $X^{1/p}$ gives

$$\big(\varphi_j, \big[[(A+\varepsilon)^p + C^p]^{1/p} - (A+\varepsilon)\big]\varphi_j\big) \geq (\varphi_j, (X-\varepsilon)\varphi_j).$$

It therefore follows from (9.6.3), followed by Fatou's Lemma [118, Lemma 1.7] applied to sums, that

$$\begin{aligned}
\operatorname{Tr} C &\geq \liminf_{\varepsilon \to 0} \sum_j \big(\varphi_j, \big[[(A+\varepsilon)^p + C^p]^{1/p} - (A+\varepsilon)\big]\varphi_j\big) \\
&\geq \sum_j \liminf_{\varepsilon \to 0} \big(\varphi_j, \big[[(A+\varepsilon)^p + C^p]^{1/p} - (A+\varepsilon)\big]\varphi_j\big) \\
&\geq \operatorname{Tr} X.
\end{aligned}$$

■

CHAPTER 10

The Dirac Operator and the Brown–Ravenhall Model

10.1 The Dirac Operator

In the previous chapters we have seen that matter is stable of the second kind if the negative particles are fermions with a fixed, finite number of 'spin states' per particle. We found, however, that stability requires a bound on the nuclear charge Z and on the fine structure constant α. The non-relativistic $\sqrt{\alpha}\,\boldsymbol{\sigma}\cdot\boldsymbol{B}(\boldsymbol{x})$ Pauli interaction of the electron spin with the magnetic field required a bound on $Z\alpha^2$ and on α, in addition to the compensating positive field energy $\int|\boldsymbol{B}|^2$. Relativistic mechanics, on the other hand, in which $|\boldsymbol{p}+\sqrt{\alpha}\boldsymbol{A}(\boldsymbol{x})|^2$ is replaced by $|\boldsymbol{p}+\sqrt{\alpha}\boldsymbol{A}(\boldsymbol{x})|$, but without the Pauli term, required a bound on $Z\alpha$ and on α.

The next chapter in the stability saga ought to be the combination of magnetic interaction together with relativistic kinetic energy. Thus, we could consider replacing the Pauli kinetic energy operator $|\boldsymbol{\sigma}\cdot(\boldsymbol{p}+\sqrt{\alpha}\boldsymbol{A}(\boldsymbol{x}))|^2$ simply by its square root $|\boldsymbol{\sigma}\cdot(\boldsymbol{p}+\sqrt{\alpha}\boldsymbol{A}(\boldsymbol{x}))|$ or, if we wish to include the electron's mass, by $\sqrt{[\boldsymbol{\sigma}\cdot(\boldsymbol{p}+\sqrt{\alpha}\boldsymbol{A}(\boldsymbol{x}))]^2+m^2}-m$. This was actually considered in the early days of quantum mechanics but was rejected for various reasons, one being that this is not a local operator[1] like the Laplacian Δ; moreover, it gives the wrong spectrum for hydrogen.

The accepted relativistic generalization of Schrödinger's kinetic energy operator $-\Delta$ was invented by Dirac in 1928 [39]. Unlike $|\boldsymbol{p}|$ it is a local operator; in fact, it is a first order differential operator. For a particle of mass m it is given,

[1] The notion of locality of an operator $\mathcal{O}$ means that the matrix elements $(\psi,\mathcal{O}\,\phi)$ vanish for any two functions ψ and ϕ of disjoint support. Hence any multiplication operator is local, including the Coulomb potential. The non-relativistic kinetic energy is local, too, but not the relativistic kinetic energy.

in our units, by

$$\boxed{D_A = \boldsymbol{\alpha} \cdot (\boldsymbol{p} + \sqrt{\alpha}\boldsymbol{A}(\boldsymbol{x})) + \beta m} \tag{10.1.1}$$

where $\boldsymbol{\alpha} = (\alpha^1, \alpha^2, \alpha^3)$ and β are 4×4 matrices, satisfying the *anti*commutation relations

$$\alpha^i \alpha^j + \alpha^j \alpha^i = 2\,\delta_{ij}\,\mathbb{I}_{\mathbb{C}^4}, \quad \alpha^i \beta + \beta \alpha^i = 0, \quad \beta^2 = \mathbb{I}_{\mathbb{C}^4}. \tag{10.1.2}$$

(We hope the reader will distinguish $\boldsymbol{\alpha}$ and α; the notation predates us.) A particular representation, which we shall use in the following, is

$$\alpha^i = \begin{pmatrix} 0 & \sigma^i \\ \sigma^i & 0 \end{pmatrix}, \qquad \beta = \begin{pmatrix} \mathbb{I}_{\mathbb{C}^2} & 0 \\ 0 & -\mathbb{I}_{\mathbb{C}^2} \end{pmatrix} \tag{10.1.3}$$

where σ^i are the Pauli matrices defined in Eq. (2.1.50). The Hilbert space for one particle has changed from $L^2(\mathbb{R}^3; \mathbb{C}^2)$ to $L^2(\mathbb{R}^3; \mathbb{C}^4)$. The operator $\boldsymbol{p}$ appearing in (10.1.1) is $\boldsymbol{p} = (p^1, p^2, p^3) = -i\nabla$, as before.

The reason D_A is like the square root of the Pauli operator T_A in (9.2.1) is easily seen by computing its square:

$$D_A^2 = \begin{pmatrix} \left[\boldsymbol{\sigma} \cdot (\boldsymbol{p} + \sqrt{\alpha}\boldsymbol{A}(\boldsymbol{x}))\right]^2 + m^2 & 0 \\ 0 & \left[\boldsymbol{\sigma} \cdot (\boldsymbol{p} + \sqrt{\alpha}\boldsymbol{A}(\boldsymbol{x}))\right]^2 + m^2 \end{pmatrix}. \tag{10.1.4}$$

The Dirac operator thus represents a sophisticated way of taking the square root of the Pauli operator.

Note that D_A is not positive or even bounded from below. In fact, the spectrum of D_A and $-D_A$ are identical, since $-D_A = U D_A U^\dagger$ with

$$U = \begin{pmatrix} 0 & \mathbb{I}_{\mathbb{C}^2} \\ -\mathbb{I}_{\mathbb{C}^2} & 0 \end{pmatrix}. \tag{10.1.5}$$

In fact, under this unitary conjugation the α^i and β change sign. Since $D_A^2 \geq m^2$ by (10.1.4), D_A has no spectrum in the interval $(-m, m)$.

The simplest case to consider is the one of just one particle without any external potential, and without any magnetic field. The Hamiltonian is then $D_0 = \boldsymbol{\alpha} \cdot \boldsymbol{p} + \beta m$ and its ground state energy is $-\infty$. To get around this problem, Dirac proposed to consider only the positive energy subspace of D_0, interpreting the negative energy states to be 'filled'.[2] That is, the one-particle space $L^2(\mathbb{R}^3; \mathbb{C}^4)$

[2] This consideration led Dirac to the prediction of antiparticles as unfilled 'holes' in the otherwise filled subspace of negative energy states. These particles, called positrons, were discovered a few years later.

splits into a direct sum $\mathcal{H}_0^+ \oplus \mathcal{H}_0^-$, where $\mathcal{H}_0^\pm$ denote the positive and negative spectral subspaces of D_0, respectively.[3] On $\mathcal{H}_0^+$, D_0 is positive by definition, and hence there is no stability issue for D_0 alone.

The following explicit representation of $\mathcal{H}_0^+$ will be useful in the following. Since the projection onto $\mathcal{H}_0^+$ is given by $(\mathbb{I} + D_0/|D_0|)/2$, a function ψ in $\mathcal{H}_0^+$ has a Fourier transform of the form

$$\widehat{\psi}(\boldsymbol{k}) = \begin{pmatrix} u(\boldsymbol{k}) \\ \frac{2\pi\sigma\cdot\boldsymbol{k}}{\sqrt{(2\pi\boldsymbol{k})^2+m^2}+m} u(\boldsymbol{k}) \end{pmatrix} \tag{10.1.6}$$

with $u \in L^2(\mathbb{R}^3; \mathbb{C}^2)$. Conversely, for any such u the function (10.1.6) is in $\mathcal{H}_0^+$. The norm and kinetic energy are given by

$$\begin{aligned} (\psi, \psi) &= \int_{\mathbb{R}^3} \langle u(\boldsymbol{k}), u(\boldsymbol{k})\rangle \left(1 + \frac{(2\pi\boldsymbol{k})^2}{\left(\sqrt{(2\pi\boldsymbol{k})^2+m^2}+m\right)^2}\right) \mathrm{d}\boldsymbol{k} \\ (\psi, D_0\psi) &= 2\int_{\mathbb{R}^3} \langle u(\boldsymbol{k}), u(\boldsymbol{k})\rangle \frac{(2\pi\boldsymbol{k})^2+m^2}{\sqrt{(2\pi\boldsymbol{k})^2+m^2}+m}\,\mathrm{d}\boldsymbol{k}. \end{aligned} \tag{10.1.7}$$

Here, $\langle\,\cdot\,,\,\cdot\,\rangle$ denotes the inner product in $\mathbb{C}^2$.

In the special case $m = 0$, these formulas simplify to

$$\begin{aligned} (\psi, \psi) &= 2\int_{\mathbb{R}^3} \langle u(\boldsymbol{k}), u(\boldsymbol{k})\rangle \mathrm{d}\boldsymbol{k} \\ (\psi, D_0\psi) &= 2\int_{\mathbb{R}^3} \langle u(\boldsymbol{k}), u(\boldsymbol{k})\rangle |2\pi\boldsymbol{k}|\,\mathrm{d}\boldsymbol{k}. \end{aligned} \tag{10.1.8}$$

We display them explicitly as they will be useful later in Section 10.5.

[3] Up to now all the models in this book could be defined via quadratic forms, and it was not necessary to talk about self-adjoint operators and their spectra. In connection with the Dirac operator, however, the notion of spectral subspace is unavoidable, at least in the presence of magnetic fields. Without magnetic fields one can just use the explicit representation (10.1.6) for the space $\mathcal{H}_0^+$, and no spectral theory is required. With magnetic fields, as discussed below, no such explicit representation is available, however. For details on operator theory, we refer to standard textbooks, e.g., [150, Vol. 1].

10.1.1 Gauge Invariance

Like the Schrödinger operator $(\boldsymbol{p}+\sqrt{\alpha}\boldsymbol{A}(\boldsymbol{x}))^2$, the Dirac operator $D_{\boldsymbol{A}}$ in (10.1.1) is **gauge invariant**. This means that for any function χ there exists a unitary operator $U(\chi)$ such that

$$D_{\boldsymbol{A}+\nabla\chi} = U(\chi) D_{\boldsymbol{A}} U(\chi)^{\dagger}.$$

In fact, $U(\chi)$ is the multiplication operator $e^{i\sqrt{\alpha}\chi(\boldsymbol{x})} \otimes \mathbb{I}_{\mathbb{C}^4}$. In particular, the spectrum of $D_{\boldsymbol{A}}$ does not change when the gradient of an arbitrary function is added to the vector potential $\boldsymbol{A}$, but the spectral subspaces *do* change. Note that χ does not affect the magnetic field $\boldsymbol{B}$, since $\operatorname{curl}\nabla\chi = 0$.

One possible choice of gauge is $\operatorname{div}\boldsymbol{A}(\boldsymbol{x}) = 0$. The following lemma shows that if the magnetic field energy is finite, one can always find an $\boldsymbol{A}$ such that $\operatorname{curl}\boldsymbol{A} = \boldsymbol{B}$ and $\operatorname{div}\boldsymbol{A} = 0$. In addition, $\boldsymbol{A}$ can be chosen to go to zero at infinity in a weak sense; more precisely $\boldsymbol{A} \in L^6(\mathbb{R}^3)$. These questions were touched upon in Chapter 2, Section 2.1.2.

Lemma 10.1 (Coulomb Gauge). *Assume that the vector field $\boldsymbol{B}$ is in $L^2(\mathbb{R}^3;\mathbb{R}^3)$, i.e., $\mathcal{E}_{\mathrm{mag}}(\boldsymbol{B}) < \infty$, and assume that* $\operatorname{div}\boldsymbol{B}(\boldsymbol{x}) = 0$ *(in the sense of distributions). Then there exists a unique $\boldsymbol{A} \in L^6(\mathbb{R}^3;\mathbb{R}^3)$ such that*

$$\operatorname{curl}\boldsymbol{A}(\boldsymbol{x}) = \boldsymbol{B}(\boldsymbol{x}) \quad \text{and} \quad \operatorname{div}\boldsymbol{A}(\boldsymbol{x}) = 0 \tag{10.1.9}$$

(again, in the sense of distributions).

Proof. If $\widehat{\boldsymbol{B}}(\boldsymbol{k})$ denotes the Fourier transform of $\boldsymbol{B}(\boldsymbol{x})$, we can take $\widehat{\boldsymbol{A}}(\boldsymbol{k}) = (2\pi i|\boldsymbol{k}|^2)^{-1}\boldsymbol{k}\wedge\widehat{\boldsymbol{B}}(\boldsymbol{k})$. Note that $\boldsymbol{k}\cdot\widehat{\boldsymbol{B}}(\boldsymbol{k}) = 0$ since $\operatorname{div}\boldsymbol{B} = 0$, whence $(2\pi i\boldsymbol{k})\wedge\widehat{\boldsymbol{A}}(\boldsymbol{k}) = \widehat{\boldsymbol{B}}(\boldsymbol{k})$, which is the same as saying that $\operatorname{curl}\boldsymbol{A}(\boldsymbol{x}) = \boldsymbol{B}(\boldsymbol{x})$. Moreover, $\boldsymbol{k}\cdot\widehat{\boldsymbol{A}}(\boldsymbol{k}) = 0$, meaning that $\operatorname{div}\boldsymbol{A}(\boldsymbol{x}) = 0$. Hence $|\boldsymbol{k}|\widehat{\boldsymbol{A}}(\boldsymbol{k}) \in L^2$.

Sobolev's inequality (2.2.4) will imply that $\boldsymbol{A} \in L^6(\mathbb{R}^3)$ if we can show that $\boldsymbol{A}$ goes to zero at infinity, in the weak sense that the measure of the set $S_\mu = \{\boldsymbol{x} : |\boldsymbol{A}(\boldsymbol{x})| \geq \mu\}$ is finite for every $\mu > 0$. For any $\varepsilon > 0$, $\widehat{\boldsymbol{f}}_\varepsilon(\boldsymbol{k}) = \widehat{\boldsymbol{A}}(\boldsymbol{k})\Theta(|\boldsymbol{k}|-\varepsilon)$ is in $L^2(\mathbb{R}^3)$, where Θ is the Heaviside step function defined in (6.6.4); moreover, $\widehat{\boldsymbol{A}}(\boldsymbol{k})\Theta(\varepsilon-|\boldsymbol{k}|)$ is in $L^1(\mathbb{R}^3)$ and its Fourier transform is uniformly bounded by $\int_{|\boldsymbol{k}|\leq\varepsilon}|\widehat{\boldsymbol{A}}(\boldsymbol{k})|\mathrm{d}\boldsymbol{k} \leq C\varepsilon^{1/2}\|\boldsymbol{B}\|_2$, using Schwarz's inequality. Choose ε small enough such that $C\varepsilon^{1/2}\|\boldsymbol{B}\|_2 < \mu/2$. Then $S_\mu \subset \{\boldsymbol{x} : |\boldsymbol{f}_\varepsilon(\boldsymbol{x})| \geq \mu/2\}$, which has finite measure since $\boldsymbol{f}_\varepsilon \in L^2$.

Finally, to show uniqueness, note that if both $\boldsymbol{A}_1$ and $\boldsymbol{A}_2$ satisfy (10.1.9), their difference is curl-free and hence must be the gradient of a function χ.

Since $\operatorname{div} \boldsymbol{A}_1 - \operatorname{div} \boldsymbol{A}_2 = \operatorname{div} \nabla \chi = \Delta \chi$ is zero, χ is a harmonic function. But the only harmonic function whose gradient goes to zero at infinity is the constant function, and hence $\boldsymbol{A}_1 = \boldsymbol{A}_2$. ■

The particular gauge given by (10.1.9) is called the **Coulomb gauge**. The condition that $\boldsymbol{A}$ goes to zero at infinity fixes $\boldsymbol{A}$ uniquely. This gauge plays a particular role in quantum electrodynamics, as we shall see in the next chapter.

We denote the set of vector potentials in the Coulomb gauge with finite magnetic field energy by $\mathcal{A}$, i.e.,

$$\mathcal{A} := \left\{ \boldsymbol{A} \in L^6,\ \operatorname{div} \boldsymbol{A} = 0, \operatorname{curl} \boldsymbol{A} \in L^2 \right\}.$$

In the Coulomb gauge (10.1.9), the field energy (2.1.16) can alternatively be expressed as

$$\mathcal{E}_{\mathrm{mag}}(\boldsymbol{B}) = \frac{1}{8\pi} \int\limits_{\mathbb{R}^3} |\nabla \boldsymbol{A}(\boldsymbol{x})|^2 \mathrm{d}\boldsymbol{x},$$

where $|\nabla \boldsymbol{A}(\boldsymbol{x})|^2$ is short for $\sum_{i=1}^3 |\nabla A^i(\boldsymbol{x})|^2$. This follows from the fact that $|\nabla \boldsymbol{A}|^2 = |\operatorname{div} \boldsymbol{A}|^2 + |\operatorname{curl} \boldsymbol{A}|^2$.

10.2 Three Alternative Hilbert Spaces

The many-body Hamiltonian for relativistic electrons in a magnetic field is formally given by

$$H = \sum_{i=1}^{N} D_{A,i} + \alpha V_C(\underline{\boldsymbol{X}}, \underline{\boldsymbol{R}}) + \mathcal{E}_{\mathrm{mag}}(\boldsymbol{B}) \tag{10.2.1}$$

in analogy with the non-relativistic case. The sophisticated reader will notice that this theory is not really relativistic. To make such a theory one would have to assign a four-dimensional space-time coordinate to each electron just as we assign a three-dimensional coordinate in non-relativistic theory. While there have been many attempts to construct such a multiple time formalism no totally satisfactory model of this kind exists. The true solution, as it is currently understood, lies in the formalism of quantum field theory, in which only *fields* exist and particles are regarded as excitations of the field. Despite these concerns we shall continue to pursue the question of the stability of the Hamiltonian described by (10.2.1).

Although it is not truly relativistic, the Hamiltonian (10.2.1) is used in one form or another in practical calculations of energies of atoms and molecules, and remarkably accurate results have been attained with its use. It can not be taken literally without some modification, however, for reasons we shall now explain.

Since H in (10.2.1) is not bounded from below, its ground state energy (defined as in the non-relativistic case as the infimum of $(\psi, H\psi)/(\psi, \psi)$ over all $\psi \neq 0$) is not a useful concept. For $N = 1$, i.e., the hydrogenic atom, the spectrum of H consists of discrete eigenvalues in the interval $(-m, m)$ together with continuous spectrum in its complement $(-\infty, m] \cup [m, \infty)$. These discrete eigenvalues can be interpreted as electron bound states; we shall discuss this further in Section 10.3 below. For $N \geq 2$, however, the spectrum of H consists of the whole real line without gaps, and eigenvalues would not normally be expected to exist. This was observed by Brown and Ravenhall in [23, 16], and results from particles in the positive energy subspace of the Dirac operator interacting with particles in the negative energy subspace.

There are several alternative ways of addressing the question of stability, three of which we shall discuss here.

10.2.1 The Brown–Ravenhall Model

A solution of the problem addressed above, proposed by Brown and Ravenhall [23] on the basis of perturbation theoretic considerations, is to allow the electron wave functions to live only in the positive energy subspace of the Dirac operator D_0. This is in accordance with Dirac's original suggestion that the negative energy states are completely filled with fermions, and hence, by the Pauli principle, only the positive energy states are available to the electrons.[4] If Λ_0^+ denotes the projection onto the positive spectral subspace of D_0, the ground

[4] The Brown–Ravenhall model is also sometimes called the 'no pair' model for the following reason. A more accurate model would have additional terms in the dynamics that cause electrons in the filled negative energy sea to move to the positive energy subspace. The resulting 'hole' in the sea is an electron 'antiparticle' called a 'positron' and this is identified, physically, as a particle. Thus, the missing terms 'create' an electron–positron pair, but the creation of a pair, even if it were included in the model, would require the expenditure of a relatively large energy ($2mc^2$). The absence of such pair production effects is a defect of the model, but it is regarded as a minor defect in the world of quantum chemistry, where $2mc^2$ is very much larger than other energies under consideration.

state energy in the Brown–Ravenhall model is thus

$$E_0 = \inf \left\{ (\psi, H\psi) \,:\, \psi \in \bigwedge^N \left[\Lambda_0^+ L^2(\mathbb{R}^3; \mathbb{C}^4)\right],\ \|\psi\|_2 = 1,\ \boldsymbol{A} \in \mathcal{A} \right\}.$$

Here, we choose to minimize over vector potentials in the Coulomb gauge $\operatorname{div} \boldsymbol{A} = 0$. Although this may seem arbitrary, it is important for the following reasons. The Brown–Ravenhall model is *not* gauge-invariant, since a gauge change can not be incorporated into the wave function in the usual way by the replacement $\psi(\boldsymbol{x}) \mapsto e^{i\sqrt{\alpha}\chi(x)}\psi(\boldsymbol{x})$, for this would generally take the wave function out of the space $\Lambda_0^+ L^2(\mathbb{R}^3; \mathbb{C}^4)$. Therefore, the energy does depend on the choice of gauge in this model. The condition $\operatorname{div} \boldsymbol{A} = 0$ (together with the restriction that A vanishes weakly at infinity) seems natural here, since without it E_0 would simply be $-\infty$. In fact, for any ψ the energy could be driven to $-\infty$ by adding $\lambda \nabla \chi$ to $\boldsymbol{A}$ for a suitable χ such that $(\psi, \boldsymbol{\alpha} \cdot \nabla \chi \psi) \neq 0$ and letting $\lambda \to \infty$.

Even with the Coulomb gauge condition, it is not obvious that the Brown–Ravenhall model is stable. *Indeed it is not* – at least not for large enough N regardless of how small α might be. This will be discussed in Section 10.5 below. *It turns out that the Coulomb potential V_C is irrelevant for the instability!*

10.2.2 A Modified Brown–Ravenhall Model

In order to define a gauge-invariant model and at the same time ensure that the kinetic energy of the electrons is always positive, one can define the relevant electron subspace to be the positive energy subspace of $D_{\boldsymbol{A}}$, not D_0. That is, the Hilbert space now depends on the choice of the vector potential. The resulting model *is* now gauge invariant. Because of this gauge invariance, it is not necessary to work in the Coulomb gauge.

For this modified Brown–Ravenhall model, the ground state energy is

$$E_0 = \inf \left\{ (\psi, H\psi) : \psi \in \bigwedge^N \left[\Lambda_{\boldsymbol{A}}^+ L^2(\mathbb{R}^3; \mathbb{C}^4)\right], \|\psi\|_2 = 1, \int |\operatorname{curl} \boldsymbol{A}|^2 < \infty \right\}.$$

Here, $\Lambda_{\boldsymbol{A}}^+$ denote the projection onto the positive energy subspace of $D_{\boldsymbol{A}}$, i.e., $\Lambda_{\boldsymbol{A}}^+ = (D_{\boldsymbol{A}} + |D_{\boldsymbol{A}}|)/(2|D_{\boldsymbol{A}}|)$. The infimum is taken over wave functions ψ and vector potentials $\boldsymbol{A}$.

The reader might object to the fact that the electron's Hilbert space depends on the magnetic vector potential $\boldsymbol{A}$, a situation that we have not encountered before. This connection between $\boldsymbol{A}$ and the state of the electron is not unnatural, however, if we interpret the field $\boldsymbol{A}$ as one that is generated by the electrons themselves. From this perspective, the elementary object is not a bare electron but an electron dressed with its magnetic field, which in some ways is more natural. We will encounter this situation again in the next chapter on quantized electromagnetic fields.

It turns out that the modified Brown–Ravenhall model is stable provided α *and* $Z\alpha$ *are both not too large.* This will be proved in Section 10.4.

10.2.3 The Furry Picture

It is possible to go one step further and include not only the magnetic vector potential in the definition of the relevant electron space but also the one-particle part of the Coulomb potential V_C, i.e., the interaction energy between the electron and the nuclei. That is, the electron Hilbert space is now the positive energy subspace of

$$D_A - \sum_{j=1}^{M} \frac{Z_j \alpha}{|\boldsymbol{x} - \boldsymbol{R}_j|}. \tag{10.2.2}$$

This is sometimes referred to as the Furry picture (after W. Furry [75]). This choice trivially leads to stability, since there are no negative terms left in the energy. If the operator (10.2.2) has eigenvalues ≤ 0, one may want to include the corresponding L^2 eigenfunctions in the electron space.

The Furry picture is often employed in calculations, and there is some justification for it [142, 176]. The Hilbert space now depends on the positions of the nuclei, however, but these are usually regarded as macroscopically determined parameters in the problem and should not, therefore, influence the fundamental Hilbert space of the microscopic particles in the system. Since stability is automatically ensured in this model, we have no need to discuss it further.[5]

[5] The stability problem has not completely disappeared, however, since one still has to be able to define the operator (10.2.2). To do so one needs that $Z_j \alpha \leq 1$ if all the $\boldsymbol{R}_j$ are distinct. Compare with footnote 6 on page 191.

10.3 The One-Particle Problem

Before discussing the general many-body problem, let us first investigate the case of one electron in the three different pictures described in the previous section. We start with considering a single particle in a magnetic field, and later discuss the hydrogenic atom.

10.3.1 The Lonely Dirac Particle in a Magnetic Field

Consider a single electron in the presence of a magnetic field with vector potential $\boldsymbol{A}$. In the Brown–Ravenhall model discussed above, the question of stability already arises in this simple system. In fact, we shall show that the question of stability depends on the value of α.

Recall that $\mathcal{H}_0^+$ denotes the positive spectral subspace of $D_0 = \boldsymbol{\alpha} \cdot \boldsymbol{p} + \beta m$.

Lemma 10.2 (Stability Criterion for a Single Particle). *There is a finite critical value of α, denoted by α_c, with $3\,\pi\,2^{-4/3} \le \alpha_c < \infty$, such that*

$$\inf_{\psi, \boldsymbol{A}} \left\{ (\psi, D_{\boldsymbol{A}}\psi) + \mathcal{E}_{\mathrm{mag}}(\boldsymbol{B}) \ : \ \psi \in \mathcal{H}_0^+, \ \|\psi\|_2 = 1, \ \boldsymbol{A} \in \mathcal{A} \right\}$$

is finite for $\alpha < \alpha_c$ and $-\infty$ for $\alpha > \alpha_c$.

Proof. The fact that $\alpha_c < \infty$ follows from scaling. We cannot, in fact, scale ψ directly since this would take us out of $\mathcal{H}_0^+$ (which depends on m in a non-trivial way). The terms $\boldsymbol{p}$, $\boldsymbol{A}$ and $\mathcal{E}_{\mathrm{mag}}(\boldsymbol{B})$ all behave like an inverse length under scaling; only m is scale independent. We can, however, use the representation (10.1.6) for $\psi \in \mathcal{H}_0^+$ and scale the corresponding u by $\lambda^{-3/2}u(\boldsymbol{p}/\lambda)$. For any fixed u, pick an $\boldsymbol{A}$ such that $(\psi, \boldsymbol{\alpha} \cdot \boldsymbol{A}\psi) < 0$ for all small m. Rescale u according to $u(\boldsymbol{p}) \to \lambda^{-3/2}u(\boldsymbol{p}/\lambda)$, $\boldsymbol{A}(\boldsymbol{x}) \to \lambda\boldsymbol{A}(\lambda\boldsymbol{x})$, and let ψ be given by (10.1.6). Then, for small enough λ, $(\psi, D_{\boldsymbol{A}}\psi) + \mathcal{E}_{\mathrm{mag}}(\boldsymbol{B})$ will be negative for α large enough. The energy can then be driven to $-\infty$ by letting λ tend to $+\infty$. Note that in the limit of large λ, $|\boldsymbol{p}|$ becomes large and hence the m in (10.1.6) becomes irrelevant.

To show that $\alpha_c > 0$, we first note that for $\psi \in \mathcal{H}_0^+$,

$$(\psi, D_0\psi) = \int_{\mathbb{R}^3} \sqrt{(2\pi \boldsymbol{k})^2 + m^2}\,|\widehat{\psi}(\boldsymbol{k})|^2 \mathrm{d}\boldsymbol{k}, \tag{10.3.1}$$

where $|\widehat{\psi}(\boldsymbol{k})|$ denotes the Euclidean $\mathbb{C}^4$ norm of the $\mathbb{C}^4$ vector $\widehat{\psi}(\boldsymbol{k})$. This can be seen by simply noting that $D_0 = \Lambda_0^+|D_0|\Lambda_0^+ - \Lambda_0^-|D_0|\Lambda_0^-$, with $\Lambda_0^\pm$ the projections onto $\mathcal{H}_0^\pm$, and using the fact that $|D_0|^2 = \boldsymbol{p}^2 + m^2$, see Eq. (10.1.4). Using (10.3.1) and Sobolev's inequality (2.2.12) we thus obtain

$$(\psi, D_0\psi) \geq S' \left(\int_{\mathbb{R}^3} |\psi(\boldsymbol{x})|^3 \mathrm{d}\boldsymbol{x} \right)^{2/3},$$

with $S' = (2\pi^2)^{1/3}$. Moreover, since $\boldsymbol{\alpha} \cdot \boldsymbol{A}(\boldsymbol{x}) \geq -|\boldsymbol{A}(\boldsymbol{x})|$,

$$(\psi, \boldsymbol{\alpha} \cdot \boldsymbol{A}(\boldsymbol{x})\psi) \geq -\int_{\mathbb{R}^3} |\boldsymbol{A}(\boldsymbol{x})||\psi(\boldsymbol{x})|^2 \mathrm{d}\boldsymbol{x} \geq -\|A\|_6 \|\psi\|_{12/5}^2 \geq -\|A\|_6 \|\psi\|_3 \|\psi\|_2,$$

where we have used Hölder's inequality in the last two steps. Finally, again using the Sobolev inequality, but this time Eq. (2.2.4) for $|\boldsymbol{p}|^2$ instead of Eq. (2.2.12) for $|\boldsymbol{p}|$,

$$\mathcal{E}_{\mathrm{mag}}(\boldsymbol{B}) = \frac{1}{8\pi} \int_{\mathbb{R}^3} |\nabla \boldsymbol{A}(\boldsymbol{x})|^2 \mathrm{d}\boldsymbol{x} \geq \frac{3}{8} \left(\frac{\pi}{4}\right)^{1/3} \|\boldsymbol{A}\|_6^2.$$

Using $\|\psi\|_2 = 1$, the energy is then bounded from below by

$$(\psi, D_{\boldsymbol{A}}\psi) + \mathcal{E}_{\mathrm{mag}}(\boldsymbol{B}) \geq (2\pi^2)^{1/3} \|\psi\|_3^2 - \sqrt{\alpha}\|\boldsymbol{A}\|_6 \|\psi\|_3 + \frac{3}{8} \left(\frac{\pi}{4}\right)^{1/3} \|\boldsymbol{A}\|_6^2,$$

which is positive for all ψ and $\boldsymbol{A}$ as long as $\alpha \leq 3\pi\, 2^{-4/3}$. ■

In the *modified* Brown–Ravenhall model, where the electron Hilbert space is the positive spectral subspace of $D_{\boldsymbol{A}}$ instead of D_0, stability of a single particle is not an issue, of course.

10.3.2 The Hydrogenic Atom in a Magnetic Field

Consider one Dirac particle in the Coulomb field of a nucleus of charge Z. For simplicity, we set $\boldsymbol{A} = 0$ for the moment. The appropriate Hamiltonian describing this system is

$$H = D_0 - \frac{Z\alpha}{|\boldsymbol{x}|}.$$

There are two ways to proceed. One is to try to find the spectrum of H on the full Hilbert space $L^2(\mathbb{R}^3; \mathbb{C}^4)$. This can actually be done explicitly, see, e.g., [16]. As long as $Z\alpha \le 1$, one finds that there are discrete eigenvalues in the interval $[0, m)$, and these look very much like the spectrum of the non-relativistic hydrogen problem with Hamiltonian $-(1/2)\Delta - Z\alpha|\boldsymbol{x}|^{-1}$, but shifted upward by m. In addition, there is continuous spectrum on $(-\infty, m] \cup [m, \infty)$.[6] As long as $Z\alpha \le 1$, there is no need to make an *a priori* restriction of the allowed function space to the positive spectral subspace of D_0. This is the approach usually taken in the analysis of one-electron atoms. It corresponds to the Furry picture discussed above. It has the conceptual disadvantage that the relevant Hilbert space changes when someone moves a nucleus (although it has to be admitted that many people do not regard this as a problem). The main point here is that there is no simple variational (i.e., energy minimization) principle for H. Even for the simple hydrogen atom with $Z\alpha \le 1$ the ground state energy can not be found by minimizing $(\psi, H\psi)$ but only by an exact calculation of the spectrum of H.[7]

An alternative approach is to consider the Brown–Ravenhall model, where one restricts the allowed wave functions to the positive energy subspace of D_0. In this way, one has a variational principle for the ground state energy. Although the spectrum of hydrogenic atoms does not come out quite as accurate as in the first approach, the latter is much more suitable for an analysis of the stability of matter. As a warm-up, let us apply this approach to hydrogenic atoms. That is, we are looking for

$$E_0 = \inf_{\psi \in \mathcal{H}_0^+,\, \|\psi\|_2 = 1} \left[(\psi, D_0\psi) - Z\alpha(\psi, |\boldsymbol{x}|^{-1}\psi) \right]. \tag{10.3.2}$$

Lemma 10.3 (Stability of Hydrogen). *There is a critical $(Z\alpha)_c$, with $2/\pi \le (Z\alpha)_c \le 4/\pi$, such that E_0 in (10.3.2) satisfies $E_0 > -\infty$ for $Z\alpha < (Z\alpha)_c$ and $E_0 = -\infty$ for $Z\alpha > (Z\alpha)_c$.*

The value of the critical $Z\alpha$ can actually be computed explicitly. It is given by $(Z\alpha)_c = 2/(\pi/2 + 2/\pi)$. Note that this number is bigger than $2/\pi$ but less

[6] As long as $Z\alpha \le 1$, H can be defined as a self-adjoint operator on a natural domain containing all infinitely differentiable functions of compact support, as was recently proved by Esteban and Loss [53]. Self-adjointness fails for $Z\alpha > 1$. The existence of a critical value of $Z\alpha$ for the Dirac Hamiltonian is analogous to the case of $|\boldsymbol{p}|$ discussed in Chapter 8 where the critical value of $Z\alpha$ was $2/\pi$.

[7] In the one-particle case, there are min-max principles for the eigenvalues of the Dirac operator in the interval $(-m, m)$. See [80, 52].

than 1. The computation of $(Z\alpha)_c$ was a challenging problem which was solved by Evans, Perry and Siedentop in [54]. Improved bounds on the energy were later given in [181], and a generalization to the case of one-electron molecules is in [5].

Proof. The fact that E_0 is bounded from below as long as $Z\alpha \leq 2/\pi$ follows immediately from Lemma 8.2 in Chapter 8, together with the observation in Eq. (10.3.1) that D_0 acts as multiplication by $\sqrt{(2\pi \boldsymbol{k})^2 + m^2}$ on the Fourier transform of functions in $\mathcal{H}_0^+$.

Consider now the case $Z\alpha > 4/\pi$. For any $E < 0$, we can find a function $\psi \in L^2(\mathbb{R}^3; \mathbb{C}^4)$ with $\|\psi\|_2 = 1$ such that

$$(\psi, |D_0|\psi) - \frac{Z\alpha}{2}(\psi, |\boldsymbol{x}|^{-1}\psi) < E.$$

This follows again from Lemma 8.2 in Chapter 8, where it was shown that $2/\pi$ is the critical coupling constant for stability of $|\boldsymbol{p}| - Z\alpha|\boldsymbol{x}|^{-1}$. Let $\psi_\pm$ denote the projection of ψ onto $\mathcal{H}_0^\pm$. We note that $\|\psi_\pm\|_2 \leq 1$. Then $(\psi, |D_0|\psi) = (\psi_+, |D_0|\psi_+) + (\psi_-, |D_0|\psi_-)$ and, by Schwarz's inequality,

$$(\psi, |\boldsymbol{x}|^{-1}\psi) \leq 2(\psi_+, |\boldsymbol{x}|^{-1}\psi_+) + 2(\psi_-, |\boldsymbol{x}|^{-1}\psi_-).$$

We thus conclude that

$$(\psi_+, |D_0| - Z\alpha|\boldsymbol{x}|^{-1}\psi_+) + (\psi_-, |D_0| - Z\alpha|\boldsymbol{x}|^{-1}\psi_-) < E.$$

One of the terms on the left side has to be less than $E/2$. Without loss of generality, we can assume it is the first term; if not, simply replace ψ by $U\psi$ with U given in (10.1.5) in order to exchange ψ_+ and ψ_-. Since E was arbitrary, this shows instability for $Z\alpha > 4/\pi$. ∎

In combination, the results of this and the previous subsection imply the following lemma concerning hydrogenic atoms in magnetic fields.

Lemma 10.4 (Hydrogen in a Magnetic Field). *For both α and $Z\alpha$ small enough,*

$$\inf_{\psi, \boldsymbol{A}} \left\{ (\psi, D_{\boldsymbol{A}}\psi) - Z\alpha(\psi, |\boldsymbol{x}|^{-1}\psi) + \mathcal{E}_{\text{mag}}(\boldsymbol{B}) \,:\, \psi \in \mathcal{H}_0^+,\ \|\psi\|_2 = 1,\ \boldsymbol{A} \in \mathcal{A} \right\} \tag{10.3.3}$$

is finite. If either α or $Z\alpha$ is too large, it is $-\infty$.

We shall not trouble the reader at this point with precise estimates on the critical parameters for the single electron but will turn our attention to the many-body problem instead. It will turn out that the Brown–Ravenhall model, as defined using the positive spectral subspace of D_0, is inherently *unstable of the first kind!* This result is not unexpected since there is no positivity condition available to control the N terms $\sum_j \boldsymbol{\alpha}_j \cdot \boldsymbol{A}(\boldsymbol{x}_j)$. Historically, the instability was first proved in [127] with the aid of the negative potential generated by M nuclei, with $M \gg 1$. It was later realized by Griesemer and Tix [81] that the nuclei were not needed! The $\boldsymbol{\alpha}_j \cdot \boldsymbol{A}(\boldsymbol{x}_j)$ terms alone can cause collapse if N is large enough. This analysis will be summarized in Section 10.5.

The modified Brown–Ravenhall model, where the electron Hilbert space is defined as the positive spectral subspace of $D_{\boldsymbol{A}}$, turns out to be stable for α and $Z\alpha$ small enough. We shall show this in the next section.

10.4 Stability of the Modified Brown–Ravenhall Model

The Hamiltonian for N electrons is

$$\boxed{H = \sum_{i=1}^{N} D_{\boldsymbol{A},i} + \alpha V_C(\underline{\boldsymbol{X}}, \underline{\boldsymbol{R}}) + \mathcal{E}_{\text{mag}}(\boldsymbol{B}),} \tag{10.4.1}$$

with $D_{\boldsymbol{A}}$ defined in (10.1.1) and V_C the total Coulomb potential given in (2.1.21). Let $\Lambda_{\boldsymbol{A}}^{+}$ denote the projection onto the positive spectral subspace of $D_{\boldsymbol{A}}$, and let $\mathcal{H}_{\boldsymbol{A}}^{+} = \Lambda_{\boldsymbol{A}}^{+} L^2(\mathbb{R}^3; \mathbb{C}^4)$ denote the corresponding subspace. For N electrons, the relevant Hilbert space is now

$$\mathcal{H} = \bigwedge_{i=1}^{N} \mathcal{H}_{\boldsymbol{A}}^{+}$$

and the ground state energy is

$$E_0 = \inf_{\psi, \boldsymbol{A}} \{(\psi, H\psi) \, : \, \psi \in \mathcal{H}, \; \|\psi\|_2 = 1, \; \boldsymbol{A} \in \mathcal{A}\} . \tag{10.4.2}$$

As in the previous chapter, we could allow for spin degrees of freedom in addition to the 4 naturally present in the one-particle space $L^2(\mathbb{R}^3; \mathbb{C}^4)$, but we shall refrain from doing so for simplicity. The choice of the Coulomb gauge, i.e., $\operatorname{div} \boldsymbol{A} = 0$, in $\mathcal{A}$ is not important because of gauge invariance, which holds in this model but which does not hold if one considers the positive spectral subspace of D_0 as the one-particle subspace, instead of the one of $D_{\boldsymbol{A}}$, as we do here.

The following theorem was proved by Lieb, Siedentop and Solovej in [127].

Theorem 10.1 (Stability of the Modified Brown–Ravenhall Model). *Let $\kappa(q, Z)$ be given by*

$$\frac{1}{\kappa(q, Z)} = \frac{Z\pi}{2} + 2.26\, q^{1/3} Z^{2/3} + 1.03\, q^{1/3}. \tag{10.4.3}$$

Then E_0 in (10.4.2) is non-negative for all N and M and nuclear positions $\boldsymbol{R}_1, \ldots, \boldsymbol{R}_M$ as long as $\alpha < \kappa(2, Z)$ and

$$\boxed{0.137\,\alpha \left(1 - \frac{\alpha^2}{\kappa(2, Z)^2}\right)^{-3/2} \leq \frac{1}{8\pi}} \tag{10.4.4}$$

For the physical value of $\alpha \approx 1/137$, this proves stability for all $Z \leq 56$. For $Z = 1$, on the other hand, stability holds for all $\alpha \leq 0.117$. It is still an open problem to show that stability holds up to the expected critical value $Z = 2\alpha^{-1}/(2/\pi + \pi/2) \approx 124$ for the physical value of α. A proof that stability holds up to $Z = 88$ is given in [91].

The quantity $\kappa(q, Z)$ in (10.4.3) is the same as in (9.5.2) in the previous chapter and it is the lower bound on the critical value of α for stability of relativistic matter obtained in Chapter 8, Theorem 8.1. As in the previous chapter on the Pauli operator, that theorem will play a crucial role in the proof of Theorem 10.1.

The following proof follows [127] closely.

Proof. As in the proof of Theorem 9.2 in the previous Chapter 9, the first step is to use the relativistic stability of matter, which states that

$$V_C(\underline{\boldsymbol{X}}, \underline{\boldsymbol{R}}) \geq -\frac{1}{\kappa} \sum_{i=1}^{N} |\boldsymbol{p}_i + \sqrt{\alpha}\boldsymbol{A}(\boldsymbol{x}_i)|$$

for all $\kappa \leq \kappa(q, Z)$ defined in (10.4.3). *A priori*, one might think that $q = 4$ here, but it is actually enough to use $q = 2$. This can be seen by the following argument, which is due to M. Loss and is given in [127, Appendix B]. It turns out that the spin-summed one-particle density matrix $\mathring{\gamma}^{(1)}$ of any $\psi \in \mathcal{H}$, defined in (3.1.32), is bounded by 2 and not only by 4 as one might naively think. In fact, since the unitary U in (10.1.5) maps the positive spectral subspace of $D_{\boldsymbol{A}}$ to the negative spectral subspace (and *vice versa*), the one-particle density matrix $\gamma^{(1)}$ of ψ satisfies $\gamma^{(1)} + U\gamma_\psi^{(1)}U^\dagger \leq \mathbb{I}$, and hence

$$\mathring{\gamma}^{(1)} = \mathrm{Tr}_{\mathbb{C}^4}\gamma^{(1)} = \frac{1}{2}\mathrm{Tr}_{\mathbb{C}^4}\left(\gamma^{(1)} + U\gamma^{(1)}U^\dagger\right) \leq 2.$$

The only place q enters in the proof of Theorem 8.1 on relativistic stability is via the kinetic energy inequality in Corollary 4.1, which depends on the largest eigenvalue of the spin-summed one-particle density matrix $\mathring{\gamma}^{(1)}$. Since this largest eigenvalue is bounded by 2 in the present case, the effective q is thus 2 and not 4.

In this way, our stability problem reduces to finding a bound on the sum of the negative eigenvalues of the one-particle operator

$$h = \Lambda_A^+ \left(D_A - \frac{\alpha}{\kappa} |\boldsymbol{p} + \sqrt{\alpha} \boldsymbol{A}| \right) \Lambda_A^+ = \Lambda_A^+ S \Lambda_A^+,$$

where we denote

$$S = |D_A| - \frac{\alpha}{\kappa} |\boldsymbol{p} + \sqrt{\alpha} \boldsymbol{A}|.$$

Like any self-adjoint operator, S can be decomposed into its positive and negative parts. That is, it can be written as $S = S_+ - S_-$, with $S_\pm$ positive operators, and $S_+ S_- = S_- S_+ = 0$. If we neglect the positive part S_+, the sum of the negative eigenvalues of $\Lambda_A^+ S \Lambda_A^+$ will only decrease, and hence

$$\operatorname{Tr} h_- \leq \operatorname{Tr} \Lambda_A^+ S_- \Lambda_A^+.$$

Note that $U S U^\dagger = S$, with U given in (10.1.5). Hence also $U S_- U^\dagger = S_-$. Moreover, since $U \Lambda_A^+ U^\dagger = \Lambda_A^- = \mathbb{I} - \Lambda_A^+$, we have

$$\operatorname{Tr} \Lambda_A^+ S_- \Lambda_A^+ = \frac{1}{2} \operatorname{Tr} S_-.$$

The next step is to use the BKS inequality, Theorem 9.4. It implies that

$$\operatorname{Tr} S_- = \operatorname{Tr} \left[|D_A| - \frac{\alpha}{\kappa} |\boldsymbol{p} + \sqrt{\alpha} \boldsymbol{A}| \right]_- \leq \operatorname{Tr} \left[|D_A|^2 - \frac{\alpha^2}{\kappa^2} |\boldsymbol{p} + \sqrt{\alpha} \boldsymbol{A}|^2 \right]_-^{1/2}.$$

Recall that $|D_A|^2 = [\boldsymbol{\sigma} \cdot (\boldsymbol{p} + \sqrt{\alpha} \boldsymbol{A})]^2 + m^2 = (\boldsymbol{p} + \sqrt{\alpha} \boldsymbol{A})^2 + \sqrt{\alpha} \boldsymbol{\sigma} \cdot \boldsymbol{B} + m^2$. For an upper bound on the square root of the negative eigenvalues, we can use $m^2 \geq 0$ as well as $\boldsymbol{\sigma} \cdot \boldsymbol{B}(\boldsymbol{x}) \geq -|\boldsymbol{B}(\boldsymbol{x})|$. Hence

$$\operatorname{Tr} \left[|D_A|^2 - \frac{\alpha^2}{\kappa^2} |\boldsymbol{p} + \sqrt{\alpha} \boldsymbol{A}|^2 \right]_-^{1/2}$$

$$\leq \operatorname{Tr} \left[\left(1 - \frac{\alpha^2}{\kappa^2} \right) |\boldsymbol{p} + \sqrt{\alpha} \boldsymbol{A}|^2 - \sqrt{\alpha} |\boldsymbol{B}(\boldsymbol{x})| \right]_-^{1/2}.$$

The LT inequality, Theorem 4.1 for $\gamma = 1/2$, implies that the latter trace is bounded from above by

$$\mathrm{Tr}\left[\left(1-\frac{\alpha^2}{\kappa^2}\right)|\boldsymbol{p}+\sqrt{\alpha}\boldsymbol{A}|^2-\sqrt{\alpha}|\boldsymbol{B}(\boldsymbol{x})|\right]_-^{1/2}$$
$$\leq 4\alpha\left(1-\frac{\alpha^2}{\kappa^2}\right)^{-3/2} L_{1/2,3}\int_{\mathbb{R}^3}|\boldsymbol{B}(\boldsymbol{x})|^2\mathrm{d}\boldsymbol{x}.$$

(The factor 4 comes from the $\mathbb{C}^4$ part of the trace.)

Recall that we still have the positive magnetic field energy at our disposal, namely $(8\pi)^{-1}\int_{\mathbb{R}^3}|\boldsymbol{B}(\boldsymbol{x})|^2\mathrm{d}\boldsymbol{x}$. Non-negativity of the ground state energy E_0 thus holds as long as

$$2\alpha\left(1-\frac{\alpha^2}{\kappa^2}\right)^{-3/2} L_{1/2,3}\leq\frac{1}{8\pi}.$$

As explained in Chapter 4, $L_{1/2,3}\leq 0.0683$. Inserting this bound for $L_{1/2,3}$ yields (10.4.4). ■

10.5 Instability of the Original Brown–Ravenhall Model

In the previous section, we showed that the modified (gauge-invariant) Brown–Ravenhall model is indeed stable of the second kind, if α is sufficiently small (independent of N). The original Brown–Ravenhall model, in contrast, is not even stable of the first kind, as we shall see in this section.

As in the previous examples of instability, it was originally thought that the negative part of the Coulomb interaction energy was necessary for instability of the first kind [127]. Later it was realized by Griesemer and Tix [81] that the nuclei were not needed. The instability is solely caused by the interaction of the electrons with the magnetic field, that is, by the $\boldsymbol{\alpha}\cdot\boldsymbol{A}$ term in the Dirac operator. What they showed is the following.

Theorem 10.2 (Instability of the First Kind for the Original Brown–Ravenhall Model). *There is a constant $C > 0$, independent of m, N and α, such that*

$$\inf_{\psi,\boldsymbol{A}}\left\{\left(\psi,\sum_{i=1}^N D_{\boldsymbol{A},i}\,\psi\right)+\mathcal{E}_{\mathrm{mag}}(\boldsymbol{B})\;:\;\psi\in\bigwedge_{i=1}^N\mathcal{H}_0^+,\ \|\psi\|=1,\ \boldsymbol{A}\in\mathcal{A}\right\}=-\infty$$

as long as $N\alpha^{3/2}\geq C$.

In particular, the original Brown–Ravenhall model is unstable for any $\alpha > 0$ and N large enough. This is true also with the inclusion of the Coulomb energy V_C, but we shall not consider this case here. The following proof of Theorem 10.2 is inspired by [81]. We shall be brief at some points and resist the temptation of giving an estimate of the constant C. It was shown in [81] that $C < 1.4 \times 10^5$.

Proof. The first point to note is that we can set $m = 0$. As discussed in the proof of Lemma 10.2, it suffices to find a $\psi \in \bigwedge^N \mathcal{H}_0^+$ (for $m = 0$) and an $\boldsymbol{A} \in \mathcal{A}$ such that $(\psi, \sum_{i=1}^N D_A \psi) + \mathcal{E}_{\mathrm{mag}}(\boldsymbol{B}) < 0$. The energy (with the m) can then be driven to $-\infty$ by scaling, since all the terms in the energy (except for m) scale like an inverse length.

The strategy now is to find N orthonormal functions in $\mathcal{H}_0^+$ such that the kinetic energy of the corresponding Slater determinant (defined in Chapter 3, Section 3.1.3) has a kinetic energy proportional to $N^{4/3}$, which is what would be expected from semiclassics of relativistic electrons in a box of size 1 (see Chapter 4, Section 4.1.1). Moreover, one tries to find a fixed $\boldsymbol{A}$ (independent of N) so that the expected value of $\sum_j \boldsymbol{\alpha}_j \cdot \boldsymbol{A}(\boldsymbol{x}_j)$ is order N. The field energy is then fixed, independent of N. If we multiply the chosen vector potentials $\boldsymbol{A}$ by λ, the total energy is thus of the order $N^{4/3} - \text{const}\,\lambda\sqrt{\alpha}N + \text{const}\,\lambda^2$. With the choice $\lambda = c\sqrt{\alpha}N$ for an appropriate $c > 0$, this becomes $N^{4/3} - \text{const}\,\alpha N^2$, which is negative for $\alpha^{3/2}N$ large enough, as claimed.

A possible choice is the following. Consider one-particle wave functions in $\mathcal{H}_0^+$ of the form

$$\phi = \begin{pmatrix} u \\ \frac{\sigma\cdot p}{|p|} u \end{pmatrix},$$

where $u(\boldsymbol{x}, \sigma) = f(\boldsymbol{x})\delta_{\sigma,1}$. We choose N orthonormal functions $f_n \in L^2(\mathbb{R}^3)$ to be the N lowest eigenfunctions of the Dirichlet Laplacian in a cubic box of length 1, multiplied by a factor $e^{i\boldsymbol{\varpi}\cdot\boldsymbol{x}}$ for some $\boldsymbol{\varpi} \in \mathbb{R}^3$. We assume that $\|f_n\|^2 = 1/2$, so that the corresponding ϕ's are normalized to 1. The kinetic energy of the Slater-determinant made up of these N functions is of the order $N^{4/3} + |\varpi|N$.

The vector potential $\boldsymbol{A}$ we choose is one that is constant (in the z-direction) inside the unit cube, and appropriately defined outside the unit cube so as to make it divergence free and have finite magnetic field energy.[8] The contribution

[8] One way to imagine such a field is to consider a finite circular pipe though which water is flowing, in such a way that the velocity field close to the portion near the central axis has a constant magnitude and then decays smoothly to zero towards the boundary of the pipe. The

of $\sum_j \boldsymbol{\alpha}_j \cdot \boldsymbol{A}(\boldsymbol{x}_j)$ is then

$$2a \sum_{n=1}^{N} \int_{\mathbb{R}^3} |\widehat{f}_n(\boldsymbol{p})|^2 \frac{p^3}{|\boldsymbol{p}|} \mathrm{d}\boldsymbol{p} \tag{10.5.1}$$

where a denotes the magnitude of $\boldsymbol{A}$ inside the unit cube. The expression $2\sum_{n=1}^{N} |\widehat{f}_n(\boldsymbol{p})|^2$ is just the momentum distribution of the system which, for large N, is approximately given by $\varrho((\boldsymbol{p}-\boldsymbol{\varpi})/N^{1/3})$ for some fixed ϱ with $\int \varrho(\boldsymbol{p})\mathrm{d}\boldsymbol{p} = 1$. If we choose $|\boldsymbol{\varpi}|$, which is at our disposal, to be equal to a large number times $N^{1/3}$, (10.5.1) is of the order aN, as desired. This concludes the implementation of our strategy outlined above, and thus the proof. ∎

10.6 The Non-Relativistic Limit and the Pauli Operator

The Hamiltonian with the Pauli kinetic energy considered in Chapter 9 is said to be the non-relativistic limit of the models considered in this chapter. 'Non-relativistic limit' in our units means that we let $m \to \infty$ with fixed α. We shall consider this limit in the three Hilbert spaces discussed in Section 10.2.

- **The Furry Picture.** For one particle in an external potential V, the Furry picture simply means that we are looking at the positive spectrum of the full Dirac operator $D_{\boldsymbol{A}} + V$. We look for a solution of the Schrödinger equation $(D_{\boldsymbol{A}} + V - E)\psi = 0$, with $\psi \in L^2(\mathbb{R}^3; \mathbb{C}^4)$. If we write

$$\psi(\boldsymbol{x}) = \begin{pmatrix} u(\boldsymbol{x}) \\ \ell(\boldsymbol{x}) \end{pmatrix} \tag{10.6.1}$$

for $u, \ell \in L^2(\mathbb{R}^3; \mathbb{C}^2)$, this equation becomes

$$\begin{aligned} \boldsymbol{\sigma} \cdot (\boldsymbol{p} + \sqrt{\alpha}\boldsymbol{A})\ell + Vu &= (E - m)u \\ \boldsymbol{\sigma} \cdot (\boldsymbol{p} + \sqrt{\alpha}\boldsymbol{A})u + V\ell &= (E + m)\ell. \end{aligned} \tag{10.6.2}$$

For energies $E \approx m + O(1)$ (for hydrogen, $E \approx m(1 - 10^{-5})$), we can ignore the term $V\ell$ in the second equation for large m, and thereby obtain $\ell \approx$

velocity field of the water will be our $\boldsymbol{A}$ field. Now we bend the pipe into a torus, but keeping part of it unbent. If the unit cube is inside the unbent part, the velocity field inside it will be uniform, as desired.

$\sigma \cdot (p + \sqrt{\alpha} A)u/(2m)$. Inserting this into the first equation yields

$$\frac{1}{2m}\left[\sigma \cdot (p + \sqrt{\alpha} A)\right]^2 u + Vu = (E - m)u, \qquad (10.6.3)$$

which is the Schrödinger equation with the Pauli kinetic energy. For a precise mathematical analysis in terms of convergence of the operators in resolvent sense (and also higher order corrections) we refer to Thaller's treatise [177, Chapter 6].

- **The Original Brown–Ravenhall Model.** For large m, the lower component of the wave function in the positive spectral subspace of D_0 is related to the upper one simply by $\ell \approx \sigma \cdot pu/(2m)$, see Eq. (10.1.6). For functions in the positive spectral subspace of D_0, the Dirac operator D_0 simply acts as $\sqrt{p^2 + m^2} \approx m + p^2/(2m)$. Hence $(\psi, D_0\psi) \approx (u, m + p^2/(2m)\, u)$. The expectation value of $\alpha \cdot A$ in (10.6.1) equals

$$\frac{1}{2m}(u, [(\sigma \cdot p)(\sigma \cdot A) + (\sigma \cdot A)(\sigma \cdot p)]\, u)$$

in this case. A simple computation shows that $(\sigma \cdot p)(\sigma \cdot A) + (\sigma \cdot A)(\sigma \cdot p) = p \cdot A + A \cdot p + \sigma \cdot B$. Hence we recover all the terms in the Pauli operator, *except* for $\alpha A^2/(2m)$.

The conclusion is that we recover the Pauli–Fierz model (discussed in the footnote on page 21) as the non-relativistic limit of the Brown–Ravenhall model, and not the true (gauge-invariant) Pauli kinetic energy. This defect is closely related to the instability of the Brown–Ravenhall model discussed in this chapter.

- **The Modified Brown–Ravenhall Model.** On the positive spectral subspace of D_A, the Pauli operator D_A acts as $|D_A| = \sqrt{[\sigma \cdot (p + \sqrt{\alpha} A)]^2 + m^2} \approx m + [\sigma \cdot (p + \sqrt{\alpha} A)]^2/(2m)$. *Hence we immediately recover the full Pauli operator (10.6.3) in the large m limit of the modified model, thus remedying the defect just mentioned.*

CHAPTER 11

Quantized Electromagnetic Fields and Stability of Matter

So far we have treated the magnetic field $\boldsymbol{B}$ as a (classical) external field, with an energy $\mathcal{E}_{\text{mag}}(\boldsymbol{B}) = (8\pi)^{-1} \int |\boldsymbol{B}(\boldsymbol{x})|^2 \mathrm{d}\boldsymbol{x}$. We would now like to broaden our horizon and treat the electromagnetic field in its proper role as a dynamical quantity. Moreover, we would like to go even further and quantize this field in the canonical way, which will automatically give us the quantization of light, as discovered by Planck in 1900 [149]. The Hilbert space for this revised quantum mechanics will be larger than the L^2 spaces considered before, and will involve **Fock space**. The stability of matter questions will become more complicated in some ways and will have to be revisited.

The ultimate version of quantum electrodynamics is even more complicated. The number of electrons is not fixed, but fluctuates. Electrons are defined by a quantized field and can be created and destroyed. We shall not venture into this realm, however, and will continue to regard the electron number N as fixed.

We begin with a review of electromagnetism, which the knowledgeable reader can safely skip. A good reference for further details is Spohn's treatise [174].

11.1 Review of Classical Electrodynamics and its Quantization

11.1.1 Maxwell's Equations

In classical electromagnetic theory, the interaction of charged particles is mediated by fields, which is to say (possibly time-dependent) functions on $\mathbb{R}^3$. There are six in fact, consisting of two (three component) vector fields $\boldsymbol{E}(\boldsymbol{x})$ and $\boldsymbol{B}(\boldsymbol{x})$, the electric field and the magnetic field. These satisfy the following equations, known as **Maxwell's equations**. With the notation $\partial f(\boldsymbol{x},t)/\partial t = \dot{f}(\boldsymbol{x},t)$, they are

$$\operatorname{div} \boldsymbol{E} = 4\pi \varrho \qquad \operatorname{curl} \boldsymbol{B} = 4\pi \boldsymbol{j} + \dot{\boldsymbol{E}} \tag{11.1.1}$$

$$\operatorname{div} \boldsymbol{B} = 0 \qquad \operatorname{curl} \boldsymbol{E} = -\dot{\boldsymbol{B}}. \tag{11.1.2}$$

The two equations (11.1.1) are the inhomogeneous Maxwell equations and the two equations (11.1.2) are the homogeneous Maxwell equations. The function ϱ is the charge density and the vector-valued function $\boldsymbol{j}$ is the electric current density. The two are related by the continuity equation

$$\dot{\varrho} + \operatorname{div} \boldsymbol{j} = 0, \tag{11.1.3}$$

which expresses the **conservation of charge**, that is, the fact that the total charge is independent of time. If (11.1.3) did not hold, Maxwell's equation would be inconsistent because $4\pi\dot{\varrho} = \operatorname{div} \dot{\boldsymbol{E}} = -4\pi \operatorname{div} \boldsymbol{j}$, since $\operatorname{div} \operatorname{curl} \boldsymbol{B} = 0$. In our case, the charges are point charges of electrons and nuclei, so ϱ is a sum of δ functions,

$$\varrho(\boldsymbol{y}) = \sum_i Q_i \delta(\boldsymbol{y} - \boldsymbol{x}_i) \quad \text{and} \quad \boldsymbol{j}(\boldsymbol{y}) = \sum_i Q_i \dot{\boldsymbol{x}}_i \delta(\boldsymbol{y} - \boldsymbol{x}_i), \tag{11.1.4}$$

where Q_i denotes the charge of particle i. Maxwell's equations have to be understood in the distributional sense (see [118]).

The fields $\boldsymbol{E}$ and $\boldsymbol{B}$ are generated by the moving charges $\boldsymbol{x}_i(t)$, according to (11.1.1). In turn, they yield the electromagnetic forces on the particles, which are

$$\boxed{\boldsymbol{F}_i = Q_i \left[\boldsymbol{E}(\boldsymbol{x}_i, t) + \dot{\boldsymbol{x}}_i \wedge \boldsymbol{B}(\boldsymbol{x}_i, t)\right].} \tag{11.1.5}$$

(The force on the particles is really ill-defined because of the self-interaction of the particles, i.e., the interacting of a particle with the field it creates. One can get around this difficulty by imagining the point charge to be smeared out a bit, but we shall ignore this problem for the moment.)

The homogeneous Maxwell equations (11.1.2) are constraints on the possible values of the fields which imply that the fields can be described economically by introducing potentials. More precisely, for a scalar (real-valued) potential

$\phi(\boldsymbol{x}, t)$ and a vector potential $\boldsymbol{A}(\boldsymbol{x}, t)$, we can define

$$\boxed{\boldsymbol{E}(\boldsymbol{x}, t) = -\nabla\phi(\boldsymbol{x}, t) - \dot{\boldsymbol{A}}(\boldsymbol{x}, t), \qquad \boldsymbol{B}(\boldsymbol{x}, t) = \operatorname{curl} \boldsymbol{A}(\boldsymbol{x}, t),} \tag{11.1.6}$$

and observe that Eqs. (11.1.2) are automatically satisfied. All fields $\boldsymbol{E}$ and $\boldsymbol{B}$ satisfying (11.1.2) can be written in this way. This follows from the fact that any divergence-free vector field can be written as a curl, and every curl-free vector field as a gradient. The equations (11.1.1) become

$$-\Delta\phi - \operatorname{div} \dot{\boldsymbol{A}} = 4\pi\varrho \tag{11.1.7}$$

and

$$\nabla \operatorname{div} \boldsymbol{A} - \Delta\boldsymbol{A} + \ddot{\boldsymbol{A}} = 4\pi \boldsymbol{j} - \nabla\dot{\phi}. \tag{11.1.8}$$

Note that $\boldsymbol{E}$ and $\boldsymbol{B}$ do not determine $\boldsymbol{A}$ and ϕ uniquely. In fact, there is a huge freedom in choosing the potentials. Namely, if $\boldsymbol{A}$ and ϕ are potentials giving rise to fields $\boldsymbol{E}$ and $\boldsymbol{B}$ via (11.1.6), so do

$$\boldsymbol{A}'(\boldsymbol{x}, t) = \boldsymbol{A}(\boldsymbol{x}, t) - \nabla\chi(\boldsymbol{x}, t), \qquad \phi'(\boldsymbol{x}, t) = \phi(\boldsymbol{x}, t) + \dot{\chi}(\boldsymbol{x}, t) \tag{11.1.9}$$

for an arbitrary function χ. One can thus impose restrictions on the potentials in order to remove this ambiguity. Such restrictions are called **gauge conditions**.

Maxwell's equations (11.1.7) and (11.1.8) simplify considerably in the **Coulomb gauge** where $\operatorname{div} \boldsymbol{A} = 0$ for all times (compare with Section 10.1.1). There is always a choice of $\boldsymbol{A}$ satisfying this gauge condition. Namely, for general $\boldsymbol{A}$ and ϕ define χ by $\Delta\chi = \operatorname{div} \boldsymbol{A}$. Then obviously $\boldsymbol{A}'$ and ϕ' in (11.1.9) give rise to the same fields $\boldsymbol{E}$ and $\boldsymbol{B}$, and $\operatorname{div} \boldsymbol{A}' = 0$ for all times.

In Coulomb gauge, the equations (11.1.7) and (11.1.8) become

$$-\Delta\phi = 4\pi\varrho \tag{11.1.10}$$

and

$$-\Delta\boldsymbol{A} + \ddot{\boldsymbol{A}} = 4\pi \boldsymbol{j} - \nabla\dot{\phi}. \tag{11.1.11}$$

The scalar potential ϕ is thus given by the instantaneous Coulomb potential

$$\boxed{\phi(\boldsymbol{x}, t) = \int_{\mathbb{R}^3} \frac{\varrho(\boldsymbol{y}, t)}{|\boldsymbol{x} - \boldsymbol{y}|} \mathrm{d}\boldsymbol{y}.} \tag{11.1.12}$$

Using the continuity equation (11.1.3), we see that

$$4\pi \boldsymbol{j} - \nabla\dot{\phi} = 4\pi\, \mathbb{P}\boldsymbol{j} \tag{11.1.13}$$

where $\mathbb{P}$ denotes the projection onto divergence-free vector fields. Explicitly,

$$\big(\mathbb{P}\boldsymbol{j}\big)(\boldsymbol{x},t) = \boldsymbol{j}(\boldsymbol{x},t) + \frac{1}{4\pi}\nabla \int\limits_{\mathbb{R}^3} \frac{\operatorname{div} \boldsymbol{j}(\boldsymbol{y},t)}{|\boldsymbol{x}-\boldsymbol{y}|}\mathrm{d}\boldsymbol{y} \tag{11.1.14}$$

or, in momentum space,

$$\big(\widehat{\mathbb{P}\boldsymbol{j}}\big)^i(\boldsymbol{k},t) = \sum_{l=1}^{3}\left(\delta_{il} - \frac{k^i k^l}{|\boldsymbol{k}|^2}\right)\widehat{j}^l(\boldsymbol{k},t). \tag{11.1.15}$$

To summarize, in Coulomb gauge the scalar potential ϕ is explicitly given by (11.1.12), and the vector potential $\boldsymbol{A}$ satisfies the equation

$$\boxed{-\Delta \boldsymbol{A} + \ddot{\boldsymbol{A}} = 4\pi\, \mathbb{P}\boldsymbol{j}.} \tag{11.1.16}$$

The fields $\boldsymbol{E}$ and $\boldsymbol{B}$ are given in terms of these potentials by (11.1.6).

Equation (11.1.16) can be interpreted as the evolution equation of coupled harmonic oscillators, at each point $\boldsymbol{x}$, driven by an external force. These harmonic oscillators are decoupled in Fourier space, where (11.1.16) becomes

$$\ddot{\widehat{\boldsymbol{A}}}(\boldsymbol{k},t) + |2\pi\boldsymbol{k}|^2\widehat{\boldsymbol{A}}(\boldsymbol{k},t) = 4\pi\, \widehat{\mathbb{P}\boldsymbol{j}}(\boldsymbol{k},t). \tag{11.1.17}$$

For every $\boldsymbol{k}$, this represents an independent harmonic oscillator with frequency $|\boldsymbol{k}|$ (or angular frequency $2\pi|\boldsymbol{k}|$). Later on we shall quantize these harmonic oscillators in the canonical way thus leading to Planck's 1900 law [149], which equates the energy of a photon of frequency $2\pi|\boldsymbol{k}|$ with the excitation energy of the corresponding harmonic oscillator. Note that the Coulomb gauge condition becomes $\boldsymbol{k}\cdot\widehat{\boldsymbol{A}}(\boldsymbol{k},t) = 0$ in momentum space. That is, $\widehat{\boldsymbol{A}}$ is perpendicular to the direction of $\boldsymbol{k}$. In particular, for each $\boldsymbol{k}$ there are only 2 independent degrees of freedom, transverse to the direction of propagation $\boldsymbol{k}$.

The electromagnetic field energy equals

$$\boxed{\frac{1}{8\pi}\int\limits_{\mathbb{R}^3}\big(\boldsymbol{E}(\boldsymbol{x},t)^2 + \boldsymbol{B}(\boldsymbol{x},t)^2\big)\,\mathrm{d}\boldsymbol{x}.} \tag{11.1.18}$$

This is the amount of work against the forces (11.1.5) to take well separated massless infinitesimal charges at rest and bring them into the stated conditions of position and velocity at time t. Using (11.1.6) and the Coulomb gauge div $\boldsymbol{A} = 0$, this energy, (11.1.18), becomes

$$\frac{1}{8\pi}\int_{\mathbb{R}^3} \left(\dot{\boldsymbol{A}}(\boldsymbol{x},t)^2 + (\operatorname{curl} \boldsymbol{A}(\boldsymbol{x},t))^2\right) \mathrm{d}\boldsymbol{x} + D(\varrho,\varrho)$$

$$= \frac{1}{8\pi}\int_{\mathbb{R}^3} \left(\dot{\widehat{\boldsymbol{A}}}(\boldsymbol{k},t)^2 + |2\pi\boldsymbol{k}|^2\, \widehat{\boldsymbol{A}}(\boldsymbol{k},t)^2\right) \mathrm{d}\boldsymbol{k} + D(\varrho,\varrho), \quad (11.1.19)$$

where $D(\varrho,\varrho)$ is the self-energy of the charge distribution ϱ, as defined in (5.1.4). The mixed term $\int \nabla\phi \cdot \dot{\boldsymbol{A}}$ disappeared by integrating by parts and using the Coulomb gauge condition. If we want to interpret (11.1.17) as the equation of motion of a harmonic oscillator of angular frequency $2\pi|\boldsymbol{k}|$ and energy given by the integrand in (11.1.19), then we have to identify $(4\pi)^{-1}\dot{\widehat{\boldsymbol{A}}}(\boldsymbol{k},t)$ as the 'momentum' corresponding to the 'position' coordinate $\widehat{\boldsymbol{A}}(\boldsymbol{k},t)$.[1] The electromagnetic field energy can thus be thought of as the energy of independent harmonic oscillators, two for every momentum $\boldsymbol{k}$ corresponding to the two transverse degrees of freedom.

11.1.2 Lagrangian and Hamiltonian of the Electromagnetic Field

The analogy with the harmonic oscillators suggests a way to derive a Hamiltonian description of the electromagnetic field, and ultimately for the field plus the matter. To do this systematically, we follow the standard route of first starting with the **Lagrangian** point of view. With $\boldsymbol{E}$ and $\boldsymbol{B}$ given as in (11.1.6), the Lagrangian is

$$\mathcal{L}(\phi, \boldsymbol{A}, \dot{\boldsymbol{A}}) = \int_{\mathbb{R}^3} \Bigg[\frac{1}{8\pi}\left(\boldsymbol{E}(\boldsymbol{x},t)^2 - \boldsymbol{B}(\boldsymbol{x},t)^2\right)$$

$$+\ \boldsymbol{j}(\boldsymbol{x},t)\cdot \boldsymbol{A}(\boldsymbol{x},t) - \varrho(\boldsymbol{x})\phi(\boldsymbol{x})\Bigg]\mathrm{d}\boldsymbol{x}. \quad (11.1.20)$$

[1] This is not to be confused with the momentum of the electromagnetic field, which is proportional to Poynting's vector $\boldsymbol{E} \wedge \boldsymbol{B}$. The words 'momentum' and 'position' are used here merely to make an analogy with an oscillating spring.

It can be readily checked that the Euler–Lagrange equations for this Lagrangian yield (11.1.1). While $\mathcal{L}$ involves the time derivative of one of the variables, $\boldsymbol{A}$, as it should, it does not involve involve the time derivative of the other variable, ϕ. Consequently the Euler–Lagrange equation for ϕ is really just a constraint equation, given by

$$-\operatorname{div}\left(\dot{\boldsymbol{A}} + \nabla\phi\right) = 4\pi\varrho, \tag{11.1.21}$$

which is the same as (11.1.7).

As already mentioned in the previous subsection, there is still some ambiguity in the choice of $\boldsymbol{A}$ and, at this point, it is convenient to work in the **Coulomb gauge** $\operatorname{div}\boldsymbol{A} = 0$. In particular, $\operatorname{div}\dot{\boldsymbol{A}} = 0$, and hence Eq. (11.1.21) becomes $\Delta\phi = -4\pi\varrho$. The scalar potential ϕ is thus given by the instantaneous Coulomb potential (11.1.12). Eliminating ϕ from the Lagrangian (11.1.20) with the aid of (11.1.12) and again using $\operatorname{div}\boldsymbol{A} = 0$, the Lagrangian becomes

$$\mathcal{L}(\boldsymbol{A}, \dot{\boldsymbol{A}}) = \int_{\mathbb{R}^3} \left[\frac{1}{8\pi}\left(\dot{\boldsymbol{A}}(\boldsymbol{x}, t)^2 - \boldsymbol{B}(\boldsymbol{x}, t)^2\right) + \boldsymbol{j}(\boldsymbol{x}, t)\cdot\boldsymbol{A}(\boldsymbol{x}, t)\right]\mathrm{d}\boldsymbol{x} - D(\varrho, \varrho). \tag{11.1.22}$$

The last expression is just the (instantaneous) electrostatic energy of the charge distribution ϱ. In case ϱ is the sum of δ functions, as in (11.1.4), this energy is really infinite owing to the infinite self-energy of a point charge. After subtracting this infinite self-energy, the electrostatic energy becomes

$$\sum_{i<j} \frac{Q_i Q_j}{|\boldsymbol{x}_i - \boldsymbol{x}_j|}. \tag{11.1.23}$$

The scalar potential ϕ has thus been eliminated from the problem, and $\boldsymbol{A}$ is the *only dynamical variable*. The *canonically conjugate momentum* to $\boldsymbol{A}$ is

$$\boldsymbol{\Pi} = \frac{\delta\mathcal{L}}{\delta\dot{\boldsymbol{A}}} = \frac{1}{4\pi}\dot{\boldsymbol{A}}.$$

The **Hamiltonian** corresponding to the Lagrangian $\mathcal{L}$ reads

$$H(\boldsymbol{A}, \boldsymbol{\Pi}) = \int_{\mathbb{R}^3} \left[2\pi\,\boldsymbol{\Pi}(\boldsymbol{x}, t)^2 + \frac{1}{8\pi}\boldsymbol{B}(\boldsymbol{x}, t)^2 - \boldsymbol{j}(\boldsymbol{x}, t)\cdot\boldsymbol{A}(\boldsymbol{x}, t)\right]\mathrm{d}\boldsymbol{x} + D(\varrho, \varrho). \tag{11.1.24}$$

In particular, we recover the magnetic field energy $\mathcal{E}_{\mathrm{mag}}(\boldsymbol{B}) = (8\pi)^{-1} \int_{\mathbb{R}^3} \boldsymbol{B}(\boldsymbol{x})^2 \mathrm{d}\boldsymbol{x}$. The reason that the additional $\boldsymbol{\Pi}^2$ term has not appeared so far is that we were always interested in minimizing the total energy, and for that purpose one can set $\boldsymbol{\Pi} = 0$, classically. This is analogous to minimizing $\boldsymbol{p}^2 + V(\boldsymbol{x})$, in which case $V(\boldsymbol{x})$ is minimized and $\boldsymbol{p} = 0$. When the field is quantized, it will be important not to drop the $\boldsymbol{\Pi}^2$ term, as we shall see, just as it is important not to drop the $\boldsymbol{p}^2$ term when minimizing the Schrödinger energy $\boldsymbol{p}^2 + V(\boldsymbol{x})$. The last term in (11.1.24) is the electrostatic energy in (11.1.23), after subtracting the infinite self-energy.

This Hamiltonian description of the electromagnetic field and its interaction with charges and currents is manifestly *not* relativistically invariant. It has to be recomputed after each Lorentz transformation. Nevertheless, it seems to be the only way to give a Hamiltonian formulation of the fields and their energies. It is used, as we said, in actual calculations of the spectra of atoms and molecules.

There does exist a relativistically invariant formulation using the 'Lorentz gauge' instead of the Coulomb gauge, in which $\phi(\boldsymbol{x})$ appears in addition to $\boldsymbol{A}(\boldsymbol{x})$. In the Lorentz gauge there are constraints between ϕ and $\boldsymbol{A}$, however, which makes the theory complicated, especially when an attempt is made to quantize it. Nevertheless, the Lorentz gauge is the most convenient one to use for the perturbation-theoretic treatment of scattering of particles. The Coulomb gauge is convenient for bound state problems, especially the calculation of the ground state energy. A good discussion of the relationship of the two gauges can be found in Heitler's book [87, Chapter 1, see §6]. An early article that established the importance of the Hamiltonian in the Coulomb gauge is Fermi's [63].

In the Coulomb gauge, which we use here, the div $\boldsymbol{A} = 0$ condition means that $\boldsymbol{A}$ is constrained to be transversal (meaning that in Fourier space $\boldsymbol{k} \cdot \widehat{\boldsymbol{A}}(\boldsymbol{k}) = 0$ for every $\boldsymbol{k}$). Hence there are only 2 degrees of freedom (the two transverse polarization vectors) at each $\boldsymbol{k}$ point. The same applies to $\boldsymbol{\Pi}(\boldsymbol{x})$. Consequently, since both $\boldsymbol{A}$ and $\boldsymbol{\Pi}$ are real-valued, we can represent them in terms of a complex valued function $a(\boldsymbol{k}, \lambda)$ as

$$\boldsymbol{A}(\boldsymbol{x}) = \sum_{\lambda=1}^{2} \int_{\mathbb{R}^3} \frac{1}{\sqrt{|\boldsymbol{k}|}} \boldsymbol{\varepsilon}_\lambda(\boldsymbol{k}) \left(e^{2\pi i \boldsymbol{k}\cdot\boldsymbol{x}} a(\boldsymbol{k}, \lambda) + e^{-2\pi i \boldsymbol{k}\cdot\boldsymbol{x}} \overline{a(\boldsymbol{k}, \lambda)} \right) \mathrm{d}\boldsymbol{k} \tag{11.1.25}$$

$$\boldsymbol{\Pi}(\boldsymbol{x}) = -\frac{i}{2} \sum_{\lambda=1}^{2} \int_{\mathbb{R}^3} \sqrt{|\boldsymbol{k}|}\, \boldsymbol{\varepsilon}_\lambda(\boldsymbol{k}) \left(e^{2\pi i \boldsymbol{k}\cdot\boldsymbol{x}} a(\boldsymbol{k}, \lambda) - e^{-2\pi i \boldsymbol{k}\cdot\boldsymbol{x}} \overline{a(\boldsymbol{k}, \lambda)} \right) \mathrm{d}\boldsymbol{k}. \tag{11.1.26}$$

The polarization vectors $\boldsymbol{\varepsilon}_1(\boldsymbol{k})$ and $\boldsymbol{\varepsilon}_2(\boldsymbol{k})$ are orthonormal and orthogonal to $\boldsymbol{k}$, i.e., $\boldsymbol{\varepsilon}_\lambda(\boldsymbol{k}) \cdot \boldsymbol{k} = 0$. Their choice is not unique, of course. They can *not* be chosen to be continuous functions of $\boldsymbol{k}$, however (because one cannot 'comb the hair on a sphere').[2] In terms of this complex valued function $a(\boldsymbol{k}, \lambda)$, the magnetic field $\boldsymbol{B} = \operatorname{curl} \boldsymbol{A}$ is expressed as

$$\boldsymbol{B}(\boldsymbol{x}) = \sum_{\lambda=1}^{2} \int_{\mathbb{R}^3} \frac{2\pi i}{\sqrt{|\boldsymbol{k}|}} \boldsymbol{k} \wedge \boldsymbol{\varepsilon}_\lambda(\boldsymbol{k}) \left(e^{2\pi i \boldsymbol{k}\cdot\boldsymbol{x}} a(\boldsymbol{k}, \lambda) - e^{-2\pi i \boldsymbol{k}\cdot\boldsymbol{x}} \overline{a(\boldsymbol{k}, \lambda)} \right) \mathrm{d}\boldsymbol{k}. \tag{11.1.27}$$

It is convenient to factor out the terms $1/\sqrt{|\boldsymbol{k}|}$ and $\sqrt{|\boldsymbol{k}|}$ in the integrands in (11.1.25) and (11.1.26), respectively, since the field energy can then be simply written as

$$\int_{\mathbb{R}^3} \left[2\pi\, \boldsymbol{\Pi}(\boldsymbol{x}, t)^2 + \frac{1}{8\pi} \boldsymbol{B}(\boldsymbol{x}, t)^2 \right] \mathrm{d}\boldsymbol{x} = \sum_{\lambda=1}^{2} \int_{\mathbb{R}^3} 2\pi |\boldsymbol{k}| |a(\boldsymbol{k}, \lambda)|^2 \mathrm{d}\boldsymbol{k}. \tag{11.1.28}$$

The fact that $\boldsymbol{\Pi}$ and $\boldsymbol{A}$ are canonically conjugate variables implies that the **Poisson brackets** between $a(\boldsymbol{k}, \lambda)$ and $\overline{a(\boldsymbol{k}, \lambda)}$ are given by

$$\begin{aligned} &\{a(\boldsymbol{k}, \lambda), \overline{a(\boldsymbol{k}', \lambda')}\} = \delta_{\lambda\lambda'} \delta(\boldsymbol{k} - \boldsymbol{k}') \\ &\{a(\boldsymbol{k}, \lambda), a(\boldsymbol{k}', \lambda')\} = \{\overline{a(\boldsymbol{k}, \lambda)}, \overline{a(\boldsymbol{k}', \lambda')}\} = 0. \end{aligned} \tag{11.1.29}$$

11.1.3 Quantization of the Electromagnetic Field

We now follow the standard rules of quantization. The canonical variables $a(\boldsymbol{k}, \lambda)$ and $\overline{a(\boldsymbol{k}, \lambda)}$ become operators $a(\boldsymbol{k}, \lambda)$ and $a^\dagger(\boldsymbol{k}, \lambda)$ on a Hilbert space, satisfying the **canonical commutation relations**

$$\begin{aligned} &[a(\boldsymbol{k}, \lambda), a^\dagger(\boldsymbol{k}', \lambda')] = \delta_{\lambda\lambda'} \delta(\boldsymbol{k} - \boldsymbol{k}') \\ &[a(\boldsymbol{k}, \lambda), a(\boldsymbol{k}', \lambda')] = [a^\dagger(\boldsymbol{k}, \lambda), a^\dagger(\boldsymbol{k}', \lambda')] = 0. \end{aligned} \tag{11.1.30}$$

Also $\boldsymbol{A}$ and $\boldsymbol{\Pi}$ become operators on this Hilbert space. They are still given by (11.1.25) and (11.1.26), with the functions $a(\boldsymbol{k}, \lambda)$ and $\overline{a(\boldsymbol{k}, \lambda)}$ replaced by the corresponding operators $a(\boldsymbol{k}, \lambda)$ and $a^\dagger(\boldsymbol{k}, \lambda)$, respectively. A natural

[2] The lack of continuity can lead to spurious technical problems. One way to restore continuity is to make use of an artificial third component $\boldsymbol{\varepsilon}_3(\boldsymbol{k})$. See [73, 120].

representation is in terms of **Fock space** [67]. It is given by

$$\mathcal{F} = \bigoplus_{N \geq 0} L^2(\mathbb{R}^3; \mathbb{C}^2)^{\otimes_s^N}$$

where $L^2(\mathbb{R}^3; \mathbb{C}^2)^{\otimes_s^N}$ stands for the N-fold *symmetric* tensor product of $L^2(\mathbb{R}^3; \mathbb{C}^2)$, i.e., the totally symmetric square integrable functions of N variables. The sum runs over all non-negative integers N, which are interpreted as *photon* numbers. For $N = 0$, $L^2(\mathbb{R}^3; \mathbb{C}^2)^{\otimes_s^0} = \mathbb{C}$ by definition. The corresponding vector is called the vacuum, denoted by Ω. It is annihilated by all the $a(\boldsymbol{k}, \lambda)$, that is,

$$a(\boldsymbol{k}, \lambda)\Omega = 0 \quad \text{for all} \quad \boldsymbol{k} \in \mathbb{R}^3 \quad \text{and} \quad \lambda \in \{1, 2\}.$$

A vector in $\mathcal{F}$ can be thought of as a sequence of functions

$$\Psi = \{\psi_0 \in \mathbb{C}, \psi_1(\boldsymbol{k}, \lambda), \psi_2(\boldsymbol{k}_1, \lambda_1, \boldsymbol{k}_2, \lambda_2), \dots\}.$$

The inner product on $\mathcal{F}$ is then

$$(\Phi, \Psi) = \sum_{N=0}^{\infty} (\phi_N, \psi_N).$$

The **annihilation operator** $a(\boldsymbol{k}, \lambda)$ maps a function of N variables to a function of $N - 1$ variables. Explicitly,

$$\big(a(\boldsymbol{k}, \lambda)\psi_N\big)(\boldsymbol{k}_1, \lambda, \dots, \boldsymbol{k}_{N-1}, \lambda_{N-1}) = \sqrt{N}\psi_N(\boldsymbol{k}_1, \lambda_1, \dots, \boldsymbol{k}_{N-1}, \lambda_{N-1}, \boldsymbol{k}, \lambda).$$

Their adjoints are called the **creation operators**, which act as

$$\begin{aligned}&\big(a^\dagger(\boldsymbol{k}, \lambda)\psi_N\big)(\boldsymbol{k}_1, \lambda_1, \dots, \boldsymbol{k}_{N+1}, \lambda_{N+1})\\ &\quad = \sqrt{N+1}\,\mathcal{S}\,\psi_N(\boldsymbol{k}_1, \lambda_1, \dots, \boldsymbol{k}_N, \lambda_N)\delta(\boldsymbol{k}_{N+1} - \boldsymbol{k})\delta_{\lambda_{N+1}\lambda}\end{aligned}$$

where $\mathcal{S}$ stands for the symmetrization operator, i.e.,

$$\mathcal{S} f(\boldsymbol{k}_1, \lambda_1, \dots, \boldsymbol{k}_N, \lambda_N) = \frac{1}{N!}\sum_{\pi} f(\boldsymbol{k}_{\pi(1)}, \lambda_{\pi(1)}, \dots, \boldsymbol{k}_{\pi(N)}, \lambda_{\pi(N)}),$$

the sum being over all permutations of the N variables. It can be readily checked that $a^\dagger(\boldsymbol{k}, \lambda)$ is the adjoint of $a(\boldsymbol{k}, \lambda)$, and that the canonical commutation relations (11.1.30) are satisfied.

Strictly speaking, $a(\boldsymbol{k}, \lambda)$ and $a^\dagger(\boldsymbol{k}, \lambda)$ are not operators but operator-valued distributions. They can be made into *bona fide* (although still unbounded) operators by smearing them out with test-functions in the usual way. Formally, this means considering operators of the form $\int_{\mathbb{R}^3} f(\boldsymbol{k})a(\boldsymbol{k},\lambda)\mathrm{d}\boldsymbol{k}$ instead of $a(\boldsymbol{k},\lambda)$, with f infinitely differentiable and of compact support. See [150, Sect. X.7] for a proper definition. Alternatively, one could confine the electromagnetic field to a large but finite box, in which case the $\boldsymbol{k}$ values would become discrete and the Fourier integral becomes a Fourier series. Then $a(\boldsymbol{k},\lambda)$ is a genuine operator.

In analogy with (11.1.28), the electromagnetic field energy is given by the operator

$$H_f = \sum_{\lambda=1}^{2} \int_{\mathbb{R}^3} 2\pi |\boldsymbol{k}|\, a^\dagger(\boldsymbol{k},\lambda)a(\boldsymbol{k},\lambda)\mathrm{d}\boldsymbol{k}. \tag{11.1.31}$$

The precise analogue of the left side of (11.1.28) would actually have $(1/2)(a^\dagger a + aa^\dagger)$ in the integrand, which differs from (11.1.31) only by an infinite constant, which is of no physical significance here. The choice (11.1.31) corresponds to an energy scale in which the vacuum has zero field energy, which is a natural condition.

In the absence of matter, the Hamiltonian (11.1.31) describes the full energy of the quantized electromagnetic field. In particular, it reproduces Planck's seminal discovery [149] that the electromagnetic radiation field is quantized in terms of photons, whose energy equals $h\nu = hc|\boldsymbol{k}| = 2\pi|\boldsymbol{k}|$ in our units, where $h = 2\pi\hbar = 2\pi$ and $c = 1$. Here, ν is the frequency and $1/|\boldsymbol{k}|$ is the wavelength. Moreover, the theory is linear in the sense that the energy of N photons is the sum of the individual energies. This follows from (11.1.31), since H_f acts on a $\Psi \in \mathcal{F}$ by multiplying ψ_N by $2\pi\sum_{i=1}^N |\boldsymbol{k}_i|$. It is noteworthy that the additivity of the energy holds even if the photons have the same value of $\boldsymbol{k}$. Different values of $\boldsymbol{k}$ refer to different, independent oscillators, so it is not surprising that the energies are additive; the additivity for photons with the same $\boldsymbol{k}$-value is a special property of the spectrum of harmonic oscillators.

The interaction of the electromagnetic field with matter is described by the $\boldsymbol{A}$ field, which formally is given by (11.1.25). In the quantum case, this is not a well-defined operator, however. This can already be seen by applying $\boldsymbol{A}(\boldsymbol{x})$ to the vacuum Ω. Each of the three components of $\boldsymbol{A}(\boldsymbol{x})\Omega$ is a one-photon state, with photon wave function $\boldsymbol{\varepsilon}_\lambda(\boldsymbol{k})|\boldsymbol{k}|^{-1/2}$. This is not a square-integrable function, however! The integral over $\boldsymbol{k}$ diverges for large $\boldsymbol{k}$, a phenomenon that is called

ultraviolet divergence. Hence $A(x)$ can not be defined as an operator on $\mathcal{F}$. Physically, this divergence is a consequence of the fact that we are looking at a fixed x, which is what we need in order to describe the interaction of the electromagnetic field with a point-particle located at x. If the particle were not a point but rather had a charge density described by $\chi(x - x_0)$, then the relevant interaction would be

$$\int_{\mathbb{R}^3} A(x)\chi(x - x_0)\mathrm{d}x = \sum_{\lambda=1}^{2} \int_{\mathbb{R}^3} \frac{\widehat{\chi}(k)}{\sqrt{|k|}} \varepsilon_\lambda(k) \Big(e^{2\pi i k\cdot x_0} a(k, \lambda) + e^{-2\pi i k\cdot x_0} a^\dagger(k, \lambda) \Big) \mathrm{d}k. \tag{11.1.32}$$

As long as $\widehat{\chi}(k)|k|^{-1/2}$ is a square-integrable function, this expression yields a well-defined operator.

The function $\widehat{\chi}(k)$ in (11.1.32) can be interpreted as an **ultraviolet cutoff**. It satisfies $\widehat{\chi}(0) = 1$ and $|\widehat{\chi}(k)| \leq 1$ (since $\chi(x) \geq 0$ and $\int \chi(x)\mathrm{d}x = 1$). In the following, it will be convenient to simply choose $\widehat{\chi}(k) = \Theta(\Lambda - |k|)$ for some $\Lambda > 0$.

11.2 Pauli Operator with Quantized Electromagnetic Field

For non-relativistic spin 1/2 particles in a quantized electromagnetic field, the Hamiltonian resembles the one in Chapter 9, Eq. (9.5.1). It is given by

$$\boxed{H = \sum_{i=1}^{N} \left[\sigma_i \cdot (p_i + \sqrt{\alpha} A(x_i))\right]^2 + \alpha\, V_C(\underline{X}, \underline{R}) + H_f,} \tag{11.2.1}$$

and acts on the Hilbert space $[\bigwedge^N L^2(\mathbb{R}^3; \mathbb{C}^2)] \otimes \mathcal{F}$, where $\mathcal{F}$ denotes the photon Fock space described in the previous section. The vector potential $A(x)$ describes the interaction between the radiation field and the particles. To simplify the notation, we take the ultraviolet cutoff function $\chi(k)$ to be the characteristic function of a ball of radius Λ. The vector potential is then

$$A(x) = \sum_{\lambda=1}^{2} \int_{|k|\leq\Lambda} \frac{1}{\sqrt{|k|}} \varepsilon_\lambda(k) \Big(e^{2\pi i k\cdot x} a(k, \lambda) + e^{-2\pi i k\cdot x} a^\dagger(k, \lambda) \Big) \mathrm{d}k. \tag{11.2.2}$$

The operators $a^\dagger(\boldsymbol{k}, \lambda)$ and $a(\boldsymbol{k}, \lambda)$ are the creation and annihilation operators of a photon of momentum $\boldsymbol{k}$ and polarization λ, satisfying the canonical commutation relations (11.1.30). Recall that the field energy H_f is defined in (11.1.31).

The reader might wonder how H_f in (11.1.31) compares with the classical field energy $\mathcal{E}_{\text{mag}}(\boldsymbol{B}) = (8\pi)^{-1} \int |\boldsymbol{B}(\boldsymbol{x})|^2 \mathrm{d}\boldsymbol{x}$. Classically, there is no $\boldsymbol{\Pi}(\boldsymbol{x})^2$ term, since we considered the $\boldsymbol{B}$ field to be time-independent. Treating the $\boldsymbol{B}$ field as a dynamical variable, the field energy has the additional term $2\pi \int |\boldsymbol{\Pi}(\boldsymbol{x})|^2 \mathrm{d}\boldsymbol{x}$. Naively one might think that the energy is now bigger because of this term, and the problem of stability of matter should become simpler. Alas, this is not the case. As remarked, $\boldsymbol{\Pi}^2$ plus $\boldsymbol{B}^2$ really gives $a^\dagger a + a a^\dagger$, which we have abbreviated to $2a^\dagger a$, thereby ignoring an infinite commutator.[3] Thus, while $a^\dagger a$ continues to be positive, H_f is not necessarily bigger (and can be less) than the classical field energy $\mathcal{E}_{\text{mag}}(\boldsymbol{B})$. *The problem of stability of matter in QED becomes more complicated because H_f can be less than $\mathcal{E}_{\text{mag}}(\boldsymbol{B})$.* An additional difficulty comes from the fact that H_f *does not commute with* $\boldsymbol{A}(\boldsymbol{x})$.

The quantized field $\boldsymbol{B}(\boldsymbol{x})$ is now an *operator* on Fock space, parametrized by $\boldsymbol{x} \in \mathbb{R}^3$, given by

$$\boxed{\boldsymbol{B}(\boldsymbol{x}) = \sum_{\lambda=1}^{2} \int_{|\boldsymbol{k}| \le \Lambda} \frac{2\pi i}{\sqrt{|\boldsymbol{k}|}} \boldsymbol{k} \wedge \boldsymbol{\varepsilon}_\lambda(\boldsymbol{k}) \left(e^{2\pi i \boldsymbol{k}\cdot\boldsymbol{x}} a(\boldsymbol{k}, \lambda) - e^{-2\pi i \boldsymbol{k}\cdot\boldsymbol{x}} a^\dagger(\boldsymbol{k}, \lambda) \right) \mathrm{d}\boldsymbol{k}.} \tag{11.2.3}$$

This $\boldsymbol{B}$ field is the quantization of the classical field in (11.1.27). As we shall see in Lemma 11.1, it *is* possible to obtain a lower bound on H_f in terms of the integral of $\boldsymbol{B}(\boldsymbol{x})^2$ over a bounded set $\Omega \subset \mathbb{R}^3$. There will be an error term that is proportional to the volume of Ω and to the fourth power of the ultraviolet cutoff Λ, however.

The operators $\boldsymbol{B}(\boldsymbol{x})$ and $\boldsymbol{A}(\boldsymbol{y})$ all commute with each other, as explicit calculation shows. A simpler way to see this is to introduce the operators $b(\boldsymbol{k}, \lambda) = a(\boldsymbol{k}, \lambda) + a^\dagger(-\boldsymbol{k}, \lambda)$, which satisfy $b^\dagger(\boldsymbol{k}, \lambda) = b(-\boldsymbol{k}, \lambda)$. Clearly, they all commute with each other. If we choose the polarization vectors to satisfy

[3] In a finite volume, where the $\boldsymbol{k}$ values are discrete, the commutator would not be infinite but proportional to the volume.

$\varepsilon_\lambda(\boldsymbol{k}) = \varepsilon_\lambda(-\boldsymbol{k})$ for all $\boldsymbol{k}$, then $\boldsymbol{B}(\boldsymbol{x})$ can be written as

$$\boldsymbol{B}(\boldsymbol{x}) = \sum_{\lambda=1}^{2} \int_{|\boldsymbol{k}|\leq\Lambda} \frac{2\pi i}{\sqrt{|\boldsymbol{k}|}} \boldsymbol{k} \wedge \varepsilon_\lambda(\boldsymbol{k}) e^{2\pi i \boldsymbol{k}\cdot\boldsymbol{x}} b(\boldsymbol{k}, \lambda) \mathrm{d}\boldsymbol{k}, \tag{11.2.4}$$

and a similar expression holds for $\boldsymbol{A}(\boldsymbol{x})$. Therefore $\boldsymbol{A}(\boldsymbol{x})$ and $\boldsymbol{B}(\boldsymbol{y})$ are linear combination of the operators $b(\boldsymbol{k}, \lambda)$ and hence commute.

The operators $q(\boldsymbol{k}, \lambda) = b(\boldsymbol{k}, \lambda) + b(-\boldsymbol{k}, \lambda)$ and $\widetilde{q}(\boldsymbol{k}, \lambda) = i(b(\boldsymbol{k}, \lambda) - b(-\boldsymbol{k}, \lambda))$ are self-adjoint and thus have real spectrum. They all commute with each other. (As mentioned above, these are strictly speaking only 'operator-valued distributions', but we can regard them as genuine operators on a Hilbert space if we discretize $\boldsymbol{k}$ so that Fourier integrals become sums.) We claim that their joint spectrum is $\mathbb{R}^2$. That is, there is a representation on Fock space in which all these operators can be regarded as ordinary real numbers without restriction (except, of course, that $q(\boldsymbol{k}, \lambda) = q(-\boldsymbol{k}, \lambda)$ and $\widetilde{q}(\boldsymbol{k}, \lambda) = -q(-\boldsymbol{k}, \lambda)$). To see this, introduce the conjugate operators $\widetilde{p}(\boldsymbol{k}, \lambda) = (1/4)[a(\boldsymbol{k}, \lambda) - a^\dagger(-\boldsymbol{k}, \lambda) - a(-\boldsymbol{k}, \lambda) + a^\dagger(\boldsymbol{k}, \lambda)]$ and $p(\boldsymbol{k}, \lambda) = (i/4)[a(\boldsymbol{k}, \lambda) - a^\dagger(-\boldsymbol{k}, \lambda) + a(-\boldsymbol{k}, \lambda) - a^\dagger(\boldsymbol{k}, \lambda)]$. The pairs p, q and $\widetilde{p}, \widetilde{q}$ each satisfy the Heisenberg commutation relations $[p(\boldsymbol{k}, \lambda), q(\boldsymbol{k}', \lambda')] = -i\delta_{\lambda,\lambda'}\delta(\boldsymbol{k} - \boldsymbol{k}')$ and they commute otherwise.

If we put all this together we see that there is a representation of Fock space (called the **Schrödinger representation** or 'Q-space', see [159] or [150, Sect. 10.7]) in which the Fourier coefficients of $\boldsymbol{A}(\boldsymbol{x})$ and $\boldsymbol{B}(\boldsymbol{x})$ are independent complex numbers. In other words, we can regard the $\boldsymbol{A}$ field and the $\boldsymbol{B}$ field as arbitrary functions (subject to the condition that their Fourier transforms vanish for $|\boldsymbol{k}| > \Lambda$) with $\boldsymbol{B} = \operatorname{curl} \boldsymbol{A}$. The field energy, however, involves $\boldsymbol{\Pi}$, which is composed of the conjugate operators, p and $\widetilde{p}$, and which do not commute with $\boldsymbol{A}$ and $\boldsymbol{B}$. That is why the case of quantized fields is different from the classical case. If the field energy were $(8\pi)^{-1} \int \boldsymbol{B}(\boldsymbol{x})^2 \mathrm{d}\boldsymbol{x}$ instead of H_f, the quantum case would simply reduce to the classical case.

The following lemma will be important when considering the question of stability of matter for the Hamiltonian (11.2.1). As mentioned above, it allows for bounding the field energy H_f from below in terms of the integral of $\boldsymbol{B}(\boldsymbol{x})^2$ over sets of finite volume. In this way, the problem of stability of matter with quantized fields can be reduced to the classical case.

Lemma 11.1 (Classical and Quantum Field Energy). *For any function* $g \in L^1(\mathbb{R}^3)$ *with* $0 \leq g(\boldsymbol{x}) \leq 1$

$$H_f \geq \frac{1}{8\pi} \int_{\mathbb{R}^3} |\boldsymbol{B}(\boldsymbol{x})|^2 g(\boldsymbol{x}) \mathrm{d}\boldsymbol{x} - 2\pi^2 \Lambda^4 \int_{\mathbb{R}^3} g(\boldsymbol{x}) \mathrm{d}\boldsymbol{x}. \tag{11.2.5}$$

This lemma was proved in [27, Lemma 3]. For a more general inequality of this type, see also [119, Lemma B.1].

Proof. Write $\boldsymbol{B}(\boldsymbol{x}) = \boldsymbol{C}(\boldsymbol{x}) + \boldsymbol{C}(\boldsymbol{x})^\dagger$, where $\boldsymbol{C}(\boldsymbol{x})$ denotes the part of $\boldsymbol{B}(\boldsymbol{x})$ in (11.2.3) containing only the annihilation operators $a(\boldsymbol{k}, \lambda)$. Then

$$\begin{aligned} \boldsymbol{B}(\boldsymbol{x})^2 &= [\boldsymbol{C}(\boldsymbol{x}) + \boldsymbol{C}(\boldsymbol{x})^\dagger]^2 \\ &= 4\boldsymbol{C}(\boldsymbol{x})^\dagger \cdot \boldsymbol{C}(\boldsymbol{x}) + 2[\boldsymbol{C}(\boldsymbol{x}), \boldsymbol{C}(\boldsymbol{x})^\dagger] + [\boldsymbol{C}(\boldsymbol{x}) - \boldsymbol{C}(\boldsymbol{x})^\dagger]^2 \\ &\leq 4\boldsymbol{C}(\boldsymbol{x})^\dagger \cdot \boldsymbol{C}(\boldsymbol{x}) + 2[\boldsymbol{C}(\boldsymbol{x}), \boldsymbol{C}(\boldsymbol{x})^\dagger]. \end{aligned}$$

The commutator $[\boldsymbol{C}, \boldsymbol{C}^\dagger]$ means $\sum_{i=1}^3 [C^i, C^{i\dagger}]$. It can be easily evaluated by using the canonical commutation relations (11.1.30). It is independent of $\boldsymbol{x}$, and given by

$$\begin{aligned} [\boldsymbol{C}(\boldsymbol{x}), \boldsymbol{C}(\boldsymbol{x})^\dagger] &= \sum_{\lambda=1}^{2} \int_{|\boldsymbol{k}| \leq \Lambda} \frac{(2\pi)^2}{|\boldsymbol{k}|} (\boldsymbol{k} \wedge \boldsymbol{\varepsilon}_\lambda(\boldsymbol{k}))^2 \mathrm{d}\boldsymbol{k} \\ &= 2(2\pi)^2 \int_{|\boldsymbol{k}| \leq \Lambda} |\boldsymbol{k}| \mathrm{d}\boldsymbol{k} = (2\pi)^3 \Lambda^4. \end{aligned}$$

Moreover,

$$\int_{\mathbb{R}^3} \boldsymbol{C}(\boldsymbol{x})^\dagger \cdot \boldsymbol{C}(\boldsymbol{x}) \mathrm{d}\boldsymbol{x} = (2\pi)^2 \sum_{\lambda=1}^{2} \int_{|\boldsymbol{k}| \leq \Lambda} |\boldsymbol{k}| a^\dagger(\boldsymbol{k}, \lambda) a(\boldsymbol{k}, \lambda) \leq 2\pi H_f.$$

Since $0 \leq g(\boldsymbol{x}) \leq 1$ the rest is obvious. ■

Recall that in the case of classical magnetic fields, stability of matter for the Pauli operator holds if and only if both α and $Z\alpha^2$ are small enough. This was shown in Theorems 9.2 and 9.3 in Chapter 9. In the case of quantized fields, a similar result holds, as we shall now describe.

The first proof of stability of the second kind for the Hamiltonian (11.2.1) was given by Bugliaro, Fröhlich and Graf in [27]. They used Lemma 11.1 in order to reduce the problem to the classical case, studied in Chapter 9. We shall follow a similar route in the proof of the following theorem about non-relativistic stability.

Let $\beta(Z)$ denote the maximal value of α such that the inequality (8.6.2) is satisfied for $q = 2$. It was shown in Lemma 8.6 that the inequality

$$\left(\psi, \sum_{i=1}^{N}\left(\frac{1}{\beta(Z)}|\boldsymbol{p}_i + \sqrt{\alpha}\boldsymbol{A}(\boldsymbol{x}_i)| - W^{\lambda}(\boldsymbol{x}_i)\right)\psi\right) \geq -\frac{Z^2}{8}\sum_{j=1}^{M}\frac{1}{D_j} \tag{11.2.6}$$

holds for all normalized, fermionic wave functions $\psi \in \bigwedge^N L^2(\mathbb{R}^3;\mathbb{C}^2)$ and all $0 < \lambda < 1$. The function W^{λ} is defined in (5.4.1), and D_j is half the distance of $\boldsymbol{R}_j$ to the nearest nucleus. Although Lemma 8.6 was formulated with classical vector potentials, it holds with quantized vector potentials as well, since these are equal to classical fields in the Schrödinger representation, as discussed above. That is, (11.2.6) holds for $\psi \in [\bigwedge^N L^2(\mathbb{R}^3;\mathbb{C}^2)] \otimes \mathcal{F}$, with $\boldsymbol{A}(\boldsymbol{x})$ the operator defined in (11.2.2).

Inequality (11.2.6) will play a crucial role in the proof of the following theorem.

Theorem 11.1 (Stability with the Pauli Operator and Quantized Fields). *Let $Z = \max_j Z_j$ and assume that*

$$\alpha^2 \leq 4.31\frac{\sqrt{3}}{8\pi^2}\beta(Z), \tag{11.2.7}$$

where $\beta(Z)$ is defined just before (11.2.6). Let H be the Hamiltonian (11.2.1) and let $\psi \in [\bigwedge^N L^2(\mathbb{R}^3;\mathbb{C}^2)] \otimes \mathcal{F}$ with $(\psi,\psi) = 1$. Then we have

$$(\psi, H\psi) \geq -N\left(\frac{\pi^2}{4\ell^2} + \frac{Z + \sqrt{2Z} + 1/2}{\ell} + \frac{\alpha^2}{\beta(Z)^2}\right) - \frac{(4\pi)^3}{3}M\Lambda^4\ell^3 \tag{11.2.8}$$

for any $\ell > 0$.

For the physical value $\alpha = 1/137$, a numerical evaluation shows that our stability criterion holds for all $Z \leq 854$. (Compare with the case of classical magnetic fields, where stability was proved in Theorem 9.2 for all $Z \leq 953$.)

Proof. Without loss of generality we can assume again that $Z_j = Z$ for all $j = 1, 2, \ldots, M$. This follows from the monotonicity of the ground state energy in the nuclear coordinates, as discussed in Section 3.2.3. In fact, the result of Proposition 3.1 applies equally well to the quantized field case.

Let χ be a nice function on $\mathbb{R}_+$, with $\chi(t) = 1$ for $t \le 1$, $\chi(t) = 0$ for $t \ge 2$ and $0 \le \chi(t) \le 1$ in-between. For $\ell > 0$ let $\eta_1(\boldsymbol{x}) = \chi(\min_j |\boldsymbol{x} - \boldsymbol{R}_j|/\ell)$ and $\eta_2(\boldsymbol{x}) = \sqrt{1 - \eta_1(\boldsymbol{x})^2}$. The function η_1 is 1 within a distance ℓ from any of the nuclei, and 0 a distance 2ℓ away from all the nuclei. Moreover, $|\nabla\eta_1|^2 + |\nabla\eta_2|^2 \le C\ell^{-2}$ for some constant $C > 0$. In fact, with the choice $\chi(t) = \cos((\pi/2)(t-1))$ for $1 \le t \le 2$ the constant C can be taken to be $C = \pi^2/4$.

Using the fact that $\eta_1^2 + \eta_2^2 = 1$, it is easy to see that[4]

$$
\begin{aligned}
\left[\boldsymbol{\sigma}\cdot(\boldsymbol{p} + \sqrt{\alpha}\boldsymbol{A}(\boldsymbol{x}))\right]^2 = {} & \eta_1(\boldsymbol{x})\left[\boldsymbol{\sigma}\cdot(\boldsymbol{p} + \sqrt{\alpha}\boldsymbol{A}(\boldsymbol{x}))\right]^2\eta_1(\boldsymbol{x}) \\
& + \eta_2(\boldsymbol{x})\left[\boldsymbol{\sigma}\cdot(\boldsymbol{p} + \sqrt{\alpha}\boldsymbol{A}(\boldsymbol{x}))\right]^2\eta_2(\boldsymbol{x}) \\
& - |\nabla\eta_1(\boldsymbol{x})|^2 - |\nabla\eta_2(\boldsymbol{x})|^2. \qquad (11.2.9)
\end{aligned}
$$

For a given $\psi \in [\bigwedge^N L^2(\mathbb{R}^3;\mathbb{C}^2)] \otimes \mathcal{F}$ and $i_j \in \{1,2\}$ for $1 \le j \le N$, let $\psi_{i_1\cdots i_N} = \prod_{j=1}^N \eta_{i_j}(\boldsymbol{x}_j)\psi$. Using (11.2.9) we can write

$$
\begin{aligned}
(\psi, H\,\psi) = {} & \sum_{i_1=1}^{2}\cdots\sum_{i_N=1}^{2}\left(\psi_{i_1\cdots i_N}, H\,\psi_{i_1\cdots i_N}\right) \\
& - \left(\psi, \textstyle\sum_{j=1}^N\left[|\nabla\eta_1(\boldsymbol{x}_j)|^2 + |\nabla\eta_2(\boldsymbol{x}_j)|^2\right]\psi\right). \qquad (11.2.10)
\end{aligned}
$$

The last term is bounded from below by $-NC\ell^{-2}(\psi,\psi)$.

We shall now derive a lower bound on $\left(\psi_{i_1\cdots i_N}, H\,\psi_{i_1\cdots i_N}\right)$ for given values of i_j, $1 \le j \le N$. We employ the improved electrostatic inequality of Theorem 5.5, which states that

$$
V_C(\underline{X}, \underline{R}) \ge -\sum_{i=1}^N W^\lambda(\boldsymbol{x}_i) + \frac{Z^2}{8}\sum_{j=1}^M \frac{1}{D_j}. \qquad (11.2.11)
$$

Our choice for λ will be the optimal value given in (5.4.3). For this value, $W^\lambda(\boldsymbol{x}) \le (Z + \sqrt{2Z} + 1/2)/\min_k |\boldsymbol{x} - \boldsymbol{R}_k|$, and hence $W^\lambda(\boldsymbol{x}_j)$ is bounded from above by $(Z + \sqrt{2Z} + 1/2)\ell^{-1}$ if $\boldsymbol{x}_j$ is in the support of η_2, i.e., if $|\boldsymbol{x}_j - \boldsymbol{R}_k| \ge \ell$ for all $1 \le k \le M$. In particular,

$$
\left(\psi_{i_1\cdots i_N}, W^\lambda(\boldsymbol{x}_j)\,\psi_{i_1\cdots i_N}\right) \le \frac{Z + \sqrt{2Z} + 1/2}{\ell}(\psi_{i_1\cdots i_N}, \psi_{i_1\cdots i_N}) \qquad (11.2.12)
$$

[4] This is known as the IMS localization formula; see [36].

if $i_j = 2$. We shall use this bound for all the particles j with $i_j = 2$. For the particles j with $i_j = 1$, we shall use (11.2.6) instead, which implies that

$$\sum_{j,\, i_j=1} W^\lambda(\boldsymbol{x}_j) \leq \frac{Z^2}{8} \sum_{k=1}^{M} \frac{1}{D_k} + \frac{1}{\beta(Z)} \sum_{j,\, i_j=1} |\boldsymbol{p}_j + \sqrt{\alpha}\boldsymbol{A}(\boldsymbol{x}_j)|. \qquad (11.2.13)$$

By combining (11.2.11)–(11.2.13) and dropping the positive kinetic energy for the particles with $i_j = 2$, we obtain the bound

$$(\psi_{i_1\cdots i_N}, H\, \psi_{i_1\cdots i_N}) \geq (\psi_{i_1\cdots i_N}, \widetilde{H}_{i_1,\cdots i_N}\, \psi_{i_1\cdots i_N}) - N\frac{Z + \sqrt{2Z} + 1/2}{\ell}(\psi_{i_1\cdots i_N}, \psi_{i_1\cdots i_N}) \qquad (11.2.14)$$

with

$$\widetilde{H}_{i_1,\cdots i_N} = \sum_{j,\, i_j=1} \left(\left[\boldsymbol{\sigma}_j \cdot (\boldsymbol{p}_j + \sqrt{\alpha}\boldsymbol{A}(\boldsymbol{x}_j))\right]^2 - \frac{\alpha}{\beta(Z)} |\boldsymbol{p}_j + \sqrt{\alpha}\boldsymbol{A}(\boldsymbol{x}_j)| \right) + H_f. \qquad (11.2.15)$$

Note that $\widetilde{H}_{i_1,\cdots i_N}$ acts non-trivially only on the variables $z_j = (\boldsymbol{x}_j, \sigma_j)$ with $i_j = 1$.

Using Lemma 11.1, with g being the characteristic function of the set

$$\Xi := \{\boldsymbol{x} \in \mathbb{R}^3 \, : \, \min_k |\boldsymbol{x} - \boldsymbol{R}_k| \leq 2\ell\},$$

we conclude that

$$H_f \geq \frac{1}{8\pi} \int_\Xi |\boldsymbol{B}(\boldsymbol{x})|^2 \mathrm{d}\boldsymbol{x} - \frac{(4\pi)^3}{3} \Lambda^4 M \ell^3.$$

After integrating out the variables z_j with $i_j = 2$, the one-particle density of $\psi_{i_1\cdots i_N}$ for the particles j with $i_j = 1$ is supported in the set Ξ. Hence we can apply Corollary 9.1 to conclude that

$$(\psi_{i_1\cdots i_N}, \widetilde{H}_{i_1,\cdots i_N}\, \psi_{i_1\cdots i_N}) \geq \left(-4N \frac{\alpha^2}{\beta(Z)^2} - \frac{(4\pi)^3}{3} \Lambda^4 M \ell^3 \right) (\psi_{i_1\cdots i_N}, \psi_{i_1\cdots i_N})$$

as long as (11.2.7) holds. This bound is valid for all values of i_j, $1 \leq j \leq N$. In combination with (11.2.10) and (11.2.14), we arrive at (11.2.8) ■

We have thus proved stability for the Pauli Hamiltonian with quantized electromagnetic field, for suitable values of α and $Z\alpha^2$. It turns out that because of the ultraviolet cutoff Λ the bounds on α and $Z\alpha^2$ are actually *not* necessary for stability of the second kind, i.e., stability holds for all values of $\alpha > 0$ and $Z > 0$. This was shown by Fefferman, Fröhlich and Graf [59] and, subsequently, by Bugliaro, Fefferman and Graf [26]. The same would be true for classical magnetic fields in the presence of an ultraviolet cutoff. The proof of this fact utilizes Lieb–Thirring inequalities for the Pauli operators which involve gradients of $\boldsymbol{B}(\boldsymbol{x})$ and which are more complicated than the ones given in Chapter 4; a partial list of references on this topic is in Section 4.4.

11.3 Dirac Operator with Quantized Electromagnetic Field

The next step is to consider the system discussed in Chapter 10, where the kinetic energy is described by the Dirac operator. The difference now is that the electromagnetic field is quantized. We expect the conclusions to be essentially the same as before, and they will be, but significant questions of principle arise because the field is not just a function on $\mathbb{R}^3$ but it is an operator on Fock space.

Formally, the Hamiltonian of the system is given by

$$\boxed{H = \sum_{i=1}^{N} D_{\boldsymbol{A},i} + \alpha V_C(\underline{\boldsymbol{X}}, \underline{\boldsymbol{R}}) + H_f} \tag{11.3.1}$$

where $D_{\boldsymbol{A},i} = D_{0,i} + \boldsymbol{\alpha}_i \cdot \boldsymbol{A}(\boldsymbol{x}_i)$ is an operator that acts on $L^2(\mathbb{R}^3; \mathbb{C}^4) \otimes \mathcal{F}$. The Dirac operator D_0 is defined in (10.1.1), and we use the standard units where $m = 1$. The Coulomb potential V_C is defined in (2.1.21) in Chapter 2, and the field energy H_f is given in (11.1.31) above. We note that the operators $D_{\boldsymbol{A},i}$ do *not* act on different factors of a Hilbert space with tensor product structure (as they did in the case of classical fields), but they still all commute. The reason for this is that $\boldsymbol{A}(\boldsymbol{x})$ and $\boldsymbol{A}(\boldsymbol{y})$ commute for all $\boldsymbol{x}$ and $\boldsymbol{y}$, as was explained after Eq. (11.2.3). Also $D_{0,i}$ clearly commutes with $\boldsymbol{A}(\boldsymbol{x}_j)$ for $i \neq j$. Because of this commutativity, it is possible to define the joint positive spectral subspace. That is to say, all the projections Λ_i^+ onto the positive spectral subspace of $D_{\boldsymbol{A},i}$ commute with each other, and hence $\prod_{i=1}^N \Lambda_i^+$, which does not depend on the order in which the product is taken, is again a projection.

It thus makes sense to restrict the Hamiltonian to the joint positive spectral subspace of all the $D_{\boldsymbol{A},i}$, given by $(\prod_{i=1}^N \Lambda_i^+)([\bigwedge^N L^2(\mathbb{R}^3; \mathbb{C}^4)] \otimes \mathcal{F})$. This subspace is not trivial and, in fact, infinite dimensional, as shown in [119,

Appendices C and D]. The resulting model is the quantized version of the modified Brown–Ravenhall model discussed in Chapter 10. The physical Hilbert space for this model no longer has a tensor product structure, since there is only one photon Hilbert space (the Fock space) which is shared by the N electrons. If each electron had its private Fock space, things would be simpler. *The electrons are intimately connected to the photons and can not be separated from them.* In other words, in this model there is no such thing, even conceptually, as an 'undressed electron'.

It is also possible to define the quantized analogue of the original Brown–Ravenhall model, where the physical Hilbert-space for the electrons is taken to be the positive spectral subspace of all the $D_{0,i}$. In this case the physical Hilbert space does have a tensor product structure, but this simplification comes with a price; the model is unstable.

The conclusions concerning the stability of these two models with quantized electromagnetic field are essentially the same as in the case of classical fields discussed in the previous chapter, except for a modification of some of the constants in the stability bounds.

The following stability criterion for the modified Brown–Ravenhall model is proved by Lieb and Loss in [119]. It should be compared with the non-quantized version in Theorem 10.1.

Theorem 11.2 (Stability of Modified B–R Model with Quantized Field). *Let* $\kappa = \max\{64.5, \pi Z\}$ *and* $\eta = \min\{4(\kappa\alpha)^2,\ 1\}$. *Assume that* $(\kappa\alpha)^2 < \eta$ *and*

$$[\,\eta - (\kappa\alpha)^2\,]^{3/2} \geq (0.060)8\pi\,\eta^2\alpha.$$

Let H *be the Hamiltonian (11.3.1). Then*

$$(\psi, H\psi) \geq \sqrt{1-\eta}\,N - \frac{18}{\pi}\Lambda\,N^{3/4}M^{1/4}\left(2\pi\frac{6\sqrt{\eta} + (\alpha/2)(\sqrt{2Z}+2.3)^2}{27}\right)^{3/4}$$

for all $\psi \in (\prod_{i=1}^{N}\Lambda_i^+)([\bigwedge^N L^2(\mathbb{R}^3;\mathbb{C}^4)]\otimes\mathcal{F})$ *with* $(\psi,\psi) = 1$. *In particular,* $Z \leq 42$ *is allowed for* $\alpha = 1/137$.

Although the proof of Theorem 11.2 in [119] does not touch upon concepts not already seen in the previous chapters, it is rather lengthy and technical. We shall not give it here, but refer the interested reader to the original work in [119] instead.

The fact that there is always instability for the original Brown–Ravenhall model, as in Theorem 10.2 in the case of classical fields, was proved in [81] by Griesemer and Tix. They were able to reduce the problem to the non-quantized case by judicious use of **coherent states** of the field. These are states in Fock space that are eigenstates of the annihilation operators. That is, for any complex valued function $\eta(\boldsymbol{k}, \lambda)$, there is a coherent state $\Psi_\eta \in \mathcal{F}$ such that $a(\boldsymbol{k}, \lambda)\Psi_\eta = \eta(\boldsymbol{k}, \lambda)\Psi_\eta$. In particular, $(\Psi_\eta, \boldsymbol{A}(\boldsymbol{x})\,\Psi_\eta) = \boldsymbol{A}_{\rm cl}(\boldsymbol{x})$ for all $\boldsymbol{x} \in \mathbb{R}^3$ and

$$\left(\Psi_\eta, H_f\,\Psi_\eta\right) = \int_{\mathbb{R}^3} \left[2\pi \boldsymbol{\Pi}_{\rm cl}(\boldsymbol{x})^2 + \frac{1}{8\pi}\boldsymbol{B}_{\rm cl}(\boldsymbol{x})^2\right] \mathrm{d}\boldsymbol{x},$$

where $\boldsymbol{A}_{\rm cl}$, $\boldsymbol{B}_{\rm cl}$ and $\boldsymbol{\Pi}_{\rm cl}$ are the classical fields defined in (11.1.25)–(11.1.27) with $\eta(\boldsymbol{k}, \lambda)$ in place of $a(\boldsymbol{k}, \lambda)$. The problem is thus reduced to the case of classical fields.

Because of the presence of the ultraviolet cutoff, however, the model is actually stable of the first kind. One way to see this is to use the operator inequality

$$\boldsymbol{A}(\boldsymbol{x})^2 \le \frac{8\Lambda}{3\pi} H_f + \frac{2}{\pi}\Lambda^2 \quad \text{for all } \boldsymbol{x} \in \mathbb{R}^3, \tag{11.3.2}$$

which was proved in [119, App. B]. Its proof is similar to the proof of Lemma 11.1 using Schwarz's inequality. Using operator monotonicity of the square root (cf. the footnote on page 87), it follows from this that

$$\sum_{i=1}^N \boldsymbol{\alpha}_i \cdot \boldsymbol{A}(\boldsymbol{x}_i) \ge -\sum_{i=1}^N |\boldsymbol{A}(\boldsymbol{x}_i)| \ge -N\sqrt{\frac{8\Lambda}{3\pi}H_f + \frac{2}{\pi}\Lambda^2}$$

and hence

$$H_f + \sqrt{\alpha}\sum_{i=1}^N \boldsymbol{\alpha}_i \cdot \boldsymbol{A}(\boldsymbol{x}_i) \ge -\Lambda\left(\frac{3}{4} + \frac{2\alpha}{3\pi}N^2\right). \tag{11.3.3}$$

The remaining part of the Hamiltonian is $\sum_{i=1}^N D_{0,i}$, which is positive by definition. Even if one adds the Coulomb potential $V_C(\underline{\boldsymbol{X}}, \underline{\boldsymbol{R}})$ the model is stable of the first kind for small enough values of α and $Z\alpha$; this follows from Theorem 10.1.

The model is always *unstable of the second kind*, however. This is our second example of stability of the first but not the second kind. (The first was non-relativistic charged bosons with $E_0 \approx -N^{5/3}$.) In fact, by proceeding as in the proof of Theorem 10.2 we see that even with an ultraviolet cutoff one can make the energy as negative as $-C\alpha\Lambda N^2$ for some constant $C > 0$. That is, the lower bound (11.3.3) is sharp except for the constant. (The fact that the bound is linear in Λ follows from dimensional analysis. The only inverse length in the problem, besides Λ, is the mass, which can without loss of generality be set equal to zero when analyzing stability, since the βm term can raise or lower the energy at most by mN.) We leave the details to the reader.

The various conclusions are summarized in the following tables borrowed from [119].

Electrons defined by projection onto the positive subspace of D_0, the free Dirac operator

	Classical or quantized field without cutoff Λ $\alpha > 0$ but arbitrarily small.	Classical or quantized field with cutoff Λ $\alpha > 0$ but arbitrarily small.
Without Coulomb potential αV_C	Instability of the first kind	Stability of the first kind. Instability of the second kind
With Coulomb potential αV_C	Instability of the first kind	Instability of the second kind. Stability of the first kind when both α and $Z\alpha$ are small enough

Electrons defined by projection onto the positive subspace of $D_{\mathbf{A}}$, the Dirac operator with field

	Classical field with or without cutoff Λ or quantized field with cutoff Λ
Without Coulomb potential αV_C	The Hamiltonian is positive
With Coulomb potential αV_C	Instability of the first kind when either α or $Z\alpha$ is too large
	Stability of the second kind when both α and $Z\alpha$ are small enough

CHAPTER 12

The Ionization Problem, and the Dependence of the Energy on N and M Separately

12.1 Introduction

The results in the preceding chapters on stability of the second kind were mostly of the form $E_0 > -C(N+M)$, where E_0 denotes the ground state energy of the system, and N and M are the number of electrons and nuclei, respectively. It is obvious, however, that if N is very large or very small compared to M the excess number of particles, positive or negative as the case may be, will float away to infinity. In other words it ought to be possible to reformulate the previous results as $E_0 > -C' \min\{N, M\}$ for a suitable C' that depends only on the nuclear-electron charge ratio Z.

In the relativistic case discussed in Chapter 8, the energy is actually non-negative for suitable α and Z, independently of N and M. From this we conclude that also the non-relativistic energy can be bounded below by $E_0 > -CN$ independent of M, as discussed in Remark 8.6. (For an alternative method, see [83, Thm. 3 of Part II].) Also the results in Chapters 9 and 10 yielded bounds of this form, since they rely in an essential way on the non-negativity of the relativistic energy in Chapter 8. This answers half the problem, namely it gives a bound on the energy of the correct form if M is larger than N.

In this chapter we shall deal with the other half of this problem, namely we shall show that for many models the energy can be bounded from below by M independently of N. More precisely, we shall show that if $N > 2Z_{\text{tot}} + M$, where $Z_{\text{tot}} = \sum_{k=1}^{M} Z_k$ is the total nuclear charge, one can remove the excess electrons without raising the energy. In other words, a system consisting of M nuclei having charges Z_k cannot bind more than $2Z_{\text{tot}} + M$ electrons; the remaining electrons will move off to infinity. Consequently a bound of the form $E_0(N, M) > -C(N+M)$ implies that $E_0(N, M) \geq E_0(2Z_{\text{tot}} + M, M) > -2C(Z_{\text{tot}} + M)$ independently of N. While the bound $2Z_{\text{tot}} + M$ is far from what is expected (namely $Z_{\text{tot}} + \text{const.}\, M$, see Section 12.3) it is adequate for our purposes.

12.2 Bound on the Maximum Ionization

By the word **ionization** we mean **negative ionization**, namely N is such that

$$\boxed{\mathcal{I} := N - Z_{\text{tot}} > 0.} \tag{12.2.1}$$

(For fixed nuclear positions and charges there is no bound on the possible positive ionization other than $-\mathcal{I} \leq Z_{\text{tot}}$, i.e., when $N = 0$.) We say that N electrons can be **bound** if $E_0(N) < E_0(N-1)$, i.e., it is necessary to use some energy to move one electron to infinity. The following upper bound on the number of electrons that can be bound by a collection of nuclei applies to several of the models considered in this book, but not all of them.

Specifically, we shall consider the cases in which the kinetic energy of the electrons is given by one of the two forms:

- **Non-relativistic**

$$T = \frac{1}{2}(\boldsymbol{p} + \sqrt{\alpha}\boldsymbol{A}(\boldsymbol{x}))^2 \tag{12.2.2}$$

- **Relativistic**

$$T = \sqrt{(\boldsymbol{p} + \sqrt{\alpha}\boldsymbol{A}(\boldsymbol{x}))^2 + m^2} - m \tag{12.2.3}$$

  for some $m > 0$.

As before, $\boldsymbol{A}$ denotes a (classical) magnetic vector potential. Different functions of $(\boldsymbol{p} + \sqrt{\alpha}\boldsymbol{A})^2$ could also be considered as kinetic energies, but we shall not do so here. The cases we do not know how to handle are the Pauli and Dirac operators studied in Chapters 9 and 10.[1]

[1] In [161] the maximal ionization of atoms described by the Pauli Hamiltonian is studied, and it is shown that a $\boldsymbol{B}$ dependent upper bound can be obtained. In contrast, the upper bound obtained here for the cases (12.2.2) and (12.2.3) is independent of $\boldsymbol{B}$.

The many-body Hamiltonian for N electrons and M nuclei is

$$\boxed{H_N = \sum_{i=1}^{N} T_i + \alpha V_C(\underline{\boldsymbol{X}}, \underline{\boldsymbol{R}}),} \tag{12.2.4}$$

with V_C given in (2.1.21). It acts on wave functions $\psi \in \bigwedge^N L^2(\mathbb{R}^3; \mathbb{C}^q)$ that are antisymmetric functions of the space-spin variables. This antisymmetry requirement is not important in the following, however, because the results in this chapter apply equally well to bosons.

We shall consider the nuclei to be point-like and fixed.[2] The bound on the number of electrons that can be bound to the nuclei will actually hold for all nuclear configurations, not merely the one minimizing the energy. Similarly, we shall consider the magnetic vector potential $\boldsymbol{A}$ to be fixed. Hence we have omitted the magnetic field energy $\mathcal{E}_{\text{mag}}(\boldsymbol{B})$ in (12.2.4), which is not relevant to the question addressed here.

The main result of this chapter is the following.

Theorem 12.1 (Bound on the Maximum Ionization). *Let $E_0(N)$ denote the ground state energy of H_N in (12.2.4). If $E_0(N) < E_0(N-1)$ then $N < 2Z_{\text{tot}} + M$.*

Part of the celebrated **HVZ Theorem**, going back to the work of Hunziker, van Winter and Zhislin [95, 183, 190], states that under the assumption $E_0(N) < E_0(N-1)$ there really is an eigenvalue at the bottom of the spectrum of H_N. That is, there is a $\psi \in L^2$ such that $H_N\psi = E_0(N)\psi$. Since it will be useful later, we state this as a separate lemma.[3]

Lemma 12.1 (Part of the HVZ Theorem). *Assume that $E_0(N) < E_0(N-1)$. Then there exists a ground state eigenfunction of H_N in (12.2.4).*

The lemma is intuitively very obvious, but a rigorous proof is lengthy and complicated. We refer to [150] or [82].[4] Lemma 12.1 is useful in the proof

---

[2] For results on the model with dynamic nuclei and/or smeared out nuclei, see [114].

[3] The full HVZ Theorem states that the essential spectrum of H_N starts at the ground state energy of H_{N-1}. See [178] or [150, Vol. 4]. In particular, if $E_0(N) < E_0(N-1)$, then $E_0(N)$ is below the essential spectrum and, therefore, is an eigenvalue.

[4] In the relativistic case, Lemma 12.1 was proved in [106]. See also [140, 143] for the case of the Brown–Ravenhall model.

of Theorem 12.1, but is not absolutely necessary. At the end of the proof of Theorem 12.1, we shall show how to avoid its use. We do this in order to give a self-contained proof of Theorem 12.1, since we do not prove Lemma 12.1 here.

Theorem 12.1 actually holds under the weaker assumption that there exists an eigenfunction corresponding to the lowest eigenvalue $E_0(N)$. That is, $N < 2Z_{\text{tot}} + M$ holds in case a ground state eigenfunction exists, even in case $E_0(N) = E_0(N-1)$. The theorem states that the maximal number of electrons that can be bound by a collection of nuclei is strictly smaller that $2Z_{\text{tot}} + M$. In particular H^{--} (a system of one proton and three electrons) does not exist (at least in the approximation in which the nucleus is static). On the other hand, it is a theorem of Zhislin [190] that if $N < Z_{\text{tot}} + 1$ then the system is bound. For completeness, we shall give a proof of this statement later in Theorem 12.2.

If we assume that $\boldsymbol{B}(\boldsymbol{x})$ decays to zero at infinity, then clearly $E_0(N) \leq E_0(N-1)$ for all N since we can always place the additional electron arbitrarily far from the nuclei with infinitesimal energy cost. The strictness of the inequality is the crucial criterion for whether the system is bound or not.

For simplicity, we shall prove Theorem 12.1 only in the case that all the nuclear charges are equal. We refer to [114] for the general case. For the application of Theorem 12.1 to the question of bounding the ground state energy E_0 from below by $-C\min\{N, M\}$, as discussed in the introduction to this chapter, we can safely consider only the case in which all the nuclear charges Z_j are equal to some common value Z. This was pointed out in Subsection 3.2.3, where it was shown that for any given configuration of nuclei positions and $Z_j \in [0, Z]$, the lowest ground state energy is obtained when, for each j, Z_j is either Z or zero.

Proof of Theorem 12.1. First, let us prove the theorem under the assumption that there is an eigenfunction ψ of H_N corresponding to the eigenvalue $E_0(N)$. This assumption is the content of Lemma 12.1 above but, since we do not include its proof here, we shall explain how to avoid the use of this assumption at the end of this proof.

The Schrödinger equation for ψ can be written as

$$\begin{aligned} 0 &= (H_N - E_0(N))\psi \qquad (12.2.5) \\ &= \left(H_{N-1} + T_N - \sum_{k=1}^{M} \frac{Z_k \alpha}{|\boldsymbol{x}_N - \boldsymbol{R}_k|} + \sum_{i=1}^{N-1} \frac{\alpha}{|\boldsymbol{x}_i - \boldsymbol{x}_N|} - E_0(N) \right) \psi. \end{aligned}$$

Pick a strictly positive function ϕ on $\mathbb{R}^3$ and multiply (12.2.5) by $\overline{\psi}(z_1, \ldots, z_N)/\phi(\boldsymbol{x}_N)$. Recall that $z = (\boldsymbol{x}, \sigma)$ denotes space-spin variables, and that ψ is antisymmetric in these variables. Since $H_{N-1} \geq E_0(N-1) > E_0(N)$, the operator $H_{N-1} - E_0(N)$ is non-negative on antisymmetric functions of the variables $z_1, \ldots, z_{N-1}$. Since it does not affect the z_N variable, this implies that

$$\int \overline{\psi}(\underline{z}) \frac{1}{\phi(\boldsymbol{x}_N)} ((H_{N-1} - E_N)\psi)(\underline{z}) \mathrm{d}\underline{z} \geq 0. \tag{12.2.6}$$

Moreover, if ϕ is **superharmonic** (that is, $\Delta\phi \leq 0$), the term with T_N is positive in view of the following lemma, which was inspired by unpublished work of Benguria (see [113, p. 632]).

Lemma 12.2 (Positivity Property of *T*). *Let T be either one of the operators (12.2.2) or (12.2.3). If ϕ is a non-negative superharmonic function, then*

$$\frac{1}{\phi} T + T \frac{1}{\phi} \geq 0$$

in the sense of quadratic forms. Equivalently, for any $\psi \in L^2(\mathbb{R}^3)$, $\mathrm{Re}\,(\psi/\phi, T\psi) \geq 0$.

Technicalities concerning the domain of definition of the operators shall be blithely ignored here. We refer the interested to reader to [114].

We postpone the proof of this lemma, and conclude the proof of Theorem 12.1 first. After integration of all the z_i we obtain from (12.2.5) and (12.2.6), with the aid of Lemma 12.2 (and the fact that the real part of zero is zero) that

$$\int |\psi(\underline{z})|^2 \frac{1}{\phi(\boldsymbol{x}_N)} \left(-\sum_{k=1}^{M} \frac{Z_k}{|\boldsymbol{x}_N - \boldsymbol{R}_k|} + \sum_{i=1}^{N-1} \frac{1}{|\boldsymbol{x}_i - \boldsymbol{x}_N|} \right) \mathrm{d}\underline{z} \leq 0.$$

Since $|\psi|^2$ is symmetric in all the variables, this can be written as

$$\int |\psi(\underline{z})|^2 \left(-\frac{1}{\phi(\boldsymbol{x}_N)} \sum_{k=1}^{M} \frac{Z_k}{|\boldsymbol{x}_N - \boldsymbol{R}_k|} \right. \\ \left. + \frac{1}{2} \sum_{i=1}^{N-1} \left(\frac{1}{\phi(\boldsymbol{x}_i)} + \frac{1}{\phi(\boldsymbol{x}_N)} \right) \frac{1}{|\boldsymbol{x}_i - \boldsymbol{x}_N|} \right) \mathrm{d}\underline{z} \leq 0. \tag{12.2.7}$$

First, consider the case of an atom, i.e., $M = 1$ and $\boldsymbol{R}_1 = 0$. We choose $\phi(\boldsymbol{x}) = 1/|\boldsymbol{x}|$. By the triangle inequality, $1/\phi(\boldsymbol{x}_i) + 1/\phi(\boldsymbol{x}_N) = |\boldsymbol{x}_i| + |\boldsymbol{x}_N| \geq$

$|\boldsymbol{x}_i - \boldsymbol{x}_N|$, and this inequality is strict except on a subset of $\mathbb{R}^3 \times \mathbb{R}^3$ of measure zero. Hence (12.2.7) yields

$$-Z + \frac{1}{2}(N-1) < 0,$$

or $N < 2Z + 1$, as claimed.

Next, consider the molecular case with all nuclear charges taken to be equal, i.e., $Z_k = Z$ for all $k = 1, \ldots, M$. We choose ϕ to be $\phi(\boldsymbol{x}) = \sum_k |\boldsymbol{x} - \boldsymbol{R}_k|^{-1}$. Then

$$\frac{1}{\phi(\boldsymbol{x}_i)} + \frac{1}{\phi(\boldsymbol{x}_N)} = \frac{\phi(\boldsymbol{x}_i) + \phi(\boldsymbol{x}_N)}{\phi(\boldsymbol{x}_i)\phi(\boldsymbol{x}_N)}$$
$$= \sum_{k=1}^{M} g_k(\boldsymbol{x}_i) g_k(\boldsymbol{x}_N) \left(|\boldsymbol{x}_i - \boldsymbol{R}_k| + |\boldsymbol{x}_N - \boldsymbol{R}_k|\right),$$

where we let $g_k(\boldsymbol{x}) = 1/(\phi(\boldsymbol{x})|\boldsymbol{x} - \boldsymbol{R}_k|)$. Again, by the triangle inequality, $|\boldsymbol{x}_i - \boldsymbol{R}_k| + |\boldsymbol{x}_N - \boldsymbol{R}_k| \geq |\boldsymbol{x}_i - \boldsymbol{x}_N|$. Using the symmetry of $|\psi|^2$, (12.2.7) then becomes

$$\int |\psi(\underline{z})|^2 \left(\sum_{k=1}^{M} \sum_{i \neq j} g_k(\boldsymbol{x}_i) g_k(\boldsymbol{x}_j) \right) \mathrm{d}\underline{z} < 2NZ.$$

The strictness of the inequality follows again from the fact that the above triangle inequality is strict except on a set of measure zero.

For given k, we write $\sum_{i \neq j} g_k(\boldsymbol{x}_i) g_k(\boldsymbol{x}_j) = |\sum_i g_k(\boldsymbol{x}_i)|^2 - \sum_i g_k(\boldsymbol{x}_i)^2$. To estimate the latter term from below, we can use the fact that $\sum_k g_k(\boldsymbol{x}_i) = 1$ to conclude that $\sum_k \sum_i g_k(\boldsymbol{x}_i)^2 \leq N$. For the first term, we use Schwarz's inequality

$$\left| \sum_i g_k(\boldsymbol{x}_i) \right|^2 \geq 2\frac{N}{M} \sum_i g_k(\boldsymbol{x}_i) - \frac{N^2}{M^2}.$$

Summing over k and using again that $\sum_k g_k(\boldsymbol{x}_i) = 1$, we conclude that

$$\sum_{k=1}^{M} \left| \sum_i g_k(\boldsymbol{x}_i) \right|^2 > \frac{N^2}{M}.$$

In combination, we have thus shown that

$$\frac{N^2}{M} - N < 2ZN,$$

or $N < M(2Z + 1)$, which is the desired bound.

The case of unequal nuclear charges is slightly more complicated, and we shall not give the details here, but refer the reader to [114]. It turns out that the natural choice $\phi(\boldsymbol{x}) = \sum_k Z_k/|\boldsymbol{x} - \boldsymbol{R}_k|$ does *not work*, however. One rather has to pick suitable positive numbers μ_k and choose $\phi(\boldsymbol{x}) = \sum_k \mu_k/|\boldsymbol{x} - \boldsymbol{R}_k|$. For a suitable choice of these numbers, one can then show that $N < 2\sum_k Z_k + M$.

As promised, we shall now relax the assumption that there exists an eigenfunction of H_N with eigenvalue $E_0(N)$. By hypothesis, $E_0(N) \leq E_0(N-1) - 3\varepsilon$ for some $\varepsilon > 0$. By definition of the ground state energy, there as a ψ such that $(\psi, H_N\psi) \leq E_0(N-1) - 2\varepsilon$. Since infinitely differentiable functions of compact support are dense in H^1 and $H^{1/2}$ [118, Thms. 7.6, 7.14 & 7.22], we can find a wave function ψ_ε supported in the set $|\boldsymbol{x}_i| \leq R_\varepsilon$ for suitable $R_\varepsilon > 0$ such that $(\psi_\varepsilon, H_N\psi_\varepsilon) < E_0(N-1) - \varepsilon$. Hence the Hamiltonian H_N restricted to such functions (that is, functions supported in a ball of radius R_ε with Dirichlet boundary conditions) has a ground state energy strictly less than $E_0(N-1)$. We can thus repeat the above proof with ψ being the ground state wave function in a box (which always exists) and arrive at the same conclusion. ■

It remains to prove Lemma 12.2, which we shall do next. The proof is taken from [114]; for an alternative proof, see [96].

Proof of Lemma 12.2. Consider first the non-relativistic case (12.2.2). For $\psi \in L^2(\mathbb{R}^3)$ and $f(\boldsymbol{x}) = \psi(\boldsymbol{x})/\phi(\boldsymbol{x})$, we can write

$$(\psi/\phi, T\,\psi) = \frac{1}{2}\int_{\mathbb{R}^3} \left(i\nabla\overline{f(\boldsymbol{x})} + \sqrt{\alpha}\boldsymbol{A}(\boldsymbol{x})\overline{f(\boldsymbol{x})}\right)$$

$$\times\left(-i\phi(\boldsymbol{x})\nabla f(\boldsymbol{x}) - if(\boldsymbol{x})\nabla\phi(\boldsymbol{x}) + \sqrt{\alpha}\boldsymbol{A}(\boldsymbol{x})\phi(\boldsymbol{x})f(\boldsymbol{x})\right)\mathrm{d}\boldsymbol{x}. \tag{12.2.8}$$

With the aid of partial integration one checks that this equals

$$\frac{1}{2}\int_{\mathbb{R}^3} \left(\phi(\boldsymbol{x})| - i\nabla f(\boldsymbol{x}) + \sqrt{\alpha}\boldsymbol{A}(\boldsymbol{x})f(\boldsymbol{x})|^2\right.$$

$$\left. - |f(\boldsymbol{x})|^2(\Delta\phi(\boldsymbol{x}) + i\sqrt{\alpha}\boldsymbol{A}(\boldsymbol{x})\nabla\phi(\boldsymbol{x}))\right)\mathrm{d}\boldsymbol{x}.$$

The last term drops out when the real part is taken, since both $\boldsymbol{A}$ and ϕ are real. The other terms are positive by our assumptions that $\phi(\boldsymbol{x}) \geq 0$ and $\Delta\phi(\boldsymbol{x}) \leq 0$. This proves that $\mathrm{Re}\,(\psi/\phi, T\,\psi) \geq 0$ in the non-relativistic case.

To prove the lemma in the relativistic case, we start with the integral representation

$$(B+1)^{1/2} - 1 = \frac{B}{\pi} \int_1^\infty (t-1)^{1/2} \frac{1}{t(t+B)} \mathrm{d}t.$$

The lemma thus follows if we can show that

$$\mathrm{Re}\,(\psi/\phi, \frac{B}{t+B}\psi) \geq 0 \tag{12.2.9}$$

for all $t \geq 1$, and with $B = (\boldsymbol{p} + \sqrt{\alpha}\boldsymbol{A})^2$ being twice the non-relativistic kinetic energy. To see (12.2.9), let $g = (t+B)^{-1}\psi$. Then

$$\left(\psi/\phi, \frac{B}{t+B}\psi\right) = t(g/\phi, B\,g) + \left(Bg, \frac{1}{\phi}\,Bg\right).$$

The last term is obviously positive, and the positivity of the real part of the first term was shown before.

For simplicity of presentation we have ignored issues concerning domains of operators here. The details are properly worked out in [114, Appendix A]. ■

12.3 How Many Electrons Can an Atom or Molecule Bind?

This question, although not directly related to the stability of matter, is a fascinating one, and we cannot resist the temptation to mention it here because, after many attempts, *it is still open!* The inequality $\mathcal{I} < Z_{\mathrm{tot}} + M$ proved above is far from optimal.

Numerical estimates and experimental observations of real atoms and molecules in nature leads one to believe that the maximum ionization, $\mathcal{I} := N - Z_{\mathrm{tot}}$ should be at most cM with c about 1, or possibly 2, irrespective of whether or not the nuclei are in their optimum locations. Atoms with $\mathcal{I} = 2$ are rare, if they exist at all.[5] No one has been able to prove anything resembling a theorem of this kind, except in the context of approximate theories, such as Hartree–Fock theory [171] or Thomas–Fermi type theories [8, 113].

This question is closely related to several others: Why are the radii of atoms (defined as the radius R such that $\int_{|x|>R} \varrho(\boldsymbol{x})\mathrm{d}\boldsymbol{x} = 1/2$, with ϱ the

[5] We are discussing atoms and molecules in vacuum, not in water, where c can be significantly larger, although still bounded.

one-particle density of the ground state) more or less independent of Z? Why is the ionization potential (given by $E_0(N = Z) - E_0(N = Z + 1) \geq 0$) more or less independent of Z? Why is the largest **ionization potential** among *all* atoms in the periodic table less than the smallest **electronegativity** (given by $E_0(N = Z - 1) - E_0(N = Z) > 0$) among *all* atoms? If this were not the case, neutral atoms with high ionization potential would be bad neighbors and would go around stealing electrons from innocent neutral neighbors.

The conjecture concerning the ionization potential and the electronegativity implies that $N \mapsto E_0(N)$ is convex at $N = Z$, i.e., $2E_0(N = Z) \leq E_0(N = Z - 1) + E_0(N = Z + 1)$. It has been further conjectured that $N \mapsto E_0(N)$ is convex for all N. This, in turn, implies the even more 'obvious', but unproved fact that if there is an eigenfunction for $E_0(N)$ then there is one for $E_0(N - 1)$. That is to say, if a nucleus can bind N electrons can it bind $N - 1$? All of these interesting questions are open ones, and we shall not discuss them further.

It is known that a theorem stating that $\mathcal{I} \sim M$ would have to take explicit account of the fact that electrons are fermions. Otherwise, if they were bosons, the ionization is as large as $\mathcal{I} \sim 0.21\, Z$ for an atom with large Z. It was shown by Benguria and Lieb [14], for an atom, that $\mathcal{I} \geq cZ$ as $Z \to \infty$ with c being determined by the solution to the Hartree equation, which is a certain non-linear differential equation. Baumgartner [9] showed numerically that this equation yields $c = 0.21$ and Solovej [170] found an upper bound which showed that $\mathcal{I} = 0.21Z$ is the correct asymptotic formula. This bosonic result shows that considerations of *electrostatics alone cannot solve the problem. The Pauli exclusion principle is essential!*

On the other hand, the situation for fermions is not entirely hopeless. Prior to the proof of Theorem 12.1 Ruskai [154] had proved that $\mathcal{I} < cZ^{6/5}$ and Sigal [163] had proved that $\mathcal{I} < 18Z$ for atoms. Later, Lieb, Sigal, Simon and Thirring [128] showed that $\mathcal{I}/Z \to 0$ as $Z \to \infty$. Despite subsequent improved quantitative estimates by Seco, Sigal and Solovej [158] and by Fefferman and Seco [61] the problem remains basically unresolved. For realistic atoms the $2Z + 1$ bound of Theorem 12.1 is still the best so far. There is a long way to go.

Another positive note, for both fermions and 'bosonic electrons', is a theorem proved by Zhislin [190] which states that the number of electrons that can be bound is at least as big as the total nuclear charge. It was originally proved in the non-relativistic case without magnetic field, but it can be easily generalized in the following way.

Theorem 12.2 (Zhislin's Binding Condition). *Assume $\mathcal{E}_{\text{mag}}(\boldsymbol{B}) < \infty$. If $N < Z_{\text{tot}} + 1$ then the ground state energy $E_0(N)$ of the Hamiltonian given in (12.2.4) satisfies $E_0(N) < E_0(N-1)$. This holds both for fermions and for bosons. It also holds for all fixed locations of the nuclei, not only for the energy minimizing configurations.*

It can be shown that the conclusion of this theorem also holds for movable nuclei. By Lemma 12.1 the ground state of the N particle problem is an eigenfunction when $N < Z_{\text{tot}} + 1$. Thus, a physical atom (with Z_{tot} an integer) can bind at least Z_{tot} electrons. A slight modification of the proof shows that the N electron atom has infinitely many eigenfunctions (i.e., bound states) if $N < Z_{\text{tot}} + 1$.

Proof in the non-relativistic case. We shall give the proof in the case of fermions; the case of bosons is completely analogous. Our proof will make use of part of the HVZ Theorem, as formulated in Lemma 12.1. It is not as easy as in the proof of Theorem 12.1 to give a proof that does not use this lemma.

We shall prove that if $N < Z_{\text{tot}} + 1$ and if there is a ground state for $N-1$ particles, then $E_0(N) < E_0(N-1)$. For $N = 2$, we merely have to show that $E_0(1) < E_0(0) = \alpha U(\underline{\boldsymbol{R}})$ in order to ensure the existence of a ground state for the one-particle problem. (This follows from Lemma 12.1 above. In the one-particle case, the proof is actually much simpler; see [118, Thm. 11.5].) This inequality follows easily from the fact that one nucleus with an arbitrarily small positive charge can bind an electron. The existence of a ground state for successively larger N then follows by induction, using Lemma 12.1 at each step.

Hence we may assume that there exists a ground state of H_{N-1}, with energy $E_0(N-1)$, which we denote by ψ. We assume that ψ is normalized, i.e., $\|\psi\|_2 = 1$. We first show that for any $R > 0$ there exists a function $\psi_R \in \bigwedge^{N-1} L^2(\mathbb{R}^3; \mathbb{C}^q)$ supported in the set $|\boldsymbol{x}_i| \leq R$ for all $i = 1, \ldots, N-1$, with energy $(\psi, H_{N-1}\,\psi) \leq E_0(N-1) + c/R^2$ for some constant $c > 0$. For this purpose, we pick a smooth function χ on $\mathbb{R}^{3(N-1)}$ that is supported in the set $|\boldsymbol{x}_i| \leq R$ for all $1 \leq i \leq N-1$. It is straightforward to check that[6]

$$\begin{aligned} &\chi(\underline{\boldsymbol{X}})^2(\boldsymbol{p}_i + \sqrt{\alpha}\boldsymbol{A}(\boldsymbol{x}_i))^2 + (\boldsymbol{p}_i + \sqrt{\alpha}\boldsymbol{A}(\boldsymbol{x}_i))^2\chi(\underline{\boldsymbol{X}})^2 \\ &\quad = 2\chi(\underline{\boldsymbol{X}})(\boldsymbol{p}_i + \sqrt{\alpha}\boldsymbol{A}(\boldsymbol{x}_i))^2\chi(\underline{\boldsymbol{X}}) - 2|\nabla_i\chi|^2 \end{aligned} \tag{12.3.1}$$

[6] This is the same as the IMS localization method of (11.2.9).

for $i = 1, \ldots, N-1$. Hence

$$E_0(N-1)(\psi, \chi^2\psi) = \frac{1}{2}(\psi, \chi^2 H_{N-1}\psi) + \frac{1}{2}(\psi, H_{N-1}\chi^2\psi)$$
$$= (\chi\psi, H_{N-1}\chi\psi) - \sum_{i=1}^{N-1}(\psi, |\nabla_i\chi|^2\psi). \qquad (12.3.2)$$

We can pick χ such that $|\nabla_i\chi|^2 \leq c/(2R^2)$. Moreover, $(\psi, \chi^2\psi) \geq 1/2$ for large enough R. This proves that $\psi_R = \chi\psi/\|\chi\psi\|$ has the desired property.

Let $\phi_R \in L^2(\mathbb{R}^3; \mathbb{C}^q)$ be radial, normalized, and supported in the shell $R < |\boldsymbol{x}| < 2R$. We claim that we can choose ϕ_R in such a way that its kinetic energy satisfies $\lim_{R\to\infty}(\phi_R, T\phi_R)R = 0$. A possible choice is $\phi_R(z) = \delta_{\sigma,1}(2\pi R)^{-1/2}\sin(\pi|\boldsymbol{x}|/R)/|\boldsymbol{x}|$ for $R < |\boldsymbol{x}| < 2R$, and zero otherwise. To bound the kinetic energy, we can simply use $(\boldsymbol{p} + \sqrt{\alpha}\boldsymbol{A})^2 \leq 2(\boldsymbol{p}^2 + \alpha\boldsymbol{A}^2)$. The $\boldsymbol{p}^2$ term yields c/R^2. For the $\boldsymbol{A}^2$ term, we can use Hölder's inequality to estimate $(\phi_R, \boldsymbol{A}^2\phi_R) \leq cR^{-1}(\int_{R\leq|\boldsymbol{x}|\leq 2R}|\boldsymbol{A}|^6)^{1/3}$. If $\boldsymbol{A} \in L^6(\mathbb{R}^3)$ the latter integral goes to zero as $R \to \infty$. It is no restriction to assume that $\boldsymbol{A} \in L^6(\mathbb{R}^3)$ since this property holds in the Coulomb gauge, as shown in Lemma 10.1, and we are free to choose this gauge. This proves the claim.

Let Υ denote the N particle wave function obtained by antisymmetrizing the function $\psi_R(z_1, \ldots, z_{N-1})\phi_R(z_N)$, namely,

$$\Upsilon(\underline{z}) = \frac{1}{\sqrt{N}}\Bigg[\psi_R(z_1, \ldots, z_{N-1})\phi_R(z_N)$$
$$- \sum_{i=1}^{N-1}\psi_R(z_1, \ldots, z_{i-1}, z_N, z_{i+1}, \ldots, z_{N-1})\phi_R(z_i)\Bigg]. \qquad (12.3.3)$$

Using the locality[7] of H_N and the fact that ϕ_R and ψ_R have disjoint support, we observe that

$$(\Upsilon, H_N\Upsilon) = (\psi_R, H_{N-1}\psi_R) + (\phi_R, T\phi_R) - Z_{\text{tot}}\int_{\mathbb{R}^3}\frac{|\phi_R(z)|^2}{|\boldsymbol{x}|}\mathrm{d}z$$
$$+ \iint_{\mathbb{R}^3\times\mathbb{R}^3}\frac{|\phi_R(z)|^2\varrho_{\psi_R}(\boldsymbol{y})}{|\boldsymbol{x}-\boldsymbol{y}|}\mathrm{d}z\mathrm{d}\boldsymbol{y}.$$

[7] For the notion of locality, see the footnote on page 181.

Here, ϱ_{ψ_R} denotes the one-particle density of ψ_R. We have assumed that R is large enough such $R > |\boldsymbol{R}_k|$ for all the nuclear positions $\boldsymbol{R}_k$. Since ϕ_R is radial and supported in $|\boldsymbol{x}| > R$, this implies that $\int |\phi_R(z)|^2/|\boldsymbol{x} - \boldsymbol{R}_k|\mathrm{d}z = \int |\phi_R(z)|^2/|\boldsymbol{x}|\mathrm{d}z$ by Newton's Theorem. Similarly, the last term equals $(N-1)\int |\phi_R(z)|^2/|\boldsymbol{x}|\mathrm{d}z$. Hence

$$(\Upsilon, H_N \Upsilon) \leq E_0(N-1) + \frac{N - 1 - Z_{\mathrm{tot}}}{2R} + (\phi_R, T\phi_R) + \frac{c}{R^2}.$$

Recall that $\lim_{R\to\infty} R(\phi_R, T\phi_R) = 0$. If $N < Z_{\mathrm{tot}} + 1$, we therefore see that $E_0(N) \leq (\Upsilon, H_N \Upsilon) < E_0(N-1)$ for large R. ■

Proof in the relativistic case. The strategy is the same as in the non-relativistic case. We shall briefly explain the main differences but leave the unilluminating details to the reader.

The localization formula (12.3.2) is more complicated in the relativistic case due to the non-locality of the kinetic energy. The localization error can still be bounded by c/R^2 independently of $\boldsymbol{A}$, as long as $R \gg 1/m$. (The heuristics behind this fact is that for momenta $|\boldsymbol{p}| \ll m$, the relativistic kinetic energy approximately equals the non-relativistic one.) This can be shown using the same strategy as in [72, Lemma B.1], where the case $m = 0$ was considered (in which case the localization error is $\sim 1/R$).

The kinetic energy of the one-particle function ϕ_R can be bounded in the same way as in the non-relativistic case, using the inequality $\sqrt{s + m^2} - m \leq s/(2m)$ for $s > 0$.

As will be pointed out in the proof of Theorem 14.2 in Chapter 14, page 265, locality of the one-body terms in the Hamiltonian is not necessary for the absence of cross-terms when computing the expectation value of the Hamiltonian in the antisymmetrized N-particle wave function. It is only needed for the two-body terms. The rest of the proof goes through without essential change. ■

CHAPTER 13

Gravitational Stability of White Dwarfs and Neutron Stars

13.1 Introduction and Astrophysical Background

Up to now we have been concerned with demonstrating the 'Triumph of Quantum Mechanics', which is that the electrostatic forces between the electrically charged particles that make up ordinary matter do not cause collapse (provided the maximum nuclear charge Z and the fine-structure constant α are not too large). The total energy is not only finite but it is proportional to the particle number. Electric and magnetic forces conspire to cancel out to a comfortable extent, but lurking in the background is a very, very much weaker force – gravity. This force is additive, however, and there can be no cancellation as there is for electric forces, but because it is so weak it becomes dominant only when N is very large – of the order of the number of particles in a star, which is about 10^{57}.

It was already realized shortly after the publication of Schrödinger's equation that a star would collapse under the influence of gravity if the kinetic energy of the particles is treated relativistically and if the number of constituent particles exceeds a certain critical value; this critical value depends on Planck's constant, h.

There are two kinds of stars to consider. One is a star made of electrically neutral particles called neutrons, and which is itself the residue of a collapsed star. Its mass is typically of the order of 1–2 solar masses, but gravity squeezes it to a radius of about 10–20 km. In contrast, the radius of our sun is roughly one million kilometers.

The second kind is an ordinary star, but which has burned out, i.e., its nuclear processes have finished. What is left is essentially a system of electrons and nuclei close to its ground state, as noted by Fowler [69]. The temperature of such stars can still be high enough for them to twinkle brightly, which is why we can see them optically and which is why they are called 'white'. But although their masses are comparable to the mass of the sun they are tiny (about the size of Earth, but still much bigger than a neutron star) which is why they are called 'dwarfs'. Anderson [3] and Stoner [175] first noted that there is a maximum possible mass of such a star now called the 'Chandrasekhar mass' after Chandrasekhar who put the calculation on a firm footing [30]. Beyond this mass the star collapses under its own gravity. Indeed, a white dwarf can slowly accrete matter from a companion star and, when its total mass exceeds about 1.4 solar masses, it collapses as a type Ia supernova.

Before studying the collapse, let us discuss, heuristically, the energy balance for non-relativistic gravitating matter. If we have N fermions of mass m with non-relativistic kinetic energy confined to a ball of radius R, the kinetic energy will be approximately $N^{5/3}/(R^2 m)$, as we saw in Chapter 4. The gravitational energy would be roughly $-\kappa N^2/R$, where

$$\kappa = Gm^2$$

is the effective interaction constant and G is Newton's gravitational constant (3.2.10). The total energy is minimized when $R \sim N^{-1/3}/(m\kappa)$. Owing to the negative exponent $-1/3$, 'a large star is smaller than a small star'. For us the important point is that the energy is always finite and there is no collapse, in the sense that the system is stable of the first kind. The system is not stable of the second kind, since the energy grows like $N^{7/3}$ for large N [105]. For neutron stars, we can set m equal to the mass of neutrons which is much larger than the mass of electrons.

The reason that white dwarfs are much larger than neutron stars is that there are two masses to consider in the former case. There is the mass μ of the nuclei (which is similar to the mass m_n of a neutron) so that $\kappa = G\mu^2$. The other is the electron mass m, so that the dominant kinetic energy in the non-relativistic regime is $N^{5/3}/(mR^2)$. This is much larger than the kinetic energy of N neutrons, $N^{5/3}/(m_n R^2)$. The balance is now when $R \sim N^{-1/3}/(Gm^2 m_n)$. In Section 13.2 we address the one-mass case. The two-mass case is discussed in Section 13.3.3.

Now let us discuss the effect of special relativity, which becomes important for such dense systems. The kinetic energy $\boldsymbol{p}^2/(2m)$ of the particles has to be replaced by $\sqrt{\boldsymbol{p}^2 + m^2} - m$. For large N, the kinetic energy of a collection of fermions confined to a ball of radius R will now only be $N^{4/3}/R$, independent of $m > 0$. The potential energy estimate given above remains the same and, therefore, the system will collapse if N is bigger than roughly $\kappa^{-3/2}$. This critical particle number depends on m only through κ and not through the kinetic energy $\sqrt{\boldsymbol{p}^2 + m^2} - m$. The value of m in the kinetic energy is important, however, for calculating various properties of stars with mass below the critical one, such as the radius.

The question we address here is whether $\kappa^{-3/2}$ is really the correct critical particle number for collapse. The collapse could conceivably occur with a smaller N, in principle, because of the short-distance singularity of the interparticle gravitational potential $-\kappa/|\boldsymbol{x}_i - \boldsymbol{x}_j|$. In fact, it does not do so. The heuristic discussion given above turns out to give the correct answer.

13.2 Stability and Instability Bounds

In this section, we consider the simplest model of gravitating particles, in which there is only one species of particles of mass m, which are either bosons or fermions. As discussed in the introduction, it will be important that the kinetic energy of the particles is relativistic. Since $|\boldsymbol{p}| - m \leq \sqrt{\boldsymbol{p}^2 + m^2} - m \leq |\boldsymbol{p}|$, the stability analysis, i.e., the question of the number of particles needed for collapse, does not depend on the factor m in the *kinetic* energy (but m is important in the potential energy), and hence we will simply use $|\boldsymbol{p}|$ as the kinetic energy of a particle of momentum $\boldsymbol{p}$ in the following. The basic Hamiltonian is then

$$\boxed{H = \sum_{i=1}^{N} |\boldsymbol{p}_i| - \kappa \sum_{1 \leq i < j \leq N} \frac{1}{|\boldsymbol{x}_i - \boldsymbol{x}_j|}} \tag{13.2.1}$$

with $\kappa = Gm^2 > 0$. By scaling, the ground state energy E_0 of H is either 0 or $-\infty$.

Theorem 13.1 (Stability of Gravitating Matter). *The ground state energy of the Hamiltonian (13.2.1) for N fermions with q spin states is, for small* κ*,*

$$\boxed{E_0 = \begin{cases} 0 & \text{if } N \leq 0.594\,\kappa^{-3/2} q^{-1/2} + o(\kappa^{-3/2}) \\ -\infty & \text{if } N \geq (9/2) 3^{3/4} \pi\, \kappa^{-3/2} q^{-1/2} + o(\kappa^{-3/2}). \end{cases}}$$

For bosons,

$$\boxed{E_0 = \begin{cases} 0 & \text{if } N \leq 1 + 4\kappa^{-1}/\pi \\ -\infty & \text{if } N \geq 1 + 128\kappa^{-1}/(15\pi). \end{cases}}$$

This theorem goes back to Lieb and Thirring [136], and the proof we give will follow a similar strategy, with some simplification using results from Chapter 8.

The theorem shows that the critical particle number scales as $\kappa^{-3/2}$ for small κ in the fermionic case, and as κ^{-1} in the bosonic case. It is actually possible to determine the precise asymptotic behavior of the critical particle number as $\kappa \to 0$. This will be explained in the next section.

Proof. Bosons. Upper Bound. For bosons, we simply take a product trial function of the form $\prod_{i=1}^{N} \phi(\boldsymbol{x}_i)$ for some $\phi \in H^{1/2}(\mathbb{R}^3)$. For such a wave function, the energy equals

$$N(\phi, |\boldsymbol{p}|\,\phi) - \kappa \frac{N(N-1)}{2} \iint\limits_{\mathbb{R}^3\times\mathbb{R}^3} \frac{|\phi(\boldsymbol{x})|^2 |\phi(\boldsymbol{y})|^2}{|\boldsymbol{x}-\boldsymbol{y}|} \mathrm{d}\boldsymbol{x}\mathrm{d}\boldsymbol{y}. \tag{13.2.2}$$

The energy is thus negative if

$$\kappa > \frac{2(\phi, |\boldsymbol{p}|\,\phi)}{N-1} \left(\iint\limits_{\mathbb{R}^3\times\mathbb{R}^3} \frac{|\phi(\boldsymbol{x})|^2 |\phi(\boldsymbol{y})|^2}{|\boldsymbol{x}-\boldsymbol{y}|} \mathrm{d}\boldsymbol{x}\mathrm{d}\boldsymbol{y} \right)^{-1}.$$

Once the energy is negative, it can be driven to $-\infty$ by scaling, as mentioned above. If we choose $\phi(\boldsymbol{x}) = \pi^{-1/2} \exp(-|\boldsymbol{x}|)$ as in the proof of Theorem 8.2, we find $(\phi, |\boldsymbol{p}|\phi) = 8/(3\pi)$ and $\int |\phi(\boldsymbol{x})|^2 |\phi(\boldsymbol{y})|^2 |\boldsymbol{x}-\boldsymbol{y}|^{-1} \mathrm{d}\boldsymbol{x}\mathrm{d}\boldsymbol{y} = 5/8$. The system is thus unstable if $\kappa > 128/(15\pi(N-1))$.

Bosons. Lower Bound. We write the Hamiltonian (13.2.1) as

$$H = \sum_{1\leq i\neq j\leq N} \left(\frac{1}{N-1} |\boldsymbol{p}_i| - \frac{\kappa}{2|\boldsymbol{x}_i - \boldsymbol{x}_j|} \right). \tag{13.2.3}$$

For given $\boldsymbol{x}_j$, $|\boldsymbol{p}_i| \geq (2/\pi)|\boldsymbol{x}_i - \boldsymbol{x}_j|^{-1}$, as shown in Chapter 8, Lemma 8.2. Hence $H \geq 0$ if $\kappa \leq (4/\pi)/(N-1)$.

Fermions. Upper Bound. As a trial function, we take the Slater determinant of the N lowest eigenfunctions of the Laplacian in a cube of side length L, with Dirichlet boundary conditions. The corresponding eigenvalues are given by $\pi^2 \boldsymbol{n}^2 L^{-2}$ with $\boldsymbol{n} \in \mathbb{N}^3$. Since $(\phi, |\boldsymbol{p}|\phi) \leq (\phi, \boldsymbol{p}^2\,\phi)^{1/2}$ for any ϕ with $(\phi, \phi) = 1$, the kinetic energy is bounded from above by

$$q \sum_{|\boldsymbol{n}|<K} |\boldsymbol{n}| \frac{\pi}{L}$$

where K is the smallest number such that the number of points $\boldsymbol{n} \in \mathbb{N}^3$ with $|\boldsymbol{n}| < K$ is bigger or equal to N/q. For large N, $K \approx (6N/\pi q)^{1/3}$, and hence

$$q \sum_{|\boldsymbol{n}|<K} |\boldsymbol{n}| \frac{\pi}{L} \approx \frac{3}{4} \left(\frac{6\pi^2}{q} \right)^{1/3} \frac{N^{4/3}}{L}.$$

(Compare with Section 4.2.) Since all the particles are inside a cubic box of side length L, $|\boldsymbol{x}_i - \boldsymbol{x}_j| \leq \sqrt{3}L$ for all $i \neq j$. Hence

$$-\kappa \sum_{1 \leq i < j \leq N} \frac{1}{|\boldsymbol{x}_i - \boldsymbol{x}_j|} \leq -\kappa \frac{N(N-1)}{2\sqrt{3}\, L}.$$

The total energy is thus negative under the assumption on N stated in the theorem.

Fermions. Lower Bound. If ψ is either symmetric or antisymmetric then, for $1 \leq L \leq N$ and $M = N - L$,

$$(\psi, H\psi) = \frac{N}{L}(\psi, \widetilde{H}\psi), \tag{13.2.4}$$

where

$$\widetilde{H} = \sum_{i=1}^{L} |\boldsymbol{p}_i| - \lambda \sum_{i=1}^{L} \sum_{j=1}^{M} \frac{1}{|\boldsymbol{x}_i - \boldsymbol{x}_j|} + \alpha \sum_{1 \leq j < k \leq M} \frac{1}{|\boldsymbol{x}_i - \boldsymbol{x}_j|}, \tag{13.2.5}$$

and where λ and α have to satisfy the constraint

$$\lambda LM - \frac{\alpha}{2} M(M-1) = \frac{\kappa}{2} L(N-1). \tag{13.2.6}$$

Equation (13.2.4) follows simply by counting the number of kinetic energy and potential energy terms, given that these terms do not depend on which particles are involved because of the symmetry assumption on the wave function ψ.

For simplicity, let us introduce the notation $\boldsymbol{R}_j = \boldsymbol{x}_{j+L}$ for $j = 1, \ldots, M$. There is no kinetic energy for the $\boldsymbol{R}$ particles in (13.2.5), hence we can regard them as fixed. We then derive a lower bound on $\widetilde{H}$ acting on totally antisymmetric functions of L variables only.

For fixed $\boldsymbol{R}_1, \ldots, \boldsymbol{R}_M$, we introduce again the Voronoi cells Γ_j, as in Section 5.2. Similarly, let $\Phi : \mathbb{R}^3 \to (0, \infty]$ denote the function which takes the value

$$\Phi(\boldsymbol{x}) = \sum_{k,\, k \neq j} \frac{1}{|\boldsymbol{x} - \boldsymbol{R}_j|} \qquad \text{for } \boldsymbol{x} \in \Gamma_j.$$

With $\mathfrak{D}(\boldsymbol{x}) = \min_j |\boldsymbol{x} - \boldsymbol{R}_j|$ as before, we can write

$$\widetilde{H} = \sum_{i=1}^{L} |\boldsymbol{p}_i| - \lambda \sum_{i=1}^{L} \left(\Phi(\boldsymbol{x}_i) + \frac{1}{\mathfrak{D}(\boldsymbol{x}_i)} \right) + \alpha \sum_{1 \leq k < l \leq M} \frac{1}{|\boldsymbol{R}_k - \boldsymbol{R}_l|}.$$

The last term here is the same as in (5.2.5) for $Z = 1$.

Pick some $0 < \varepsilon < 1$ and split the kinetic energy as $|\boldsymbol{p}| = \varepsilon|\boldsymbol{p}| + (1-\varepsilon)|\boldsymbol{p}|$. On the space of antisymmetric functions, the operator

$$\sum_{i=1}^{L} \big((1-\varepsilon)|\boldsymbol{p}_i| - \lambda \Phi(\boldsymbol{x}_i) \big)$$

is bounded from below by q times the sum of the negative eigenvalues of $(1-\varepsilon)|\boldsymbol{p}| - \lambda\Phi(\boldsymbol{x})$. From the relativistic LT inequality in Theorem 4.2, together with the bound on the optimal constant in Eq. (4.1.23), it follows that

$$\sum_{i=1}^{L} \big((1-\varepsilon)|\boldsymbol{p}_i| - \lambda \Phi(\boldsymbol{x}_i) \big) \geq -0.0257 \frac{\lambda^4 q}{(1-\varepsilon)^3} \int_{\mathbb{R}^3} |\Phi(\boldsymbol{x})|^4 \mathrm{d}\boldsymbol{x}.$$

To bound the latter integral, we first note that for $\boldsymbol{x} \in \Gamma_j$ and $D_j = \min_{k \neq j} |\boldsymbol{R}_j - \boldsymbol{R}_k|/2$

$$\Phi(\boldsymbol{x}) = \sum_{k,\, k \neq j} \int_{\mathbb{S}^2} \frac{\mathrm{d}\boldsymbol{\Omega}}{|\boldsymbol{x} - \boldsymbol{R}_k - D_k \boldsymbol{\Omega}|}$$

where $\boldsymbol{\Omega}$ is a vector of length 1 and $\mathrm{d}\boldsymbol{\Omega}$ denotes the normalized surface measure of the unit sphere $\mathbb{S}^2$. (This is just Newton's theorem.) Hence, by Schwarz's inequality, and for $\boldsymbol{x} \in \Gamma_j$,

$$\Phi(\boldsymbol{x})^2 \leq (M-1) \sum_{k,\, k \neq j} \int_{\mathbb{S}^2} \frac{\mathrm{d}\boldsymbol{\Omega}}{|\boldsymbol{x} - \boldsymbol{R}_k - D_k \boldsymbol{\Omega}|^2} \leq M \sum_{k=1}^{M} \int_{\mathbb{S}^2} \frac{\mathrm{d}\boldsymbol{\Omega}}{|\boldsymbol{x} - \boldsymbol{R}_k - D_k \boldsymbol{\Omega}|^2}.$$

The last term is independent of j and is an upper bound valid in all of $\mathbb{R}^3$. The integral of $|\Phi(\boldsymbol{x})|^4$ is thus bounded by

$$\int_{\mathbb{R}^3} |\Phi(\boldsymbol{x})|^4 \mathrm{d}\boldsymbol{x} \le M^2 \sum_{k,l} \int_{\mathbb{R}^3} \mathrm{d}\boldsymbol{x} \int_{\mathbb{S}^2} \frac{\mathrm{d}\boldsymbol{\Omega}}{|\boldsymbol{x} - \boldsymbol{R}_k - D_k \boldsymbol{\Omega}|^2} \int_{\mathbb{S}^2} \frac{\mathrm{d}\boldsymbol{\Omega}'}{|\boldsymbol{x} - \boldsymbol{R}_l - D_l \boldsymbol{\Omega}'|^2}$$

$$= \pi^3 M^2 \sum_{k,l} \iint_{\mathbb{S}^2 \times \mathbb{S}^2} \frac{\mathrm{d}\boldsymbol{\Omega}\, \mathrm{d}\boldsymbol{\Omega}'}{|\boldsymbol{R}_k - \boldsymbol{R}_l - D_k \boldsymbol{\Omega} + D_l \boldsymbol{\Omega}'|}. \tag{13.2.7}$$

The last equality uses Eq. (5.1.7). For $k \neq l$, the last integral is just $|\boldsymbol{R}_k - \boldsymbol{R}_l|^{-1}$ by Newton's Theorem (since $|\boldsymbol{R}_k - \boldsymbol{R}_l| \ge D_k + D_l$). Similarly, for $k = l$ it is D_k^{-1}, which can be trivially bounded by $2\sum_{j,\, j\neq k} |\boldsymbol{R}_j - \boldsymbol{R}_k|^{-1}$. Hence

$$\int_{\mathbb{R}^3} |\Phi(\boldsymbol{x})|^4 \mathrm{d}\boldsymbol{x} \le 6\pi^3 M^2 \sum_{1\le k<l\le M} \frac{1}{|\boldsymbol{R}_k - \boldsymbol{R}_l|}.$$

We are left with the task of finding a lower bound to

$$\sum_{i=1}^{L} \left(\varepsilon |\boldsymbol{p}_i| - \frac{\lambda}{\mathfrak{D}(\boldsymbol{x}_i)} \right). \tag{13.2.8}$$

Assume that $\varepsilon > \pi\lambda/2$. It was proved in Chapter 8, Lemma 8.5, that (13.2.8) is bounded from below by

$$-1.514 \frac{\lambda^4 q}{(\varepsilon - \pi\lambda/2)^3} \sum_{j=1}^{M} \frac{1}{D_j}.$$

Again noting that $D_j^{-1} \le 2\sum_{k,\, j\neq k} |\boldsymbol{R}_j - \boldsymbol{R}_k|^{-1}$, this proves that

$$\tilde{H} \ge \left[\alpha - \lambda^4 \frac{0.0257\, q}{(1-\varepsilon)^3} 6\pi^3 M^2 - \lambda^4 \frac{6.056\, q}{(\varepsilon - \pi\lambda/2)^3} \right] \sum_{1\le k<l\le M} \frac{1}{|\boldsymbol{R}_k - \boldsymbol{R}_l|}. \tag{13.2.9}$$

We shall choose M large and λ and ε small. In this case, the third term in square parentheses in (13.2.9) is negligible compared to the second term. We choose $\alpha = \sigma\lambda^4 M^2$ with $\sigma = 0.0257 \times 6\pi^3 q$ and $\lambda M = (N/2\sigma)^{1/3}$, with $\lambda \ll 1$ and $1 \ll M \ll N$. To leading order, the equation (13.2.6) then becomes

$$\kappa = \frac{3}{2}(2\sigma)^{-1/3} N^{-2/3}.$$

In other words, we have shown that $H \geq 0$ if $N \leq (2\sigma)^{-1/2}(3/2\kappa)^{3/2}$ plus terms of lower order for small κ. ■

13.3 A More Complete Picture

13.3.1 Relativistic Gravitating Fermions

As stated in the Introduction, a more accurate model uses the Hamiltonian

$$H = \sum_{i=1}^{N} \left(\sqrt{|\boldsymbol{p}_i|^2 + m^2} - m \right) - \kappa \sum_{1 \leq i < j \leq N} \frac{1}{|\boldsymbol{x}_i - \boldsymbol{x}_j|} \tag{13.3.1}$$

with $m > 0$. As pointed out above, the critical value of N for stability does not depend on m, since $|\boldsymbol{p}| - m \leq \sqrt{|\boldsymbol{p}_i|^2 + m^2} - m \leq |\boldsymbol{p}|$. Other properties for subcritical N, like the size of the star, for instance, will depend on m. In the limit of small κ and large N one would expect that to leading order the energy and particle density is exactly given by the **semiclassical approximation**, by which we mean the following.

First, the gravitational energy is approximated by the classical energy of a particle distribution described by $\varrho(\boldsymbol{x})$, given by

$$-\frac{1}{2}\kappa \iint_{\mathbb{R}^3 \times \mathbb{R}^3} \frac{\varrho(\boldsymbol{x})\varrho(\boldsymbol{y})}{|\boldsymbol{x} - \boldsymbol{y}|} \mathrm{d}\boldsymbol{x}\mathrm{d}\boldsymbol{y} = -\kappa D(\varrho, \varrho),$$

where the same notation as in (5.1.3) is used. The lowest kinetic energy for a given density function ϱ in the semiclassical approximation is

$$\int_{\mathbb{R}^3} j(\varrho(\boldsymbol{x}))\mathrm{d}\boldsymbol{x}$$

where j is defined as

$$j(t) = \frac{q}{(2\pi)^3} \int_{|\boldsymbol{k}| < (6\pi^2 t/q)^{1/3}} \left(\sqrt{|\boldsymbol{k}|^2 + m^2} - m \right) \mathrm{d}\boldsymbol{k}.$$

(Compare with the discussion of the semiclassical approximation in Section 4.1.1.) This integral can be evaluated explicitly. It is

$$j(t) = \frac{q}{16\pi^2} \left[\eta(2\eta^2 + m^2)\sqrt{\eta^2 + m^2} - m^4 \ln\left(\frac{\eta + \sqrt{\eta^2 + m^2}}{m} \right) \right] - tm,$$

with $\eta = (6\pi^2 t/q)^{1/3}$. For large t, $j(t) \approx (3/4)(6\pi^2/q)^{1/3} t^{4/3}$, whereas for small t, $j(t) = (3/10m)(6\pi^2/q)^{2/3} t^{5/3}$, which is exactly what we would expect in these limits.

Altogether, the semiclassical energy functional is thus defined to be

$$\boxed{\mathcal{E}^{\text{classical}}(\varrho) = \int_{\mathbb{R}^3} j(\varrho(\boldsymbol{x}))\mathrm{d}\boldsymbol{x} - \kappa D(\varrho, \varrho).} \tag{13.3.2}$$

Its infimum over all non-negative densities ϱ with $\int_{\mathbb{R}^3} \varrho(\boldsymbol{x})\mathrm{d}\boldsymbol{x} = N$ is the semiclassical energy $E_0^{\text{classical}}(N, \kappa, m)$. It has the scaling property

$$E_0^{\text{classical}}(N, \kappa, m) = N^{4/3} E_0^{\text{classical}}(1, \kappa N^{2/3}, m N^{-1/3}).$$

From this one immediately deduces that, within the semiclassical approximation, there is a critical value of $\kappa N^{2/3}$ for stability of the system. In other words, the critical value of N, which we denote by N_c, equals a constant times $\kappa^{-3/2}$, in agreement with our bounds in Theorem 13.1.

There are two questions one would like to ask.

- Does the quantum-mechanical problem (13.3.1) lead to this semiclassical minimization problem to leading order as $\kappa \to 0$ and $N \to \infty$, with $N\kappa^{3/2}$ fixed?
- Is there a minimizing ϱ for $\mathcal{E}^{\text{classical}}$ and is it unique (up to translations, of course)?

It is easy to show that if a minimizing ϱ for $\mathcal{E}^{\text{classical}}$ exists, it satisfies the variational equation

$$j'(\varrho(\boldsymbol{x})) = \left[\kappa \int_{\mathbb{R}^3} \frac{\varrho(\boldsymbol{y})}{|\boldsymbol{x} - \boldsymbol{y}|}\mathrm{d}\boldsymbol{y} - \mu\right]_+ \tag{13.3.3}$$

where $[\,\cdot\,]_+$ denotes the positive part, and $\mu \geq 0$ is some constant, adjusted so that $\int \varrho(\boldsymbol{x})\mathrm{d}\boldsymbol{x} = N$.

The answer to both these questions is affirmative, as proved by Lieb and Yau in [137]. The most difficult part in the proof of the correctness of $E_0^{\text{classical}}$ is the lower bound on the ground state energy of H in (13.3.1), and this follows the method in our Theorem 13.1 except that the estimates are done much more carefully in [137].

With the aid of standard techniques in the calculus of variations it is not hard to prove that there is a minimizer of $\mathcal{E}^{\text{classical}}$ as long as $N < N_c$. The proof of the uniqueness of a solution of the equation (13.3.3) for $N < N_c$ is somewhat non-standard and turns out to be rather involved.[1] When $N = N_c$, there is also a minimizer of $\mathcal{E}^{\text{classical}}$ if we set $m = 0$, i.e., replace $j(t)$ by $(3/4)(6\pi^2/q)^{1/3}t^{4/3}$. The solution of the corresponding variational equation (known as the Lane–Emden equation) turns out to be unique up to translations and rescaling; that is, one can replace $\varrho(\boldsymbol{x})$ by $\lambda^3\varrho(\lambda\boldsymbol{x} + \boldsymbol{y})$ for any $\lambda > 0$ and $\boldsymbol{y} \in \mathbb{R}^3$.

13.3.2 Relativistic Gravitating Bosons

The situation for bosons is similar to the fermionic case discussed in the previous subsection. The effective functional for small κ and large N is now the Hartree functional

$$\mathcal{E}^{\text{Hartree}}(\varrho) = \left(\sqrt{\varrho}, \left(\sqrt{|\boldsymbol{p}|^2 + m^2} - m\right)\sqrt{\varrho}\right) - \kappa D(\varrho, \varrho). \tag{13.3.4}$$

It results by assuming that all the particles are in the same one-particle state $\phi(\boldsymbol{x}) = \sqrt{\varrho(\boldsymbol{x})/N}$,[2] compare with Eq. (13.2.2). That is, the many-body wave function is of the form $\psi(\boldsymbol{x}_1, \ldots, \boldsymbol{x}_N) = \prod_{i=1}^N \phi(\boldsymbol{x}_i)$. The Hartree energy, which is the infimum of (13.3.4) over all non-negative ϱ with $\int_{\mathbb{R}^3} \varrho(\boldsymbol{x})\mathrm{d}\boldsymbol{x} = N$, has the scaling property

$$E^{\text{Hartree}}(N, \kappa, m) = N E^{\text{Hartree}}(1, \kappa N, m N^{-1}).$$

This shows that the critical value of N, denoted again by N_c, now equals a constant times κ^{-1}, in accordance with our bounds in Theorem 13.1.

It was proved in [137] that the Hartree functional (13.3.4) correctly describes the ground state energy of (13.3.1) for bosons in the limit $\kappa \to 0$ and $N \to \infty$

[1] Equation (13.3.3) is treated in many physics textbooks, see, e.g., [30] or [186, Sect. II.3.4], by converting it into a second order differential equation in the radial variable $|\boldsymbol{x}|$. It is an elementary fact that the resulting equation has a unique solution if the value of ϱ at the origin is specified, and textbooks tend to leave it at that. The difficult part of the proof in [137] is to show that there can only be *one* such value for a given N.

[2] It is not a restriction to assume that ϕ is non-negative, as we do here, since the representation (3.2.17) shows that replacing ϕ by $|\phi|$ can only lower the kinetic energy.

with κN fixed. Moreover, there is a minimizing $\phi(\boldsymbol{x}) = \sqrt{\varrho(\boldsymbol{x})}$ for $\mathcal{E}^{\text{Hartree}}$ as long as $N < N_c$, which satisfies the Hartree equation

$$\left(\sqrt{|\boldsymbol{p}|^2 + m^2} - m\right)\phi(\boldsymbol{x}) - \kappa \int_{\mathbb{R}^3} \frac{|\phi(\boldsymbol{y})|^2}{|\boldsymbol{x} - \boldsymbol{y}|}\phi(\boldsymbol{x}) = -\mu\phi(\boldsymbol{x}).$$

Uniqueness of solutions to this equation is still an open problem, in general. For small values of $\kappa \int |\phi(\boldsymbol{x})|^2 \mathrm{d}\boldsymbol{x}$ uniqueness was proved by Lenzmann [104].[3]

13.3.3 Inclusion of Coulomb Forces

In Section 13.2 we obtained bounds on the critical number of particles for collapse of a gravitating system of either bosons or fermions. In the model considered there, there was only one kind of particle and the gravitational attraction among them was proportional to $\kappa = Gm^2$, where m is the mass of the particle. This model is appropriate for a neutron star (neglecting general relativistic effects and nuclear forces, etc), in which case m equals the mass of a neutron m_{n}, which is approximately equal to the mass of a proton, namely ≈ 2000 in our units (where the electron mass equals 1). The main conclusion was that the critical particle number N_c is proportional to $\kappa^{-3/2}$ since neutrons are fermions.

For the physical values of m_{n} and G the resulting critical particle number is roughly $N_c \approx 10^{57}$, and the mass of the neutron star, which is $N_c m_{\text{n}}$, is of the same order as the mass of our sun, but the neutron star is much smaller in size (by a factor of 10^5). The main reason for this size difference, as mentioned in Section 13.1, is the lack of electrons in neutron stars, which are very light but produce a big pressure. If neutrons were bosons, N_c would be proportional to κ^{-1}, in which case $N_c \approx 10^{38}$, and $N_c m_{\text{n}} \approx$ the mass of a mountain instead of the mass of a star.

White dwarfs, on the other hand, can be thought of as consisting of a gas of nuclei and electrons, and hence not simply of one species of electrically neutral gravitating particles. The main gravitational energy in a white dwarf comes from the nuclei, since they are much heavier than the electrons. The Pauli principle for the electrons is responsible for the pressure that keeps the star from collapsing, however, even in case the nuclei are bosons. An approximate model to consider

[3] For the non-relativistic analogue of the Hartree equation, existence and uniqueness of solutions was proved in [109].

is the Hamiltonian (13.2.1) for the electrons, but with $\kappa = G(\mu/Z)^2$, however, where μ and Z are the mass and charge of the nuclei. While this model is the one usually employed, it ignores the fact that there are really two kinds of charged particles which make different contributions to the energy.

In this section we show that a model in which there are electrons and nuclei with both electrostatic and gravitational interactions is stable in essentially the same parameter region as the simpler Hamiltonian (13.2.1) for just one species of (neutral) particles, with effective coupling constant that is proportional to the square of the mass per unit charge of the heavier particles. The presentation is similar to the one in [136]. No attempt is made to evaluate the precise constants as we did earlier.

The Hamiltonian for this model is given by

$$\boxed{\begin{aligned} H = {}& \sum_{i=1}^{N}\left(\sqrt{|\boldsymbol{p}_i|^2+m}-m\right)+\sum_{k=1}^{M}\left(\sqrt{|\boldsymbol{P}_k|^2+\mu}-\mu\right) \\ &+\sum_{1\le i<j\le N}\frac{\alpha-Gm^2}{|\boldsymbol{x}_i-\boldsymbol{x}_j|}+\sum_{1\le k<l\le M}\frac{Z^2\alpha-G\mu^2}{|\boldsymbol{R}_k-\boldsymbol{R}_l|}-\sum_{i=1}^{N}\sum_{k=1}^{M}\frac{Z\alpha+Gm\mu}{|\boldsymbol{x}_i-\boldsymbol{R}_k|}, \end{aligned}} \tag{13.3.5}$$

where $\boldsymbol{P}_k$ and $\boldsymbol{R}_k$ denote the momenta and positions of the nuclei, which could either be bosons or fermions. The charge and mass of the nuclei are Z and μ, respectively. The electron mass $m = 1$ in our units, but for transparency we shall retain it in our notation.

For stability, we have to assume that $Z\alpha + Gm\mu \le 2/\pi$, which, for physical values, is satisfied for all nuclei with $Z \le 87$.

We shall also assume in the following that $Z^2\alpha > G\mu^2$. This condition is amply satisfied in practice, since $G\mu^2/(Z^2\alpha) \approx 10^{-36}$ for all nuclei in the periodic table. (In fact, μ/Z is roughly constant throughout the periodic table, since atomic nuclei contain roughly as many neutrons as protons.)

For convenience, we define an **effective nuclear charge** to be

$$Z_b = \sqrt{Z^2 - \frac{G\mu^2}{\alpha}}$$

and an **effective electron charge** to be

$$Z_e = \frac{Z\alpha + Gm\mu}{Z_b\alpha}.$$

Then H can be written as

$$H = \sum_{i=1}^{N} \left(\sqrt{|\boldsymbol{p}_i|^2 + m} - m \right) + \sum_{k=1}^{M} \left(\sqrt{|\boldsymbol{P}_k|^2 + \mu} - \mu \right)$$
$$+ \sum_{1 \le i < j \le N} \frac{\alpha Z_e^2 - \kappa}{|\boldsymbol{x}_i - \boldsymbol{x}_j|} + \sum_{1 \le k < l \le M} \frac{\alpha Z_b^2}{|\boldsymbol{R}_k - \boldsymbol{R}_l|} - \sum_{i=1}^{N} \sum_{k=1}^{M} \frac{\alpha Z_e Z_b}{|\boldsymbol{x}_i - \boldsymbol{R}_k|}, \tag{13.3.6}$$

with

$$\kappa = G \left(m + \frac{\mu}{Z} \right)^2 \left(1 - \frac{G\mu^2}{Z^2 \alpha} \right)^{-1}.$$

In practice, $2000 \lesssim \mu/(Zm) \lesssim 6000$, and hence $\kappa \approx G\mu^2/Z^2$.

We choose some $0 < \eta < 1$ and write $H = H^C + H^G$, where

$$H^C = (1 - \eta) \sum_{i=1}^{N} \left(\sqrt{|\boldsymbol{p}_i|^2 + m} - m \right) + \sum_{k=1}^{M} \left(\sqrt{|\boldsymbol{P}_k|^2 + \mu} - \mu \right)$$
$$+ \sum_{1 \le i < j \le N} \frac{\alpha Z_e^2}{|\boldsymbol{x}_i - \boldsymbol{x}_j|} + \sum_{1 \le k < l \le M} \frac{\alpha Z_b^2}{|\boldsymbol{R}_k - \boldsymbol{R}_l|} - \sum_{i=1}^{N} \sum_{k=1}^{M} \frac{\alpha Z_e Z_b}{|\boldsymbol{x}_i - \boldsymbol{R}_k|} \tag{13.3.7}$$

and

$$H^G = \eta \sum_{i=1}^{N} \left(\sqrt{|\boldsymbol{p}_i|^2 + m} - m \right) - \sum_{1 \le i < j \le N} \frac{\kappa}{|\boldsymbol{x}_i - \boldsymbol{x}_j|}.$$

The Hamiltonian H^C contains only Coulomb forces. For appropriate values of η, α, Z_e and Z_b it is bounded from below by $-(1 - \eta)Nm$, as was shown in Chapter 8, Theorem 8.1. The nuclear kinetic energy is not needed for this bound, and the statistics of the nuclei is irrelevant. If the nuclei happen to be fermions, the conclusion would hold even for $\eta = 1$ if Z is not too big, by simply exchanging the role of the nuclei and the electrons in Theorem 8.1.

The remaining part H^G contains only gravitating particles; it is equal to the Hamiltonian studied in Section 13.2, with an effective coupling constant κ/η. In particular, this shows that *the full Hamiltonian H in (13.3.5) is stable as long as $\kappa \le$ const.$\,N^{-2/3}$ (for appropriate values of Z and α). This* *result holds for fermionic electrons and nuclei of either statistics, boson or fermion.*

It remains an open problem to study the semiclassical limit $N, M \to \infty$, $\kappa \to 0$ with $N\kappa^{3/2}$ fixed. In this limit, one expects the system to be described by a semiclassical functional, which is more complicated than (13.3.2) because of exchange terms resulting from the Coulomb interaction potentials (see Chapter 6).

CHAPTER 14

The Thermodynamic Limit for Coulomb Systems

14.1 Introduction

In the previous chapters we established the fact that the ground state energy is bounded below by a constant times the total particle number for a variety of models of particles interacting via electrostatic and magnetic forces. The natural next question would be whether it is strictly proportional to the particle number for large particle number, that is, whether the limit of the energy per particle exists. For the ground state energy it is actually easy to see that this is the case, as we shall demonstrate in Section 14.2.

A more interesting question is what happens if we confine a large number of particles to a large box, with the number of particles per unit volume, i.e., the density ϱ, fixed. This would describe, for instance, a gas or a liquid or even a solid in a container. Again we expect the energy, in the limit of large system size, to be equal to the particle number times some function of the density, independent of the volume or the shape of the box. To make things even more realistic, one can discuss the very same question at a positive temperature $T > 0$, in which case the relevant quantity to look at is the free energy, i.e., the energy minus T times the entropy. This general question of the existence of the thermodynamic limit will be addressed in Section 14.3.

In the previous proofs of the stability of matter the main concern was the short distance $|\boldsymbol{x}|^{-1}$ singularity of the Coulomb potential, and to show that it does not cause the system to implode and have a very negative energy that grows faster than the particle number. In contrast, it is the long distance $|\boldsymbol{x}|^{-1}$ nature of the Coulomb potential that is the source of difficulty for proving the existence of a thermodynamic limit. By confining the particles to a box they are prevented from escaping from each other, and the danger is that the energy could be positive and grow faster than the particle number. For instance, if we put only negatively charged particles in a box of diameter L, the particles would stick to the walls of the container, and the energy would grow like $N^2/L = N^{5/3}\varrho^{1/3}$ for large N and fixed ϱ. Clearly, *charge neutrality will be an essential input here, whereas it was not before.*

Another concern might be that even if the energy is bounded from above and below by some constants times the particle number, the energy per particle still might oscillate and not converge as N goes to infinity. Although this is physically ridiculous it is a mathematical possibility that has to be addressed.

In the physically realistic case there are several kinds of particles to be considered. While there is only one negative particle of importance (the electron[1]), there are several kinds of nuclei, some of which might be bosons and some fermions. To simplify the discussion here we shall just take one kind of nucleus of charge $+Ze$ and mass μ. We emphasize, however, that the proof given here can easily be generalized to include several species of particles, see [116].

The discussion begins with the thermodynamic limit for particles of indeterminate density in their ground state, i.e., with no confining box and free to be where they like in $\mathbb{R}^3$. This would be appropriate for the discussion of a solid at very low temperature. Following that we shall give a quick review of the essential principles of statistical mechanics needed for the formulation of the thermodynamic limit at positive temperature and fixed density. The main theorem concerns the existence of the thermodynamic limit of the free energy of neutral Coulomb systems. The proof we shall give follows closely the original work by Lebowitz and Lieb [103, 116]. Only the Hamiltonians without magnetic field will be discussed. We remark that the corresponding question for matter interacting with the quantized electromagnetic field is still an open problem. Partial results in that direction were obtained in [121]. In Section 14.7 the 'jellium' model is briefly discussed. In this model the nuclei are replaced by a fixed, uniformly charged background, chosen to make the system charge neutral.

The astute reader will notice an important fact about all the systems under consideration. Quantum mechanics plays no role in the proof except to provide a lower bound to the energy that satisfies stability of the second kind. Thus, the proofs go through for classical (i.e., non-quantum) statistical mechanics if there is such a lower bound. Indeed, in the jellium model this is the case. We shall briefly explain classical statistical mechanics in Section 14.4. For the model with electrons and nuclei considered in the rest of this book, quantum mechanics is

[1] There are other kinds of negative fermions, like the muon, but they are unstable and have a very short lifetime.

not needed, as shown by Onsager [145], if we impose a hard core condition that $|\boldsymbol{x}_i - \boldsymbol{R}_j| \geq a > 0$ for all i, j. Quantum mechanics does have an effect on the final energy, however.

14.2 Thermodynamic Limit of the Ground State Energy

Before considering the general problem of the thermodynamic limit for a given density and temperature we first consider the simpler problem of the ground state energy in an infinite volume. In the previous chapters, we have shown that this ground state energy is bounded from below by a constant times the total particle number. Here, we shall supplement this result by showing that the ground state energy per particle actually has a limit as the number of particles goes to infinity.

In contrast to the general case discussed in the next section, charge neutrality is not needed here. It was also irrelevant for the question of stability discussed in the previous chapters.

For simplicity, we shall only consider systems without magnetic fields. We shall also restrict our attention to just one species of nuclei. The extension to more general situations is straightforward.

The Hamiltonian under consideration will be the same as in Chapters 7 and 8, except that we allow for a finite nuclear mass and also take the nuclear kinetic energy into account. It is given by

$$\boxed{H = \sum_{i=1}^{N} T_m(\boldsymbol{p}_i) + \sum_{j=1}^{M} T_\mu(\boldsymbol{P}_j) + \alpha V_C(\underline{\boldsymbol{X}}, \underline{\boldsymbol{R}}).} \tag{14.2.1}$$

The electrons have positions $\boldsymbol{x}_i$ and momenta $\boldsymbol{p}_i$, while the nuclei have positions $\boldsymbol{R}_j$ and momenta $\boldsymbol{P}_j$. They can be either bosons or fermions. The kinetic energy of the electrons is either of the form

$$T_m(\boldsymbol{p}) = \frac{\boldsymbol{p}^2}{2m} \qquad \text{or} \qquad T_m(\boldsymbol{p}) = \sqrt{\boldsymbol{p}^2 + m^2} - m$$

with $m > 0$. The same choices apply to the kinetic $T_\mu(\boldsymbol{P})$ of the nuclei, with mass μ instead of mass m. A special choice is $\mu = \infty$, which means that the nuclei do not move, but are fixed in some minimum energy configuration. The Coulomb potential $V_C(\underline{\boldsymbol{X}}, \underline{\boldsymbol{R}})$ is defined in (2.1.21), with $Z_j = Z$ for $j = 1, 2, \ldots, M$.

Let $E_0(N, M)$ denote the ground state energy of (14.2.1). We assume, as usual, that the electrons are fermions. If they were bosons the ground state energy would grow as $-N^{7/5}$, which implies that the thermodynamic limit exists for bosons only in the sense that the energy per particle is $-\infty$ in this limit. Real electrons have $q = 2$, but our conclusions hold for any fixed, finite q. In the relativistic case, we shall assume that $Z\alpha$ and α are appropriately chosen for stability of matter to hold. Sufficient conditions for this are given in Theorem 8.1.

Theorem 14.1 (Thermodynamic Limit of the Ground State Energy). *Let $E_0(N, M)$ denote the ground state energy of (14.2.1). Then E_0 is* **subadditive**, *that is*

$$E(N_1 + N_2, M_1 + M_2) \leq E(N_1, M_1) + E(N_2, M_2). \tag{14.2.2}$$

Now let N_j and M_j for $j = 1, 2, \ldots$ be a sequence such that $N_j + M_j \to \infty$ as $j \to \infty$ and such that $N_j/(N_j + M_j)$ converges to some number η with $0 < \eta < 1$. Then there is a function $e(\eta)$ defined on $(0, 1)$ such that

$$\boxed{\lim_{j\to\infty} \frac{E_0(N_j, M_j)}{N_j + M_j} = e(\eta),} \tag{14.2.3}$$

regardless of the sequences N_j and M_j (provided $N_j/(N_j + M_j) \to \eta$). Moreover e is bounded and convex on $(0, 1)$, i.e.,

$$e(\lambda\eta_1 + (1-\lambda)\eta_2) \leq \lambda e(\eta_1) + (1-\lambda)e(\eta_2) \tag{14.2.4}$$

for $0 \leq \lambda \leq 1$.

Proof. The subadditivity (14.2.2) is a simple consequence of our being able to use, as a variational function for the problem with $N_1 + N_2$ electrons and $M_1 + M_2$ nuclei, a product function of the form $\psi^1 \otimes \psi^2_{\boldsymbol{y}}$, where ψ^1 is a function for the N_1 electrons and M_1 nuclei, and similarly for ψ^2. The subscript $\boldsymbol{y} \in \mathbb{R}^3$ indicates a translation of ψ^2, i.e. $\psi^2_{\boldsymbol{y}}(\boldsymbol{x}_1, \boldsymbol{x}_2, \ldots) \equiv \psi^2(\boldsymbol{x}_1 + \boldsymbol{y}, \boldsymbol{x}_2 + \boldsymbol{y}, \ldots)$. The expectation value of the Hamiltonian H converges, as $\boldsymbol{y} \to \infty$, to $(\psi^1, H\psi^1) + (\psi^2, H\psi^2)$. There is a small technical problem associated with the product function $\psi^1 \otimes \psi^2_{\boldsymbol{y}}$, namely it will not necessarily have the right permutation symmetry properties. This however can easily be corrected by appropriately symmetrizing or antisymmetrizing this function, which will not affect the limiting energy as $|\boldsymbol{y}| \to \infty$.

The assertion (14.2.3) is a direct consequence of (14.2.2) together with the stability bound $E_0(N, M) \geq A(N + M)$. To prove it we first note that there is also an upper bound of the form $E_0(N, M) \leq B(N + M)$ for some finite number B. In our case we can obviously take $B = 0$ but this upper bound can be deduced from (14.2.2) alone by noting that $E_0(N, M) \leq NE_0(1, 0) + ME_0(0, 1)$.

Since $E_0(N, M)/(N + M)$ is bounded, there are sequences (by passing to subsequences if necessary) so that the limit in (14.2.3) exists. Now suppose that there are two sequences (N_j, M_j) and (N'_j, M'_j), both with the same limiting η, such that $\lim_{j\to\infty} E_0(N_j, M_j)/(N_j + M_j) < \lim_{j\to\infty} E_0(N'_j, M'_j)/(N'_j + M'_j)$. By going to a subsequence of (N'_j, M'_j) and appropriately relabeling the elements, we can assume that

$$N'_j = L_j N_j + r_j$$

for some sequence of integers L_j that goes to $+\infty$ as $j \to \infty$, and $r_j < N_j$. If we write

$$M'_j = L_j N_j + \delta_j$$

then $\lim_{j\to\infty} \delta_j/(L_j N_j) = 0$ (since both N_j/M_j and N'_j/M'_j converge to the same ratio). By subadditivity and the upper bound on E_0,

$$E_0(N'_j, M'_j) \leq L_j E_0(N_j, M_j) + B(r_j + \delta_j)$$

and hence

$$\lim_{j\to\infty} \frac{E_0(N'_j, M'_j)}{N'_j + M'_j} \leq \lim_{j\to\infty} \frac{E_0(N_j, M_j)}{N_j + M_j}.$$

This contradicts the assumption that the left side be strictly bigger than the right side, and hence the two limits agree.

As for the convexity, we shall prove weak convexity, namely $e(\frac{1}{2}\eta_1 + \frac{1}{2}\eta_2) \leq \frac{1}{2}e(\eta_1) + \frac{1}{2}e(\eta_2)$. This, together with the boundedness of $e(\eta)$ easily implies convexity (cf. [86, Thm. 1.11]). Let (N^1_j, M^1_j) and (N^2_j, M^2_j) be sequences converging to η_1 and η_2, respectively. By (14.2.2),

$$\begin{aligned} E_0(N^1_j(N^2_j + M^2_j) + N^2_j(N^1_j + M^1_j),\ M^1_j(N^2_j + M^2_j) + M^2_j(N^1_j + M^1_j)) \\ \leq (N^2_j + M^2_j)E_0(N^1_j, M^1_j) + (N^1_j + M^1_j)E_0(N^2_j, M^2_j). \end{aligned}$$

After we divide this inequality by $2(N^1_j + M^1_j)(N^2_j + M^2_j)$, the left side converges to $e(\frac{1}{2}\eta_1 + \frac{1}{2}\eta_2)$, and the right side converges to $\frac{1}{2}e(\eta_1) + \frac{1}{2}e(\eta_2)$. ■

14.3 Introduction to Quantum Statistical Mechanics and the Thermodynamic Limit

In the previous section we have discussed the existence of the thermodynamic limit of the ground state energy of systems in infinite volume. We will now turn our attention to systems that are confined to finite volumes and have positive temperature. The Hamiltonian under consideration will be the same as in the previous section. We shall consider only the neutral case here, i.e., $N = MZ$.

In order to define an operator restricted to functions that are supported on some set $\Omega \subset \mathbb{R}^3$, we consider the quadratic form[2]

$$\mathcal{E}(\psi) = \int \left(\sum_{i=1}^{N} T_m(2\pi \boldsymbol{k}_i) + \sum_{j=1}^{M} T_\mu(2\pi \boldsymbol{K}_j) \right) |\widehat{\psi}(\underline{\boldsymbol{k}}, \underline{\boldsymbol{K}})|^2 \mathrm{d}\underline{\boldsymbol{k}}\, \mathrm{d}\underline{\boldsymbol{K}}$$

$$+ \alpha \int V_C(\underline{\boldsymbol{X}}, \underline{\boldsymbol{R}}) |\psi(\underline{\boldsymbol{X}}, \underline{\boldsymbol{R}})|^2 \mathrm{d}\underline{\boldsymbol{X}}\, \mathrm{d}\underline{\boldsymbol{R}} \qquad (14.3.1)$$

for $\psi \in H^1(\mathbb{R}^{3(N+M)})$ (in the non-relativistic case) or $\psi \in H^{1/2}(\mathbb{R}^{3(N+M)})$ (in the relativistic case) with the property that $\psi(\underline{\boldsymbol{X}}, \underline{\boldsymbol{R}}) = 0$ if either $\boldsymbol{x}_i \notin \Omega$ for some i or $\boldsymbol{R}_j \notin \Omega$ for some j.

Under the assumption that $\mathcal{E}(\psi) \geq -C(\psi, \psi)$ for all ψ, i.e., stability of the first kind, there is a standard method (the Friedrichs extension) of defining a corresponding operator H_Ω such that $\mathcal{E}(\psi) = (\psi, H_\Omega \psi)$. We refer to [150, Thm. X.23] for details. In the non-relativistic case, the operator H_Ω is simply given by (14.2.1) with **Dirichlet boundary conditions** on the boundary of Ω, i.e., $\psi = 0$ on the boundary. Stability of the first kind always holds in the non-relativistic case (Theorem 7.1) and it holds in the relativistic case as long as α and $Z\alpha$ are not too large, as proved in Theorem 8.1 in Chapter 8.

For bounded Ω, the operator H_Ω has discrete spectrum and a complete set of eigenfunctions.[3] Moreover, the eigenvalues of H_Ω (including multiplicity) increase fast enough so that $e^{-\beta H_\Omega}$ is trace-class for any $\beta > 0$, that is, we can

[2] We shall slightly differ from the notation in previous chapters and use $\underline{\boldsymbol{k}}$ for $(\boldsymbol{k}_1, \ldots, \boldsymbol{k}_N)$ and $\underline{\boldsymbol{K}}$ for $(\boldsymbol{K}_1, \ldots, \boldsymbol{K}_M)$.

[3] Spectral theory is not really needed here. We can define the partition function also by a variational principle, see Remark 14.2.

define the **partition function** $Z(\beta, \Omega, N, M)$ as

$$Z(\beta, \Omega, N, M) := \text{Tr } e^{-\beta H_\Omega} = \sum_E e^{-\beta E}, \tag{14.3.2}$$

where the sum runs over all eigenvalues E of H_Ω in the appropriate symmetry class (Bose or Fermi). The positive parameter β equals $1/k_B T$, where k_B equals Boltzmann's constant and $T > 0$ is the temperature.

Remark 14.1 (Spin and Symmetry). If the electrons have q spin states then the electron coordinates should really be $z = (\boldsymbol{x}, \sigma)$ as before. The spin plays no essential role, however, and there is the following easy way to think about spin in order to be able to concentrate only on the spatial variables $\boldsymbol{x}_i$. Just pretend that there are q kinds of electrons and that each kind has $q = 1$. That is, there are N^i particles of type i (with $i = 1, \dots, q$) and the allowed wave functions are functions of $N^1, N^2, \dots, N^q$ spatial variables that are separately antisymmetric in each of the N^i variables. We can compute the partition function with given numbers $N^1, \dots, N^q$ of particles of each type. After that the partition function given in (14.3.2) is computed by summing over all choices of the N^i such that $\sum_{i=1}^q N^i = N$. No combinatorial coefficients are needed – as there would be if one tried to compute the partition function by summing over the various symmetry types of the $\boldsymbol{x}$ variables that can arise from antisymmetry in the $(\boldsymbol{x}, \sigma)$ variables. We thus see that it is only necessary to think of q species of 'spinless' fermions. For simplicity of notation we shall restrict our attention to $q = 1$ in the sequel.

The following variational principle is essential for our proof of the thermodynamic limit [153, Prop. 2.5.4].

Lemma 14.1 (Variational Principle for the Partition Function). *Let H be a self-adjoint operator that is bounded from below, with $e^{-\beta H}$ trace class. Let $\{\psi_1, \psi_2 \dots\}$ denote a set (finite or infinite) of orthonormal functions. Then*

$$\text{Tr } e^{-\beta H} \geq \sum_i \exp\left[-\beta(\psi_i, H\,\psi_i)\right].$$

In particular, Tr $e^{-\beta H}$ *equals the supremum of* $\sum_i \exp[-\beta(\psi_i, H\,\psi_i)]$ *over all (finite) sets of orthonormal functions.*

Proof. By convexity of the exponential function, $(\psi, e^{-\beta H}\psi) \geq \exp[-\beta(\psi, H\,\psi)]$ for any normalized function ψ. To see this, let E_i and ϕ_i denote the eigenvalues and corresponding orthonormal eigenfunctions of H. Then

$$(\psi, e^{-\beta H}\psi) = \sum_i e^{-\beta E_i} |(\phi_i, \psi)|^2 \geq \exp\left[-\beta \sum_i E_i |(\phi_i, \psi)|^2\right]$$
$$= \exp[-\beta(\psi, H\,\psi)],$$

where we have used Jensen's inequality[4] for the convex exponential function, as well as $\sum_i |(\phi_i, \psi)|^2 = (\psi, \psi) = 1$. Since $\text{Tr}\, e^{-\beta H} \geq \sum_i (\psi_i, e^{-\beta H}\psi_i)$ if the ψ_i are orthonormal (because the trace of a trace-class operator can be computed in *any* orthonormal basis), the claim is proved. ∎

Remark 14.2. The variational principle for the partition function can be written as

$$\boxed{\text{Tr}\, e^{-\beta H} = \sup \sum_i \exp\left[-\beta \mathcal{E}(\psi_i)\right]}$$

where the supremum is over all (finite) sets of orthonormal functions. Hence it is not necessary to introduce the Hamiltonian *operator* in order to define the partition function. The Hermitian quadratic form $\mathcal{E}(\psi)$ (defined in (14.3.1) for the model under consideration) suffices.

The **free energy** for a finite domain Ω is defined as

$$\boxed{F(\beta, \Omega, N, M) := -\frac{1}{\beta} \ln Z(\beta, \Omega, N, M).} \tag{14.3.3}$$

The **specific free energy** is the free energy per unit volume, i.e.,

$$f(\beta, \Omega, N, M) := F(\beta, \Omega, N, M)/|\Omega|. \tag{14.3.4}$$

Here and in the following, $|\cdot|$ denote the volume or Lebesgue measure of a set in $\mathbb{R}^d$.

For the Hamiltonians (14.2.1), the specific free energy is bounded from below, independent of the shape or the volume of the domain Ω. (In the relativistic case,

[4] See the footnote on page 135.

we have to assume that $Z\alpha < 2/\pi$ and α is small enough.) This can be seen from the following consideration. From the stability of matter, Theorem 7.1 in the non-relativistic case and Theorem 8.1 in the relativistic case, we see that, for $0 < \varepsilon < 1$,

$$(1-\varepsilon)\sum_{i=1}^{N} T_m(\boldsymbol{p}_i) + \alpha V_C(\underline{\boldsymbol{X}}, \underline{\boldsymbol{R}}) \geq -C(N+M)$$

for some $C > 0$ and, in the relativistic case, for ε small enough. Hence $H \geq \varepsilon \sum_{i=1}^{N} T_m(\boldsymbol{p}_i) + \sum_{j=1}^{M} T_\mu(\boldsymbol{P}_j) - C(N+M)$. In particular, from the variational principle stated in Lemma 14.1, $Z \leq e^{\beta(N+M)C} \operatorname{Tr} \exp[-\beta\varepsilon \sum_{i=1}^{N} T_m(\boldsymbol{p}_i) - \beta \sum_{j=1}^{M} T_\mu(\boldsymbol{P}_j)]$. It is easy to see that the latter trace is bounded from above by $e^{D(N+M)}$ for some $D > 0$, see [153] or any other standard textbook on statistical mechanics.

The **thermodynamic limit** concerns the behavior of $f(\beta, \Omega, N, M)$ as Ω tends to $\mathbb{R}^3$ in a suitable sense, and N and M tend to infinity as well, with $N/|\Omega|$ and $M/|\Omega|$ converging to a finite non-zero value. In order to define what is meant by 'suitable sense', we need the following definition.

Definition 14.1 (Regular Sequence). If $\Omega \subset \mathbb{R}^d$ is an open set and $h > 0$, define $V(h; \Omega)$ to be the measure of the set of points $\boldsymbol{x} \in \Omega$ such that the distance of $\boldsymbol{x}$ to Ω^c, the complement of Ω, is less than or equal to h.

For $h < 0$, let $V(h; \Omega)$ denote the measure of the points $y \in \Omega^c$ whose distance to Ω is less than or equal to $-h$.

A sequence of bounded open sets $\Omega_j \subset \mathbb{R}^d$, $j \in \mathbb{N}$, is said to be a **regular sequence** if the following two conditions hold.

(1) The sequence is a **Van Hove sequence**, i.e., $\lim_{j\to\infty} |\Omega_j| = \infty$ and $\lim_{j\to\infty} V(h; \Omega_j)/|\Omega_j| = 0$ for all fixed $h \in \mathbb{R}$.
(2) The sequence satisfies the **ball condition**, i.e., if $\mathcal{B}_j$ is the smallest ball containing Ω_j then $\liminf_{j\to\infty} |\Omega_j|/|\mathcal{B}_j| > 0$.

The following theorem, which is the main result of this chapter, shows that for a regular sequence of domains, the thermodynamic limit of the specific free energy exists, and is independent of the sequence. It was proved in [103, 116].

Theorem 14.2 (Thermodynamic Limit of the Free Energy). *Fix some $\beta > 0$ and $\varrho > 0$, and consider a regular sequence of domains $\Omega_j \in \mathbb{R}^d$, and a sequence of integers N_j such that $N_j/|\Omega_j| \to \varrho$ as $j \to \infty$. With $M_j = N_j/Z$ (i.e., in the* *neutral case), the limit*

$$\lim_{j\to\infty} f(\beta, \Omega_j, N_j, M_j) =: \bar{f}(\beta, \varrho) \tag{14.3.5}$$

exists and is finite. The limit function $\bar{f}$ is independent of the sequence. It is a convex function of ϱ and a concave function of $1/\beta$.

The convexity/concavity is important for it is the basis of **thermodynamic stability**. Convexity in ϱ means that the pressure (defined to be $-f + \varrho\partial f/\partial\varrho$) is a non-decreasing function of ϱ. (Squeezing a container increases the pressure.) If this were not true the system would spontaneously implode.

Likewise, concavity in $T = 1/(k_B\beta)$ means that the temperature increases as the energy per unit volume, defined to be $f - T\partial f/\partial T$, increases. If this were not so then hot bodies would suck energy away from cold bodies.

Remark 14.3 (Spin and Symmetry, redux). As explained in Remark 14.1, spin can be taken into account by considering the specific free energy of q species of 'spinless electrons', which will then be a function of q densities $\varrho^i = N^i/|\Omega|$, $i = 1, \ldots, q$. In the thermodynamic limit, this function will be a jointly convex[5] function of $(\varrho^1, \ldots, \varrho^q)$, and this can be proved in a similar way as convexity of $\bar{f}(\beta, \varrho)$ in ϱ in Theorem 14.2. It is also obviously a symmetric function of the q variables $\varrho^1, \ldots, \varrho^q$, which we denote by $\bar{f}(\beta, \varrho^1, \ldots, \varrho^q)$. It is easy to see that

$$\bar{f}(\beta, \varrho) = \inf_{\sum_i \varrho^i = \varrho} \bar{f}(\beta, \varrho^1, \ldots \varrho^q).$$

This follows from the fact that in the sum of the partition functions over different ways of splitting N into q integers, only the largest summand survives in the thermodynamic limit. In fact, the error that we make in $\bar{f}$ when considering only the largest contribution is at most $q(\beta|\Omega|)^{-1}\ln N$, since there are less than N^q summands. In particular, from convexity and symmetry it follows that

$$\bar{f}(\beta, \varrho) = \bar{f}(\beta, \varrho/q, \ldots, \varrho/q).$$

Several generalizations of Theorem 14.2 are possible, as shown in [116], but we shall not give the details here. These concern the following.

[5] A function of q variables is *jointly convex* if $f(\lambda\varrho^1 + (1-\lambda)\mu^1, \ldots, \lambda\varrho^q + (1-\lambda)\mu^q) \leq \lambda f(\varrho^1, \ldots, \varrho^q) + (1-\lambda) f(\mu^2, \ldots, \mu^q)$ for $0 \leq \lambda \leq 1$.

- **Non-neutral systems.** The system does not have to be strictly neutral, but the net charge has to be small enough. More precisely, as long as the net charge Q is much less than $N^{2/3}$, the resulting electrostatic energy does not contribute in the thermodynamic limit. If $Q = \bar{Q}N^{2/3}$ it is necessary to make the sequence of domains all have the same shape, e.g., a ball or an ellipsoid. Then the limiting specific free energy equals $\bar{f}(\beta, \varrho) + \varrho^{4/3}\bar{Q}^2/(2C)$ where C is the electrostatic capacity [118, Sect. 11.15] for the domain of the given shape and with unit volume. If $Q/N^{2/3} \to \infty$ then the specific free energy is $+\infty$ in the thermodynamic limit.
- **Multiple species.** Instead of having just one species of negative fermions and one species of positive nuclei, multiple species can be accommodated. For stability it is important that either all the negative or all the positive particles be fermions.
- **Short-range forces.** In addition to the Coulomb forces among the particles, one can include short-range potentials, meaning potentials of the form $V(\boldsymbol{x}_i - \boldsymbol{x}_j)$ which are integrable at infinity. They may or may not include hard cores. In this way one could take ionized atoms as the fundamental particles, or take explicitly the form factor of the nuclei into account.

Historically, the problem of the thermodynamic limit was studied mainly for systems with purely short range forces, such as hard sphere particles. Many names are associated with this development, including Onsager, Van Hove, Bogoliubov, Lee, Yang, van Kampen, Wils, Mazur, van der Linden, Griffiths, Dobrushin, Sinai and especially Ruelle [153] and Fisher [65], who evolved what came to be the canonical technique for proving the existence of the thermodynamic limit in quantum and classical statistical mechanics. The main idea is to prove that the specific free energy decreases with increasing volume, except for terms of lower order. The thermodynamic limit is then established because decreasing sequences have limits. Stability of the second kind makes its appearance in the assertion that this limit is finite because the sequence is bounded below.

The canonical technique in which big cubic boxes were packed with smaller cubic boxes would not work for the long range Coulomb force. The necessary modification was given in [116] where cubes were replaced by balls, and the long-range interaction was ameliorated by the electrostatic screening given by Newton's theorem, Theorem 5.2. The replacement of cubes by balls led to the geometric problem of how to pack large balls by smaller balls efficiently. The solution of the problem is presented in Section 14.5. Apart from this change from cubes to balls, and the resulting analytic complication this change induces, the basic strategy remains that of the canonical method mentioned above.

Recently an alternative proof of the thermodynamic limit (in the grand-canonical ensemble) for Coulomb systems was constructed by Hainzl, Lewin and Solovej [83]. The strategy there is different from the canonical approach; it is shown that the specific free energy *increases* with volume, except for negligible terms.

The proof of Theorem 14.2 will be given in Section 14.6. It will make essential use of the 'cheese theorem' discussed in Section 14.5. Before explaining the proof of Theorem 14.2 let us remind the reader of the definition of the partition function in classical statistical mechanics.

14.4 A Brief Discussion of Classical Statistical Mechanics

As explained in Section 2.1, a classical mechanical system is described by the Hamiltonian, which is now a real-valued function of the various momenta $\underline{\boldsymbol{P}} = (\boldsymbol{p}_1, \ldots, \boldsymbol{p}_N)$ and positions $\underline{\boldsymbol{X}} = (\boldsymbol{x}_1, \ldots, \boldsymbol{x}_N)$ of the particles. If there are N^j particles of species j, with $N = \sum_j N^j$, the classical partition function for particles in a domain Ω is

$$e^{-\beta F} = Z = \frac{1}{\prod_j N^j!} h^{-3N} \int_{\mathbb{R}^{3N}} \mathrm{d}\underline{\boldsymbol{P}} \int_{\Omega^N} \mathrm{d}\underline{\boldsymbol{X}}\, e^{-\beta H(\underline{\boldsymbol{P}},\underline{\boldsymbol{X}})}, \tag{14.4.1}$$

where h is Planck's constant. It plays the role of a normalization constant which makes Z dimensionless. Formula (14.4.1) is suggested by taking a semiclassical approximation of the quantum partition function (compare with the discussion in Section 4.1.1). The factorial factors result from the quantum-mechanical restriction to either symmetric or antisymmetric functions (bosons or fermions) which reflects the indistinguishability of particles within the same species.

If H is of the form $\sum_i \boldsymbol{p}_i^2/(2m_i) + V(\underline{\boldsymbol{X}})$, then the $\underline{\boldsymbol{P}}$ integration can be carried out and one obtains

$$Z = \prod_{i=1}^{N} \left(\frac{2\pi m_i}{\beta h^2} \right)^{3/2} Q \tag{14.4.2}$$

where

$$Q := e^{-\beta|\Omega| f^{\text{conf}}} := \frac{1}{\prod_j N^j!} \int_{\Omega^N} \mathrm{d}\underline{X}\, e^{-\beta V(\underline{X})} \tag{14.4.3}$$

is called the **configurational partition function**,[6] and f^{conf} is the configurational part of the free energy per unit volume.

For an ideal gas, where $V = 0$, $Q = |\Omega|^N / \prod_j N^j!$, and hence, using Stirling's formula, the specific free energy $F/|\Omega|$ becomes

$$\bar{f} = \frac{1}{\beta} \sum_j \varrho^j \left(\ln \frac{\beta^{3/2} h^3 \varrho^j}{(2\pi m^j)^{3/2}} - 1 \right)$$

in the thermodynamic limit, where m^j denotes the mass of the particles of species j, and ϱ^j is the density of these particles. For interacting systems, the existence of the thermodynamic limit of the classical specific free energy $f = F/|\Omega|$ is equivalent to the existence of the thermodynamic limit of the configurational part f^{conf}.

Let us now consider the Coulomb systems discussed in the rest of this book, where the interaction potential is given in (2.1.21). For such systems, $Q = +\infty$, because of the attractive $1/|\boldsymbol{x}|$ potential between electrons and nuclei. This means that classically there is no stability of the first kind. On the other hand, if we modify the potential slightly stability of the second kind can be obtained. The way to do this was discovered by Onsager [145] who introduced a *hard-core* condition. Let us imagine that there is some fixed radius $a > 0$ such that $V = +\infty$ unless $|\boldsymbol{x}_i - \boldsymbol{R}_j| \geq a$ for all i, j. For this V, Q is finite and stability of the second kind holds. This can be easily seen as follows. If $|\boldsymbol{x}_i - \boldsymbol{R}_j| \geq a$ for all i and j, we can replace the point charges at $\boldsymbol{x}_i$ and $\boldsymbol{R}_j$ by smeared out charges uniformly distributed over spheres of radius $a/2$. The electron–nucleus attraction does not change under this replacement, whereas the interelectron and internuclear repulsions decrease in case the particles are closer to each other than a distance a. The total Coulomb energy (including the self-energy of all the particles) is positive. Therefore V is bounded below by the negative of the self-energy of the charge distributions, namely $-(N + M)/a$.

[6] The configurational partition function is often defined with a prefactor $|\Omega|^{-N}$ or 1 instead of $1/\prod_j N^j!$.

The existence of the thermodynamic limit for this classical hard-core Coulomb gas can be shown in essentially *the same way* as the one in quantum case, which we shall discuss in the sequel. To make the following proof work without modifications one has to introduce also a hard-core constraint with respect to the boundary of Ω; i.e., the electrons and nuclei have to stay away a distance $a/2$ from the boundary of Ω. This does not affect the value of the specific free energy in the thermodynamic limit, of course.

Another Coulomb system for which stability of the second kind holds both classically and quantum mechanically is the jellium model, which we shall discuss in Section 14.7. There we shall also explain in some detail the differences between the proofs in the classical and quantum cases.

14.5 The Cheese Theorem

Our goal in this section is to prove that one can find a sequence of balls B_j, $j = 0, 1, 2 \ldots$, in $\mathbb{R}^d$ of geometrically increasing radii, $R_j = (1+p)^j$ with $p \in \mathbb{N}$, such that every ball B_j can be filled with $n_k = p^{k-1}(1+p)^{k(d-1)}$ *disjoint* copies of balls of radius R_{j-k}, with $k = 1, 2, \ldots, j$. If we can do this then the filling of the jth ball will be exponentially quick. More precisely, if $V_j = \sigma_d R_j^d$ denotes the volume of B_j then the unfilled volume fraction of B_j is

$$\begin{aligned} 1 - V_j^{-1} \sum_{k=1}^{j} n_k V_{j-k} &= 1 - \sum_{k=1}^{j} p^{k-1}(1+p)^{k(d-1)} \left(\frac{R_{j-k}}{R_j} \right)^d \\ &= 1 - \sum_{k=1}^{j} p^{k-1}(1+p)^{k(d-1)}(1+p)^{-kd} = \left(\frac{p}{1+p} \right)^j . \end{aligned} \tag{14.5.1}$$

This sequence of balls will be called the **standard sequence** and will be used not only to define a thermodynamic limit but also as a standard to show that an *arbitrary sequence* of suitably behaved domains has the *same* thermodynamic limit.

The next theorem shows that this sequence can be constructed if p is large enough ($p \geq 9$ for $d = 2$ and $p \geq 26$ for $d = 3$). For convenience of exposition we reverse the order by starting with a ball of radius 1 and then filling it with balls of radii $(1+p)^{-j}$ for all $j \geq 1$. The theorem has come to be known as the 'cheese theorem' because it shows that a ball of unit radius can have its volume annihilated by carving out balls of ever smaller radius. After a few balls have been cut out, the remaining domain looks like Swiss Emmenthaler cheese.

Theorem 14.3 (Cheese Theorem). *Let p be a positive integer satisfying*

$$1 + p \geq (2^d - 1)2\sqrt{d} + 2^d \sigma_d^{-1} \tag{14.5.2}$$

(where σ_d denotes the volume of the unit ball in $\mathbb{R}^d$). Let

$$r_j = (1+p)^{-j} \qquad \text{and} \qquad n_j = p^{j-1}(1+p)^{j(d-1)}.$$

It is possible to pack the unit ball in $\mathbb{R}^d$ with the disjoint union $\bigcup_{j=1}^{\infty}$ (n_j balls of radius r_j). It is also possible to pack the cube in $\mathbb{R}^d$ of volume σ_d with the same set of balls.

Proof. The *fundamental observation* is this: Recall the definition of $V(h;\Omega)$ in Definition 14.1. Let $\lambda\mathbb{Z}^d$ denote the hypercubic lattice in $\mathbb{R}^d$ with side length λ. The points in this lattice define hypercubes in $\mathbb{R}^d$ of volume λ^d, with the property that the vertices of these hypercubes are contained in the lattice.

The number of hypercubes that are contained entirely in Ω is at least as large as $\lambda^{-d}(|\Omega| - V(\lambda\sqrt{d};\Omega))$, where $|\Omega|$ is the measure of Ω. To see this, note that the number of hypercubes entirely contained in Ω equals λ^{-d} times their total volume. The total volume occupied by these hypercubes is at least $|\Omega|$ minus $V(\lambda\sqrt{d};\Omega)$.

A simple corollary of this observation is that we can pack at least $\lambda^{-d}(|\Omega| - V(\lambda\sqrt{d};\Omega))$ disjoint balls of diameter λ in Ω.

We prove the theorem for the unit ball; the proof for packing the cube of volume σ_d is essentially the same. Starting with Ω equal to the unit ball, we easily calculate $V(h;\Omega) = \sigma_d(1-(1-h)^d)$ for $0 \leq h \leq 1$, which is bounded by $V(h;\Omega) \leq \sigma_d((1+h)^d - 1) \leq \sigma_d h(2^d - 1)$. Setting $\lambda_1 = 2/(1+p)$, we have to check that $n_1 = (1+p)^{d-1}$ is less than or equal to $\lambda_1^{-d}(|\Omega| - V(\lambda_1\sqrt{d};\Omega))$, which follows immediately from (14.5.2) using the upper bound on V just derived.

The next step is to insert n_2 balls of radius r_2 on a hypercubic lattice of size $\lambda_2 = 2/(1+p)^2$, and use the 'fundamental observation' to make sure that they will fit. Having done so, we then insert n_3 balls of radius r_3 on a hypercubic lattice of size $\lambda_3 = 2/(1+p)^3$ and so forth. At each stage we have to calculate $V(\lambda_j\sqrt{d};\Omega_j)$, where Ω_j is the cheese left after removing the balls in the previous $j-1$ steps. That is, Ω_j is of the form $\Omega \setminus \bigcup_i \mathcal{B}_i$, where the $\mathcal{B}_i$ form a finite collection of disjoint balls, namely n_k balls of radius r_k with $k = 1, \ldots, j-1$.

We shall overestimate this volume by adding up the individual contributions to $V(\lambda_j\sqrt{d};\Omega_j)$ from Ω and the balls $\mathcal{B}_i$, i.e.,

$$V(\lambda_j\sqrt{d};\Omega_j) \le V(\lambda_j\sqrt{d};\Omega) + \sum_{k=1}^{j-1} n_k r_k^d \, V(-\lambda_j\sqrt{d}\, r_k^{-1};\Omega).$$

As before, we use $V(h;\Omega) \le V(-h;\Omega) \le \sigma_d h(2^d - 1)$ for $0 \le h \le 1$. This yields

$$\begin{aligned} V(\lambda_j\sqrt{d};\Omega_j) &\le \sigma_d(2^d - 1)\lambda_j\sqrt{d}\left(1 + \sum_{k=1}^{j-1} n_k r_k^{d-1}\right) \\ &= \sigma_d(2^d - 1)\lambda_j\sqrt{d}\left(1 + \frac{p^{j-1} - 1}{p - 1}\right). \end{aligned}$$

Moreover, $|\Omega_j| = \sigma_d(1 - \sum_{k=1}^{j-1} n_k r_k^d) = \sigma_d(p/(1+p))^{j-1}$. The inequality $n_j \le \lambda_j^{-d}(|\Omega_j| - V(\lambda_j\sqrt{d};\Omega_j))$ thus holds if

$$1 \le 2^{-d}\sigma_d\left(1 + p - (2^d - 1)2\sqrt{d}\frac{1 + p^{1-j}(p-2)}{p-1}\right).$$

Since $p \ge 2$ and $p^{1-j} \le 1$, this inequality holds if and only if it holds for $j = 1$, that is,

$$1 \le 2^{-d}\sigma_d\left(1 + p - (2^d - 1)2\sqrt{d}\right)$$

which holds true by the hypothesis (14.5.2). ■

As stated before, we want to consider a general sequence of domains and show that the thermodynamic limit of the free energy is the same as for the standard sequence. The following corollary of the cheese theorem will be important for this purpose.

Corollary 14.1 (Filling General Domains by Balls). *Let $\Omega \subset \mathbb{R}^d$ be a bounded open set. For $p \in \mathbb{N}$ satisfying (14.5.2), let $R_j = (1+p)^j$ and $n_j = p^{j-1}(1+p)^{j(d-1)}$. Fix some $J \in \mathbb{N}$. Then the following two properties hold.*

(1) Let m denote the largest integer less than or equal to $[\,|\Omega| - V((1+p)^J \sigma_d^{1/d}\sqrt{d};\Omega)]/(\sigma_d(1+p)^{Jd})$. Assume that $m > 0$. Then Ω contains the disjoint union $\bigcup_{j=0}^{J-1}(m\, n_{J-j}$ balls of radius $R_j)$.

(2) *Let $\mathcal{B}$ be any domain containing Ω, and let $\Lambda := \mathcal{B} \setminus \Omega$. Furthermore, let μ denote the largest integer less than or equal to $[\,|\Lambda| - V(-(1+p)^J \sigma_d^{1/d}\sqrt{d};\Omega) - V((1+p)^J \sigma_d^{1/d}\sqrt{d};\mathcal{B})]/(\sigma_d(1+p)^{Jd})$. Assume that $\mu > 0$. Then Λ contains the disjoint union $\bigcup_{j=0}^{J-1}(\mu\, n_{J-j}$ balls of radius $R_j)$.*

The first part of the corollary says that as long as the surface to volume ratio of a domain is small, the domain can be packed very efficiently with balls from our standard sequence, in fact with the same efficiency as the unit ball can be packed with smaller balls. The second part says, in particular, that the complement of the domain that lies inside a circumscribed ball can also be packed in a similar fashion.

Proof. By the 'fundamental observation' explained in the proof of Theorem 14.3, the domain Ω contains m disjoint cubes of volume $\sigma_d(1+p)^{Jd}$. According to the last statement in the Cheese Theorem 14.3, these can be packed with balls as stated.

The second part of the Corollary is proved in exactly the same way, noting that $V(h;\Lambda) \leq V(-h;\Omega) + V(h;\mathcal{B})$ for $h > 0$. ■

Corollary 14.1 will be applied to large domains Ω that are sufficiently regular in the sense of Definition 14.1.

14.6 Proof of Theorem 14.2

To avoid unenlightening complications, we shall assume that Z is an integer, as it is in nature.

14.6.1 Proof for Special Sequences

Pick an integer p for which the conclusion of the cheese theorem holds. For each fixed density $\varrho > 0$ we choose a special sequence of domains Ω_j of increasing volume, with $j = 0, 1, \ldots$. This sequence will be the balls described in Section 14.5, except that we choose the radii such that the number of nuclei in the volumes are integers. Explicitly, Ω_j is a ball with volume $|\Omega_j| = Z(1+p)^{3j}/\varrho$, and $M_j = (1+p)^{3j}$ and $N_j = Z(1+p)^{3j}$. We denote the radius of Ω_0 by r_0, i.e., $(4\pi r_0^3)/3 = Z/\varrho$.

For $j > 0$, we decompose Ω_j into a finite set of balls according to Theorem 14.3. That is, Ω_j contains $n_k = p^{k-1}(1+p)^{2k}$ balls of radius $r_{j-k} =$

$(1+p)^{j-k}r_0$ with $k = 1, \ldots, j$. Let λ_k denote the fraction of the volume occupied by the balls of size r_{j-k}, i.e.,

$$\lambda_k = n_k \frac{r_{j-k}^3}{r_j^3} = \frac{1}{p}\left(\frac{p}{1+p}\right)^k. \tag{14.6.1}$$

Note that $\sum_{k=1}^{\infty} \lambda_k = 1$.

The fundamental inequality we would like to show is the following.

Lemma 14.2 (Recursion Relation for the Free Energy). *Let $f_j = f(\beta, \Omega_j, N_j, M_j)$ and let λ_k be given in (14.6.1). Then*

$$f_j \leq \sum_{k=1}^{j} \lambda_k f_{j-k} + c(1+p)^3 \lambda_{j+1} \tag{14.6.2}$$

for some constant $c > 0$ (depending only on ϱ).

Let us defer the proof of this lemma for the moment, and show how (14.6.2) implies the existence of the limit of f_j as $j \to \infty$. Let d_j be defined as

$$d_j = f_j - \sum_{k=1}^{j} \lambda_k f_{j-k}. \tag{14.6.3}$$

Note that (14.6.2) says that $d_j \leq c(1+p)^3 \lambda_{j+1}$. The solution to the set of equations (14.6.3) is

$$f_j = d_j + \lambda_1 \sum_{k=0}^{j-1} d_k, \tag{14.6.4}$$

which the reader can easily verify by induction, starting with $j = 1$. If we write

$$\sum_{k=0}^{j-1} d_k = \sum_{k=0}^{j-1} \left[d_k - c(1+p)^3 \lambda_{k+1}\right] + c(1+p)^3 \sum_{k=0}^{j-1} \lambda_{k+1},$$

we can see that the limit as $j \to \infty$ of this sum exists. In fact, the limit of the last sum on the right exists (and is finite), and the limit of the first sum on the right must exist since all the summands are non-positive. Moreover, it must be finite since f_j is bounded from below, uniformly in j. This implies that $\lim_{j\to\infty} d_j = 0$, and hence also f_j in (14.6.4) has a limit, which is finite. It is at this point that the stability of matter enters to show that the limit is finite instead of $-\infty$.

We are left with the task of proving (14.6.2), which is the heart of the matter and which is the reason for using balls as our fundamental domains.

Proof of Lemma 14.2. We shall employ the variational principle for the partition function in Lemma 14.1. The set of orthonormal functions will consist of functions Ψ that are products of functions ψ_α supported entirely in individual balls, or in the complement of all the balls. The functions ψ_α will be of the right symmetry type. Strictly speaking, Ψ has to be appropriately (anti)symmetrized among particles in different balls, but this will not have any effect, as we shall see. The key idea is to place the particles in the individual balls in such a way that the resulting state is invariant under rotations, in which case *Newton's theorem (Theorem 5.2 in Chapter 5) will imply that there is no average interaction between different balls*, as will be shown. If there were no interaction between balls then (14.6.2) would be an equality.

We write Ω_j as the union of balls labeled by $b = 1, 2, \ldots,$ and their complement in Ω_j, which we denote by Γ_j. There are $n_k = p^{k-1}(1+p)^{2k}$ balls of radius $r_{j-k} = (1+p)^{j-k} r_0$. Let N^b be equal to ϱ times the volume of the ball number b (not to be confused with N_j), and $M^b = N^b/Z$. The number of particles in Γ_j, the complement of the balls, is then $N^\gamma = N_j - \sum_b N^b$ and $M^\gamma = N^\gamma/Z$.

For every ball b, let ψ_α^b denote a set of orthonormal functions of $N^b + M^b$ particles supported in the ball b, with the correct permutation symmetry type. Let $\vec{\alpha}$ denote the multi-index $\vec{\alpha} = (\alpha^1, \alpha^2, \ldots)$. As N_j-particle wave functions on Ω_j we choose products of the form

$$\Psi_{\vec{\alpha}} = \psi^\gamma \prod_b \psi_{\alpha^b}^b,$$

where ψ^γ is a fixed normalized function of $N^\gamma + M^\gamma$ particles supported in the complement of all the balls. Let $\mathcal{S}$ denote the operator that projects onto the correct permutation symmetry type for *all* the N_j particles. We claim that $(\mathcal{S}\Psi_{\vec{\alpha}}, \mathcal{S}\Psi_{\vec{\alpha}'}) = 0$ if $\vec{\alpha} \neq \vec{\alpha}'$, and

$$\frac{(\mathcal{S}\Psi_{\vec{\alpha}}, H\,\mathcal{S}\Psi_{\vec{\alpha}})}{(\mathcal{S}\Psi_{\vec{\alpha}}, \mathcal{S}\Psi_{\vec{\alpha}})} = (\Psi_{\vec{\alpha}}, H\,\Psi_{\vec{\alpha}})$$

for the Hamiltonian (14.2.1). This claim follows from the fact that a permutation (of either the electrons or the nuclei) either leaves *all* the particles in the same domain, or it must move at least *two* particles out of their original domain. In the latter case, the one-particle terms in the Hamiltonian have vanishing matrix elements between the original and the permuted function. The same is true for the two-particle terms, simply because they are local.[7] (For the one-particle terms, locality is *not* needed, which is important in the relativistic case.)

In particular, the variational principle in Lemma 14.1 implies that

$$\exp\left[-\beta|\Omega_j|f_j\right] \geq \sum_{\vec{\alpha}} \exp[-\beta(\Psi_{\vec{\alpha}}, H\,\Psi_{\vec{\alpha}})].$$

The expectation value of H can be written as[8]

$$(\Psi_{\vec{\alpha}}, H\,\Psi_{\vec{\alpha}}) = \sum_b (\psi^b_{\alpha^b}, H\,\psi^b_{\alpha^b}) + \frac{\alpha}{2}\sum_{b\neq b'} \iint_{\mathbb{R}^3\times\mathbb{R}^3} \frac{Q^b_{\alpha^b}(\boldsymbol{x})Q^{b'}_{\alpha^{b'}}(\boldsymbol{y})}{|\boldsymbol{x}-\boldsymbol{y}|}\mathrm{d}\boldsymbol{x}\,\mathrm{d}\boldsymbol{y}$$

$$+(\psi^\gamma, H\,\psi^\gamma) + \alpha\sum_b \iint_{\mathbb{R}^3\times\mathbb{R}^3} \frac{Q^b_{\alpha^b}(\boldsymbol{x})Q^\gamma(\boldsymbol{y})}{|\boldsymbol{x}-\boldsymbol{y}|}\mathrm{d}\boldsymbol{x}\,\mathrm{d}\boldsymbol{y}, \qquad (14.6.5)$$

where Q^b_α denote the charge density of ψ^b_α, defined in Eq. (3.1.8) in Chapter 3, and likewise for Q^γ. We note that $\int Q^b_\alpha(\boldsymbol{x})\mathrm{d}\boldsymbol{x} = \int Q^\gamma(\boldsymbol{x})\mathrm{d}\boldsymbol{x} = 0$ because of charge neutrality in all the balls.

Jensen's inequality[9] implies that

$$\sum_{\vec{\alpha}} \exp[-\beta(\Psi_{\vec{\alpha}}, H\,\Psi_{\vec{\alpha}})]$$

$$\geq e^{-\beta\alpha W}\sum_{\vec{\alpha}} \exp\left[-\beta\sum\nolimits_b(\psi^b_{\alpha^b}, H\,\psi^b_{\alpha^b}) - \beta(\psi^\gamma, H\,\psi^\gamma)\right],$$

where W is the average value of the interaction between different domains, given by

$$W = \sum_{\vec{\alpha}} \xi_{\vec{\alpha}}\left(\frac{1}{2}\sum_{b\neq b'}\iint \frac{Q^b_{\alpha^b}(\boldsymbol{x})Q^{b'}_{\alpha^{b'}}(\boldsymbol{y})}{|\boldsymbol{x}-\boldsymbol{y}|}\mathrm{d}\boldsymbol{x}\,\mathrm{d}\boldsymbol{y} + \sum_b \iint \frac{Q^b_{\alpha^b}(\boldsymbol{x})Q^\gamma(\boldsymbol{y})}{|\boldsymbol{x}-\boldsymbol{y}|}\mathrm{d}\boldsymbol{x}\,\mathrm{d}\boldsymbol{y}\right) \qquad (14.6.6)$$

with

$$\xi_{\vec{\alpha}} = \exp\left[-\beta\sum\nolimits_b(\psi^b_{\alpha^b}, H\,\psi^b_{\alpha^b})\right]\left(\sum_{\vec{\alpha}}\exp\left[-\beta\sum\nolimits_b(\psi^b_{\alpha^b}, H\,\psi^b_{\alpha^b})\right]\right)^{-1}.$$

Note that $\sum_{\vec{\alpha}} \xi_{\vec{\alpha}} = 1$.

---

[7] For the notion of locality, see the footnote on page 181.

[8] The symbols H in this equation *et seq* refer to different particle numbers, which should be clear from the context.

[9] See the footnote on page 135.

Given the set of orthonormal functions ψ_α^b for a given ball b, we are at liberty to replace them by a new set which is obtained from the original set by rotation $\mathcal{R}^b$ of all the particle coordinates around the center of the ball. For this new set, the expectation value of H stays the same, while all the $Q_\alpha^b(\boldsymbol{x})$ change to $Q_\alpha^b(\mathcal{R}^b\boldsymbol{x})$. The only effect of these replacements (one for each ball b) will be to replace $Q^b_{\alpha^b}(\boldsymbol{x})$ by $Q^b_{\alpha^b}(\mathcal{R}^b\boldsymbol{x})$ for all $\vec{\alpha}$ and all b in (14.6.6), while $\xi_{\vec{\alpha}}$ remains unchanged.

The important point is that there is a choice of $\mathcal{R}^b$ such that $W \leq 0$. This follows from the fact that the average of W over all rotations of all the balls equals zero. This is a consequence of Newton's theorem, since all the terms in (14.6.6) vanish for radial Q^b. Note that for this argument it is not necessary for Q^γ to be radial.

We conclude that

$$\exp\left[-\beta|\Omega_j|f_j\right] \geq \exp\left[-\beta(\psi^\gamma, H\,\psi^\gamma)\right]\prod_b\left(\sum_\alpha \exp\left[-\beta(\psi_\alpha^b, H\,\psi_\alpha^b)\right]\right).$$

Since this inequality holds for any choice of the orthonormal functions ψ_α^b, the variational principle in Lemma 14.1 implies that

$$\exp\left[-\beta|\Omega_j|f_j\right] \geq \exp\left[-\beta(\psi^\gamma, H\,\psi^\gamma) - \beta\sum_{k=1}^{j} n_k|\Omega_{j-k}|f_{j-k}\right].$$

Since $\lambda_k = n_k|\Omega_{j-k}|/|\Omega_j|$ according to (14.6.1), this implies (14.6.2) with

$$c = \frac{1}{(1+p)^3\lambda_{j+1}|\Omega_j|}(\psi^\gamma, H\,\psi^\gamma).$$

It is left to derive a uniform upper bound on c. The number of particles N^γ in the complement Γ_j equals $\varrho|\Omega_j|(1+p)\lambda_{j+1}$. Moreover, according to the cheese theorem Γ_j contains $n_{j+1} = p^j(1+p)^{2(j+1)}$ disjoint balls of radius $(1+p)^{-1}r_0$. We place the particles in these balls, with an average number of $\varrho(4\pi/3)(1+p)^{-2}r_0^3$ per ball. By the same argument as above, we can arrange the states such that the interaction between balls is negative. The energy $(\psi^\gamma, H\,\psi^\gamma)$ is then bounded above by a constant (depending on ϱ) times N^γ which, for convenience, we write as a constant times $N^\gamma(1+p)^2 = |\Omega_j|(1+p)^3\lambda_{j+1}$. This completes the proof. ■

14.6.2 Proof for General Domains

We will now extend the analysis in the previous subsection to an arbitrary regular sequence of domains Ω_j. We choose the particle numbers M_j to be as close to $\varrho|\Omega_j|/Z$ as possible, i.e., $|M_j - \varrho|\Omega_j|/Z| \leq 1$. Moreover, $N_j = M_j Z$ because of neutrality. For this choice of particle numbers we shall show that $f(\beta, \Omega_j, N_j, M_j)/|\Omega_j|$ converges to the same number as for the standard sequence of domains.

Pick an $\varepsilon > 0$, and choose J large enough such that for $j \geq J$, f_j, the specific free energy of the standard sequence defined in the previous section, is within ε of its limiting value as $j \to \infty$, which we denote by $\bar{f}$ (and whose existence was proved above). Let $K > J$. According to Corollary 14.1, item (1), we can pack $m\, n_{K-j}$ balls of radius $(1+p)^j r_0$ into the domain Ω_j, $j = J, \ldots, K-1$, where m is the largest integer less than or equal to $\varrho(1+p)^{-3K}[|\Omega_j| - V((1+p)^K(4\pi/3)^{1/3}\sqrt{3}; \Omega_j)]$.

The volume that is occupied by these balls equals

$$V_j = \frac{m}{\varrho}(1+p)^{3K}\left(1 - \left(\frac{p}{p+1}\right)^{K-J}\right).$$

Because of the assumption that Ω_j is a Van Hove sequence, $m/|\Omega_j|$ converges to $\varrho(1+p)^{-3K}$ as $j \to \infty$. We thus conclude that $\lim_{j\to\infty} V_j/|\Omega_j| = 1 - (p/(1+p))^{K-J}$ for fixed K and J.

Proceeding as in the previous subsection and using that $f_j \leq \bar{f} + \varepsilon$ for $J \leq j \leq K$, we have

$$f(\beta, \Omega_j, N_j, M_j) \leq \frac{V_j}{|\Omega_j|}(\bar{f} + \varepsilon) + \frac{C}{|\Omega_j|}(N_j - \varrho V_j)$$

for some $C > 0$ (depending only on ϱ and p). The last term is the contribution from the $N_j - \varrho V_j$ particles outside the balls. Note that $N_j - \varrho V_j$ equals ϱ times the volume of the domain plus or minus one particle. By putting these particles into appropriate smaller balls, as explained at the end of the previous subsection, their contribution to the free energy is easily seen to be proportional to the total particle number in this domain.

We conclude that

$$\limsup_{j\to\infty} f(\beta, \Omega_j, N_j, M_j) \leq (\bar{f} + \varepsilon)\left(1 - \left(\frac{p}{p+1}\right)^{K-J}\right) + C\varrho\left(\frac{p}{p+1}\right)^{K-J}.$$

Since K can be made arbitrarily large and ε is arbitrarily small, this proves that

$$\limsup_{j\to\infty} f(\beta, \Omega_j, N_j, M_j) \le \bar{f}.$$

For the lower bound, we proceed in a similar way, this time using item (2) of Corollary 14.1. For given j, we choose $\mathcal{B}_j$ to be the smallest ball containing Ω_j and whose radius equals $(1+p)^n r_0$ for some integer n. For fixed $J < K$ chosen as before, such that $|f_j - \bar{f}| \le \varepsilon$ for all $j \ge J$, $\mathcal{B}_j \setminus \Omega_j$ can be packed with μn_{K-j} balls of radius $(1+p)^j r_0$, with $j = J, \ldots, K-1$, where μ equals the largest integer less than or equal to $\varrho(1+p)^{-3K}[\,|\mathcal{B}_j \setminus \Omega_j| - V(-(1+p)^K(4\pi/3)^{1/3}\sqrt{3}; \Omega_j) - V((1+p)^K(4\pi/3)^{1/3}\sqrt{3}; \mathcal{B}_j)]$. The total volume of these balls is now

$$V_j = \frac{\mu}{\varrho}(1+p)^{3K}\left(1 - \left(\frac{p}{p+1}\right)^{K-J}\right).$$

We consider the free energy for the domain $\mathcal{B}_j$ with $\varrho|\mathcal{B}_j|$ particles. Again by putting the particles appropriately into the balls, we conclude that

$$|\mathcal{B}_j|(\bar{f} - \varepsilon) \le |\Omega_j| f(\beta, \Omega_j, N_j, M_j) + V_j(\bar{f} - \varepsilon) + C\left(\varrho|\mathcal{B}_j| - N_j - \varrho V_j\right).$$

The last term comes from particles outside Ω_j and all the balls. It is important that these particles also be placed appropriately in balls in order to make sure that their Coulomb interaction with particles in the other domains vanishes.

The regularity of the sequence Ω_j implies that $(|\mathcal{B}_j| - V_j)/|\Omega_j| \to 1 - (p/(1+p))^{K-J}$ as $j \to \infty$. Moreover, $(|\mathcal{B}_j| + V_j)/|\Omega_j| \le C'$ for some constant C' independent of j. Hence

$$\liminf_{j\to\infty} f(\beta, \Omega_j, N_j, M_j) \ge \bar{f}\left(1 - \left(\frac{p}{p+1}\right)^{K-J}\right) - \varepsilon C' - C\varrho\left(\frac{p}{p+1}\right)^{K-J}.$$

Since K and ε are arbitrary, this concludes the proof of the fact that

$$\lim_{j\to\infty} f(\beta, \Omega_j, N_j, M_j) = \bar{f}$$

for any regular sequence of domains Ω_j and appropriately chosen particle numbers satisfying $|M_j - \varrho|\Omega_j|/Z| \le 1$.

14.6.3 Convexity

For each fixed β and ϱ, the specific free energy has a thermodynamic limit, which does not depend on the chosen sequence of domains. We denote it by $\bar{f}(\beta, \varrho)$. In this subsection, we shall show that $\bar{f}$ is a convex function of ϱ and a concave function of $1/\beta$.

We first show convexity in ϱ. Fix $0 < \lambda < 1$. We choose a sequence of domains Ω_j, such that every Ω_j consists of two disjoint balls Ω_j^1 and Ω_j^2 of volume V_j^1 and V_j^2, respectively, with the property that

$$\frac{V_j^1}{V_j^1 + V_j^2} = \lambda \quad \text{for all } j.$$

This sequence of domains satisfies the regularity condition of Definition 14.1. Choose M_j^1 and M_j^2 such that $|M_j^1 - \varrho_1 V_j^1/Z| \leq 1/2$ and $|M_j^2 - \varrho_2 V_j^2/Z| \leq 1/2$. Then

$$\left| \frac{M_j^1 + M_j^2}{V_j^1 + V_j^2} - \lambda\varrho_1 - (1-\lambda)\varrho_2 \right| \leq 1.$$

By locating N_j^1 particles in the ball Ω_j^1 and N_j^2 particles in the ball Ω_j^2, we conclude (using the variational principle and Newton's theorem) that

$$\begin{aligned} f(\beta, \Omega_j, N_j^1 + N_j^2, M_j^1 + M_j^2) \leq{}& \lambda f(\beta, \Omega_j^1, N_j^1, M_j^1) \\ &+ (1-\lambda) f(\beta, \Omega_j^1, N_j^1, M_j^1). \end{aligned} \tag{14.6.7}$$

All the three quantities converge in the limit $j \to \infty$ to $\bar{f}$ with the respective densities. That is, in the limit $j \to \infty$ (14.6.7) becomes

$$\bar{f}(\beta, \lambda\varrho_1 + (1-\lambda)\varrho_2) \leq \lambda \bar{f}(\beta, \varrho_1) + (1-\lambda)\bar{f}(\beta, \varrho_2).$$

This proves convexity of $\bar{f}$ in ϱ. In particular, $\bar{f}$ is a continuous function of ϱ, since all bounded convex functions are continuous.

The concavity of $\bar{f}$ in $1/\beta$ follows easily from the variational principle, Lemma 14.1. Since $\ln \sum_i \exp[-\beta(\phi_i, H\phi_i)]$ is convex in β, $\ln Z$ is also convex in β. Hence $-\beta^{-1} \ln Z$ is concave in $1/\beta$, for every given domain and particle number. The pointwise limit of concave functions is concave.

14.6.4 General Sequences of Particle Numbers

In order to conclude the proof of Theorem 14.2, we still have to show that $f(\beta, \Omega_j, N_j, M_j)$ converges to $\bar{f}(\beta, \varrho)$ for arbitrary sequences of particle numbers satisfying $\lim_{j\to\infty} N_j/|\Omega_j| = \varrho$. Our main tool will be the monotonicity of the free energy in the size of the domain.

For a given Ω and $t > 0$, let $t\,\Omega$ denote the scaled domain. The function $t \mapsto F(\beta, t\,\Omega, N, M)$ is clearly monotone decreasing in t, by the variational principle, Lemma 14.1. In particular, since $\varrho_j := N_j/|\Omega_j| \leq \varrho(1+\varepsilon)$ for $\varepsilon > 0$ and large enough j, we have $F(\beta, \Omega_j, N_j, M_j) \leq F(\beta, [\varrho_j/(\varrho(1+\varepsilon))]\Omega_j, N_j, M_j)$. (Note that $N_j/[\varrho_j/(\varrho(1+\varepsilon))|\Omega_j|] = \varrho(1+\varepsilon)$, independently of j.) The latter quantity, divided by $|\Omega_j|$, converges to $(1+\varepsilon)^{-1}\bar{f}(\beta, \varrho(1+\varepsilon))$ as $j \to \infty$.

In particular,

$$\limsup_{j\to\infty} \frac{1}{|\Omega_j|} F(\beta, \Omega_j, N_j, M_j) \leq \frac{1}{1+\varepsilon} \bar{f}(\beta, \varrho(1+\varepsilon)).$$

In the same way one shows that

$$\liminf_{j\to\infty} \frac{1}{|\Omega_j|} F(\beta, \Omega_j, N_j, M_j) \geq \frac{1}{1-\varepsilon} \bar{f}(\beta, \varrho(1-\varepsilon)).$$

Since ε is arbitrary and $\bar{f}$ is continuous in ϱ, this proves the claim. Recall that the continuity of $\bar{f}$ follows from its convexity, which was shown in the previous subsection. ■

14.7 The Jellium Model

The jellium model was invented by Wigner [187] as a simplified model of a solid. In this model there is only one kind of particle and they move in a fixed, uniform background of electric charge of the opposite sign. In Wigner's interpretation of the model the particles were nuclei and the background was formed by the sea of electrons in a solid. In the absence of quantum mechanics and the uncertainty principle the nuclei will locate themselves at positions that minimize the total energy. Presumably they will form a regular lattice, but this has never been proved. In the quantum case the particles can be expected to form a lattice only when the density is small, so that the kinetic energy, $\int |\nabla\psi|^2$, will not get in the way of localization. A phase transition is supposed to occur from a smeared out state at high density to a localized lattice state at low density.

Here, we shall continue with our previous convention and call the moveable particles electrons. The background will have uniform positive charge density ϱ in some domain Ω. The number of electrons is $N = \varrho\,|\Omega|$ for charge neutrality, and they are confined to the domain Ω. The Hamiltonian is

$$\boxed{H = \sum_{i=1}^{N}\left(\frac{1}{2}\boldsymbol{p}_i^2 - \alpha W(\boldsymbol{x}_i)\right) + \alpha I(\underline{X}) + \alpha C,} \tag{14.7.1}$$

where

$$\begin{aligned} W(\boldsymbol{x}) &= \varrho \int_{\Omega} \frac{1}{|\boldsymbol{x}-\boldsymbol{y}|}\mathrm{d}\boldsymbol{y} \\ I(\underline{X}) &= \sum_{1\le i<j\le N} \frac{1}{|\boldsymbol{x}_i-\boldsymbol{x}_j|}, \\ C &= \frac{1}{2}\varrho^2 \int_{\Omega}\int_{\Omega} \frac{1}{|\boldsymbol{x}-\boldsymbol{y}|}\mathrm{d}\boldsymbol{x}\mathrm{d}\boldsymbol{y} = \frac{1}{2}\varrho \int_{\Omega} W(\boldsymbol{x})\mathrm{d}\boldsymbol{x}. \end{aligned} \tag{14.7.2}$$

This Hamiltonian is appropriate both in the classical and the quantum cases. In the latter $\boldsymbol{p}^2$ is to be interpreted as $-\Delta$ with Dirichlet boundary conditions on Ω.

Note that $-\Delta W(\boldsymbol{x}) = 4\pi\varrho$ for $\boldsymbol{x} \in \Omega$ while $-\Delta W(\boldsymbol{x}) = 0$ for $\boldsymbol{x} \notin \Omega$. If Ω is a ball of radius R we can compute W. Namely, $W(\boldsymbol{x}) = \pi\varrho(6R^2 - 2|\boldsymbol{x}|^2)/3$ for $|\boldsymbol{x}| \le R$, while $W(\boldsymbol{x}) = 4\pi R^3\varrho/(3|\boldsymbol{x}|)$ for $|\boldsymbol{x}| \ge R$.

Even if we omit the kinetic energy term there is stability of the second kind. This can be seen by replacing each charge $-\sqrt{\alpha}$ at $\boldsymbol{x}_i$ by the same charge uniformly distributed in a small ball of radius r centered at $\boldsymbol{x}_i$. (Compare with the discussion of the Onsager hard-core gas in Section 14.4.) By Newton's theorem the interaction of the small balls is smaller than $|\boldsymbol{x}_i - \boldsymbol{x}_j|^{-1}$. The interaction of the ball with the background is a little less negative than it is for a point, but only a little. The reader can easily work out that the optimum value of r is $r = (3/4\pi\varrho)^{1/3}$ and the energy is bounded from below by $-0.9\,\alpha N(4\pi\varrho/3)^{1/3}$. (For details of the calculation see [124].)

The existence of the thermodynamic limit of the free energy, in analogy with our Theorem 14.2, was proved by Lieb and Narnhofer [124]. We shall content ourselves with just outlining the essential new feature that has to be considered for this system. This concerns the distribution of the particles into the subdomain of a big ball that is not covered by smaller balls.

Before we do so, let us point out that this system is *not* thermodynamically stable in the following sense. Theorem 14.2 states that the limiting free energy is convex in ϱ and concave in $1/\beta$ and we explained the physical significance of these properties. While the concavity in $1/\beta$ will continue to hold in the jellium model, the convexity in ϱ will not hold, generally. The reason for lack of ϱ convexity is that the background is held fixed like the membrane of a drum. If it were allowed to adjust itself by changing its size, and thereby its charge density, in order to minimize f, then convexity would be restored.

Now we turn to the proof of the existence of the thermodynamic limit of the specific free energy, the main point of which is to prove the limit, at fixed ϱ, for the standard increasing sequence of balls. For the standard sequence, we shall give the complete proof in the classical case and remark on the necessary modification in the quantum case at the end of this section.

Let us first discuss the classical case. We take a ball B_j and fill it with balls $B_{j-1}, \ldots, B_0$ as before. There will be a certain part of B_j that is not covered by the smaller balls and we call this region D_j. In each subdomain we put exactly the right number of electrons so that the density is the given ϱ, namely the number of electrons equals ϱ times the volume of the subdomain. This means that in doing the configurational integral for the partition function, we integrate only over that subset of Ω^N where the number of particles in each subdomain is precisely ϱ times the volume of the ball. This obviously gives a lower bound on Q. The number of ways we can distribute the particles in the subdomains is the multinomial coefficient $N!$ divided by the product of the factorials of the individual particle numbers in the subdomains.

The peculiar region D_j also contains electrons at density ϱ. We have to construct the filling this way, for if we had tried to put all the electrons into the small balls, with none in D_j, the electrostatic energy of such a configuration would be significantly too high, even for large j where the fraction of the volume covered by the balls is very close to 1. In the proof of Theorem 14.2, we confined these particles to smaller balls of size B_{-1}. In the jellium case we can not do that; the resulting Coulomb energy in D_j would be too big since the charge distribution would then be far from uniform.

If we could eliminate the Coulomb interaction among the different subdomains we would obtain the following upper bound on the configurational free energy:

$$f_j^{\text{conf}} \leq \sum_{k=1}^{j} \lambda_k f_{j-k}^{\text{conf}} + p\,\lambda_j\,\phi_j \tag{14.7.3}$$

where f_j^{conf} is the configurational specific free energy for the domain B_j, and ϕ_j is the configurational specific free energy of D_j. Note that $p\,\lambda_j = 1 - \sum_{k=1}^{j} \lambda_k = |D_j|/|B_j|$. The factorials of the particle numbers coming from the permutations of the particles exactly compensate the denominators appearing in the definition (14.4.3) of the configurational partition function.

We claim that Eq. (14.7.3) is valid even in the presence of the Coulomb interactions between the different subdomains. The integral over configurations of particles within each ball contains an average over rotations of any given configuration. By Jensen's inequality and convexity of the exponential function (see the footnote on page 135) the average of the exponential function can be bounded from below by the exponential of the average. Since all the subdomains except for one (namely, D_j) are neutral balls, the average over rotations of the interaction energy between different subdomains is zero by Newton's theorem. In this way, we arrive at the fundamental inequality (14.7.3). We note that the argument given here is the classical version of the argument in the quantum case given in Lemma 14.2.

It remains to find an upper bound to ϕ_j. We shall show that (for large j)

$$\phi_j \leq \frac{\varrho}{\beta}(\ln \varrho - 1). \tag{14.7.4}$$

This follows again from Jensen's inequality, this time applying it to the integration over *all* configurations of the electrons in D_j. The average Coulomb energy within D_j is negative when all the electrons are smeared out uniformly over D_j (namely the number of electrons in D_j times the negative of the self-energy of a single electron smeared out over D_j), and hence the integral of the exponential factor is bounded from above by $|D_j|^{\varrho|D_j|}/(\varrho|D_j|)!$. An application of Stirling's formula yields (14.7.4).

By our previous analysis of inequalities of the type (14.7.3), we know that a limit exists for f_j^{conf} as $j \to \infty$. This finishes the proof in the classical case for the standard sequence. The proof for general sequences of domains and particle numbers parallels that in Section 14.6. For details we refer the reader to [124].

The proof in the quantum case has the same structure. The only new difficulty comes from estimating ϕ_j, the specific free energy of D_j. We can not just smear the electrons over the domain D_j, as we did in the classical case, for two reasons. One is the Dirichlet boundary conditions on D_j which prevents the particles from getting too close to the boundary of D_j, and the other is the antisymmetry requirement on the allowed wave functions, which also raises the kinetic energy. These are technical rather than conceptual problems and are overcome in [124, Sect. 5.1].

List of Symbols

$\boldsymbol{E}(\boldsymbol{x})$	Electric field	200
$\mathcal{E}(\psi)$	Energy functional	14
$\mathcal{E}_{\mathrm{mag}}(\boldsymbol{B})$	Magnetic field energy	11
$\mathcal{E}^{\mathrm{TF}}(\varrho)$	Thomas–Fermi functional	128
$f(\beta, \Omega, N, M)$	Specific free energy	254
$F(\beta, \Omega, N, M)$	Free energy	254
$\boldsymbol{F}$	Force	9
$\mathcal{F}$	Fock space	208
g	Gyromagnetic ratio	54
G	Gravitational constant	55
$G_e(\boldsymbol{x})$	Green's functions for $-\Delta + e$	76
h	Planck's constant	15
$\hbar$	$h/(2\pi)$	15
H	Hamiltonian	15
H_f	Quantized electromagnetic field energy	209
$H(\boldsymbol{x}, \boldsymbol{p})$	Hamilton function	10
$H^1(\mathbb{R}^d)$	Set of functions f with $f \in L^2(\mathbb{R}^d)$ and $\nabla f \in L^2(\mathbb{R}^d)$	16
$\mathcal{H}_{\boldsymbol{A}}^+$	Positive spectral subspace of $D_{\boldsymbol{A}}$	193
$I(\underline{\boldsymbol{X}})$	Electron–electron repulsion	13
I_0	Foldy's constant	137
$\mathcal{I}$	Negative ionization	222
$\boldsymbol{j}$	Current density	201
K	Constant in the kinetic energy inequality	71
$L_{\gamma,d}$	Constants in the Lieb–Thirring inequality	66
$L_{\gamma,d}^{\mathrm{cl}}$	Classical Lieb–Thirring constants	64
$L^2(\mathbb{R}^d)$	Space of square-integrable functions	16
$L^2(\mathbb{R}^d; \mathbb{C}^q)$	Space of square-integrable functions with values in $\mathbb{C}^q$	18
$L^p(\mathbb{R}^d)$	Space of pth power integrable functions	26
$\mathcal{L}$	Lagrangian	204
M	Number of nuclei	13
N	Number of electrons	13
N_e	Number of eigenvalues less than or equal to $-e$	77
$\boldsymbol{p}$	Canonical momentum	10
P_N	N particle density	105
$\mathbb{P}$	Projection onto divergence-free vector fields	203
q	Number of spin states	34
Q	Configurational partition function	259
$Q_\psi(\boldsymbol{x})$	Charge density	32
$\underline{\boldsymbol{R}}$	Nuclear coordinates $\boldsymbol{R}_1, \ldots, \boldsymbol{R}_M$	13

$\mathcal{S}$	Symmetrization operator	208
$T(\boldsymbol{p})$	Kinetic energy	10
T_ψ	Expectation value of the kinetic energy	14
Tr	Trace	39
$\mathrm{Tr}^{(N-k)}$	Partial trace over $N-k$ variables	43
$U(\underline{\boldsymbol{R}})$	Nucleus–nucleus repulsion	13
$V_C(\underline{\boldsymbol{X}}, \underline{\boldsymbol{R}})$	Coulomb potential energy	13
V_ψ	Expectation value of the potential energy	15
$V(h;\Omega)$	Measure of points a distance h from the boundary of Ω	255
$W(\underline{\boldsymbol{X}}, \underline{\boldsymbol{R}})$	Electron–nucleus attraction	13
$\boldsymbol{x}_i$	Position of particle i	9
$\underline{\boldsymbol{X}}$	Coordinates of N particles $\boldsymbol{x}_1, \ldots, \boldsymbol{x}_N$	31
$\underline{z}$	Space-spin coordinates of N particles $z_1, \ldots, z_N$, with $z_i = (\boldsymbol{x}_i, \sigma_i)$	34
$\underline{Z}$	Nuclear charges $Z_1, \ldots, Z_M$	13
Z	Maximal nuclear charge $Z = \max\{Z_1, \ldots, Z_M\}$	52
Z_{tot}	Total nuclear charge $\sum_{k=1}^{M} Z_k$	221
$Z(\beta, \Omega, N, M)$	Partition function	253
α	Fine-structure constant	1
$\boldsymbol{\alpha}, \beta$	Dirac matrices	182
Δ	Laplacian $= \boldsymbol{\nabla}^2$	15
$\boldsymbol{\varepsilon}_\lambda(\boldsymbol{k})$	Polarization vectors	206
$\gamma^{(k)}$	k-particle reduced density matrix	42
$\mathring{\gamma}^{(1)}$	Spin-summed one-particle density matrix	44
Γ	Density matrix	39
Γ_j	Voronoi cell	92
κ	$Gm^2/(\hbar c)$, where m is the particle mass	55
λ_C	Compton wavelength	22
Λ_A^+	Projection onto the positive spectral subspace of D_A	187
Λ	Ultraviolet cutoff	210
μ_0	Bohr magneton	20
ψ	Wave function	14
ϕ	Scalar potential	202
$\boldsymbol{\Pi}$	Canonically conjugate momentum to $\boldsymbol{A}$	205
$\varrho_\psi(\boldsymbol{x})$	One-particle density	32
σ_d	Volume of the unit ball in $\mathbb{R}^d$	260
$\boldsymbol{\sigma}$	Pauli matrices	19
$\underline{\sigma}$	Spins of N particles $\sigma_1, \ldots, \sigma_N$	34
Θ	Heaviside step function	64

Bibliography

Note: The page numbers {pages xxx} are the pages in the text where the reference is quoted.

[1] C. Adam, B. Muratori, and C. Nash, *Zero modes of the Dirac operator in three dimensions*, Phys. Rev. D **60**, 125001 (1999). {page 167.}

[2] M. Aizenman and E. H. Lieb, *On Semi-Classical Bounds for Eigenvalues of Schrödinger Operators*, Phys. Lett. A **66**, 427–429 (1978). {page 68.}

[3] W. Anderson, *Über die Grenzdichte der Materie und der Energie*, Z. Phys. **56**, 851–856 (1929). {page 234.}

[4] H. Araki, *On an inequality of Lieb and Thirring*, Lett. Math. Phys. **19**, 167–170 (1990). {page 85.}

[5] A. A. Balinsky and W. D. Evans, *Stability of one-electron molecules in the Brown–Ravenhall model*, Commun. Math. Phys. **202**, 481–500 (1999). {page 192.}

[6] A. A. Balinsky and W. D. Evans, *On the zero modes of Pauli operators*, J. Funct. Anal. **179**, 120–135 (2001). *On the zero modes of Weyl–Dirac operators and their multiplicity*, Bull. London Math. Soc. **34**, 236–242 (2002). {page 167.}

[7] J. Bardeen, L. N. Cooper, and J. R. Schrieffer, *Microscopic Theory of Superconductivity*, Phys. Rev. **106**, 162–164 (1957). {page 136.}

[8] B. Baumgartner, *On the degree of ionization in the TFW theory*, Lett. Math. Phys. **7**, 439–441 (1983). {page 228.}

[9] B. Baumgartner, *On Thomas–Fermi–von Weizsäcker and Hartree energies as functions of the degree of ionization*, J. Phys. A: Math. Gen. **17**, 1593–1602 (1984). {page 229.}

[10] R. J. Baxter, *Inequalities for potentials of particle systems*, Ill. J. Math. **24**, 645–652 (1980). {page 98.}

[11] E. Beckenbach and R. Bellman, *Inequalities*, Springer-Verlag (1961). {page 86.}

[12] W. Beckner, *Pitt's inequality and the uncertainty principle*, Proc. Amer. Math. Soc. **123**, 1897–1905 (1995). {page 143.}

[13] R. Benguria, *The Lane–Emden equation revisited*, in: *Advances in differential equations and mathematical physics*, Birmingham, AL, 2002, Am. Math. Soc., Providence, RI, Contemp. Math. **327**, 11–19 (2003). {page 137.}

[14] R. Benguria and E. H. Lieb, *Proof of the Stability of Highly Negative Ions in the Absence of the Pauli Principle*, Phys. Rev. Lett. **50**, 1771–1774 (1983). {page 229.}

[15] R. Benguria and M. Loss, *A simple proof of a theorem of Laptev and Weidl*, Math. Res. Lett. **7**, 195–203 (2000). {page 67.}

[16] H. A. Bethe and E. E. Salpeter, *Quantum mechanics of one- and two-electron atoms*, Academic Press (1957). {pages 186, 191.}

[17] R. Bhatia, *Matrix Analysis*, Springer (1997). {pages 88, 178.}

[18] M. Sh. Birman, *The spectrum of singular boundary problems*, Math. Sb. **55**, 124–174 (1961). English transl. in Amer. Math. Soc. Transl. Ser. 2, **53**, 23–80 (1966). {pages 76, 80.}

[19] M. Sh. Birman, L. S. Koplienko, and M. Z. Solomyak, *Estimates for the spectrum of the difference between fractional powers of two self-adjoint operators*, Soviet Mathematics **19**, 1–6 (1975). Translation of Izvestiya Vysshikh Uchebnykh Zavedenii. Matematika **19**, 3–10 (1975). {page 178.}

[20] M. Sh. Birman and M. Z. Solomyak, *Spectral Asymptotics of nonsmooth elliptic operators II*, Trans. Moscow Math. Soc. **28**, 1–32 (1973). Translation of Trudy Moskov. Mat. Obšč. {pages 171, 175, 178.}

[21] N. N. Bogoliubov, *On the theory of superfluidity*, J. Phys. (U.S.S.R.) **11**, 23 (1947). {page 136.}

[22] A. Bohm, *Quantum Mechanics: Foundations and Applications*, Springer Texts and Monographs in Physics, third edition (1993). {page xiv.}

[23] G. Brown and D. Ravenhall, *On the interaction of two electrons*, Proc. Roy. Soc. London A **208**, 552–559 (1951). {page 186.}

[24] D. Brydges and P. Federbush, *A note on energy bounds for boson matter*, J. Math. Phys. **17**, 2133–2134 (1976). {page 133.}

[25] L. Bugliaro, C. Fefferman, J. Fröhlich, G. M. Graf, and J. Stubbe, *A Lieb–Thirring bound for a magnetic Pauli Hamiltonian*, Commun. Math. Phys. **187**, 567–582 (1997). {page 82.}

[26] L. Bugliaro, C. Fefferman, and G. M. Graf, *A Lieb–Thirring bound for a magnetic Pauli Hamiltonian, II*, Rev. Mat. Iberoamericana **15**, 593–619 (1999). {pages 82, 217.}

[27] L. Bugliaro, J. Fröhlich, and G. M. Graf, *Stability of quantum electrodynamics with nonrelativistic matter*, Phys. Rev. Lett. **77**, 3494–3497 (1996). {pages 171, 213.}

[28] H. Cavendish, *Experiments to determine the density of the earth*, Phil. Trans. R. Soc. **88**, 469–526 (1798). {page 55.}

[29] G. K.-L. Chan and N. C. Handy, *Optimized Lieb–Oxford bound for the exchange-correlation energy*, Phys. Rev. A **59**, 3075–3077 (1999). {page 111.}

[30] S. Chandrasekhar, *The Maximum Mass of Ideal White Dwarfs*, Astrophysical Jour. **74**, 81–82 (1931); see also *On stars, their evolution and their stability*, Rev. Mod. Phys. **56**, 137–147 (1984). {pages 234, 242.}

[31] A. J. Coleman, *Structure of Fermion Density Matrices*, Rev. Mod. Phys. **35**, 668–686 (1963). {page 46.}

[32] J. G. Conlon, *The ground state energy of a classical gas*, Commun. Math. Phys. **94**, 439–458 (1984). {pages 4, 157.}

[33] J. G. Conlon, *A new proof of the Cwikel–Lieb–Rosenbljum bound*, Rocky Mountain J. Math. **15**, 117–122 (1985). {page 67.}

[34] J. G. Conlon, E. H. Lieb, and H.-T. Yau, *The $N^{7/5}$ Law for Charged Bosons*, Commun. Math. Phys. **116**, 417–448 (1988). {pages 130, 135, 137, 138.}

[35] M. Cwikel, *Weak type estimates for singular values and the number of bound states of Schrödinger operators*, Ann. of Math. **106**, 93–100 (1977). {page 67.}

[36] H. L. Cycon, R. G. Froese, W. Kirsch, and B. Simon, *Schrödinger operators with application to quantum mechanics and global geometry*, Springer Texts and Monographs in Physics (1987). {page 215.}

[37] I. Daubechies, *An uncertainty principle for fermions with generalized kinetic energy*, Commun. Math. Phys. **90**, 511–520 (1983). {pages 69, 70, 154.}

[38] I. Daubechies and E. H. Lieb, *One-electron relativistic molecules with Coulomb interaction*, Commun. Math. Phys. **90**, 497–510 (1983). {pages 57, 140, 141, 157, 158.}

[39] P. A. M. Dirac, *The quantum theory of the electron*, Proc. Roy. Soc. London A **117**, 610–624 (1928). *Part II, ibid.* **118**, 351–361 (1928). {pages 4, 181.}

[40] P. A. M. Dirac, *Note on exchange phenomena in the Thomas atom*, Proc. Camb. Phil. Soc. **26**, 376–385 (1930). {page 108.}

[41] J. Dolbeault, A. Laptev, and M. Loss, *Lieb–Thirring inequalities with improved constants*, J. Eur. Math. Soc. **10**, 1121–1126 (2008). {page 67.}

[42] W. F. Donoghue, *Monotone Matrix Functions and Analytic Continuation*, Springer (1974). {page 88.}

[43] F. J. Dyson, *Ground-state energy of a finite system of charged particles*, J. Math. Phys. **8**, 1538–1545 (1967). {pages 3, 135, 137, 138.}

[44] F. J. Dyson and A. Lenard, *Stability of matter I*, J. Math. Phys. **8**, 423–434 (1967); *II, ibid.* **9**, 1538–1545 (1968). {pages 3, 3, 6, 62, 121, 130, 133.}

[45] D. M. Elton, *New examples of zero modes*, J. Phys. A **33**, 7297–7303 (2000). *The local structure of zero mode producing magnetic potentials*, Commun. Math. Phys. **229**, 121–139 (2002). {page 167.}

[46] H. Epstein, *Remarks on two theorems of E. Lieb*, Commun. Math. Phys. **31**, 317–325 (1973). {page 85.}

[47] L. Erdős, *Magnetic Lieb–Thirring inequalities*, Commun. Math. Phys. **170**, 629–668 (1995). {page 82.}

[48] L. Erdős and J. P. Solovej, *The kernel of Dirac operators on $\mathbb{S}^3$ and $\mathbb{R}^3$*, Rev. Math. Phys. **13**, 1247–1280 (2001). {page 167.}

[49] L. Erdős and J. P. Solovej, *Magnetic Lieb–Thirring inequalities with optimal dependence on the field strength*, J. Stat. Phys. **116**, 475–506 (2004). {page 82.}

[50] L. Erdős and J. P. Solovej, *Uniform Lieb–Thirring inequality for the three-dimensional Pauli operator with a strong non-homogeneous magnetic field*, Ann. Henri Poincaré **5**, 671–741 (2004). {page 82.}

[51] H. Eschrig, *The Fundamentals of Density Functional Theory*, second edition, Edition am Gutenbergplatz Leipzig (2003). {page 105.}

[52] M. J. Esteban, M. Lewin, and E. Séré, *Variational methods in relativistic quantum mechanics*, Bull. Amer. Math. Soc. **45**, 535–593 (2008). {page 191.}

[53] M. J. Esteban and M. Loss, *Self-adjointness for Dirac operators via Hardy–Dirac inequalities*, J. Math. Phys. **48**, 112107-1–8 (2007); see also *Self-adjointness via partial Hardy-like inequalities*, in *Mathematical Results in Quantum Mechanics*, proceedings of QMath 10, Moeciu, Romania, Beltita *et al.* eds., pp. 41–47, World Scientific (2008). {page 191.}

[54] W. D. Evans, P. Perry, and H. Siedentop, *The spectrum of relativistic one-electron atoms according to Bethe and Salpeter*, Commun. Math. Phys. **178**, 733–746 (1996). {page 192.}

[55] P. Federbush, *A new approach to the stability of matter problem*, J. Math. Phys. **16**, 347–351 (1975); *ibid.* 706–709 (1975). {page 130.}

[56] C. Fefferman, *The uncertainty principle*, Bull. Amer. Math. Soc. **9**, 129–206 (1983). {pages 67, 130.}

[57] C. Fefferman, *Stability of Coulomb systems in a magnetic field*, Proc. Natl. Acad. Sci. USA **92**, 5006–5007 (1995). {pages 4, 171.}

[58] C. Fefferman, *On electrons and nuclei in a magnetic field*, Adv. Math. **124**, 100–153 (1996). {pages 4, 171.}

[59] C. Fefferman, J. Fröhlich, and G. M. Graf, *Stability of ultraviolet-cutoff quantum electrodynamics with non-relativistic matter*, Commun. Math. Phys. **190**, 309–330 (1997). {page 217.}

[60] C. Fefferman and R. de la Llave, *Relativistic Stability of Matter. I.*, Rev. Mat. Iberoamericana **2**, 119–215 (1986). {page 157.}

[61] C. L. Fefferman and L. A. Seco, *Asymptotic neutrality of large ions*, Commun. Math. Phys. **128**, 109–130 (1990). {page 229.}

[62] E. Fermi, *Un metodo statistico per la determinazione di alcune priorieta dell'atomo*, Atti Acad. Naz. Lincei, Rend. **6**, 602–607 (1927). {page 3.}

[63] E. Fermi, *Quantum theory of radiation*, Rev. Mod. Phys. **4**, 87–132 (1932). {page 206.}

[64] M. Fierz and W. Pauli, *On relativistic wave equations for particles of arbitrary spin in an electromagnetic field*, Proc. Roy. Soc. London A **173**, 211–232 (1939). {page 21.}

[65] M. E. Fisher, *The free energy of a macroscopic system*, Arch. Rat. Mech. Anal. **17**, 377–410 (1964). {page 257.}

[66] M. E. Fisher and D. Ruelle, *The Stability of Many-Particle Systems*, J. Math. Phys. **7**, 260–270 (1966). {page 3.}

[67] V. Fock, *Konfigurationsraum und zweite Quantelung*, Z. Phys. **75**, 622–647 (1932). {page 208.}

[68] L. L. Foldy, *Charged boson gas*, Phys. Rev. **124**, 649–651 (1961). *Errata ibid.* **125**, 2208 (1962). {page 137.}

[69] R. H. Fowler, *On dense matter*, Monthly Notices Royal Astr. Soc. **87**, 114–122 (1926). {page 233.}

[70] R. L. Frank and A. Laptev, *Spectral inequalities for Schrödinger operators with surface potentials*, In: *Spectral theory of differential operators*, T. Suslina and D. Yafaev, eds., Amer. Math. Soc. Transl. Ser. 2 **225**, 91–102 (2008). {page 70.}

[71] R. L. Frank, E. H. Lieb, and R. Seiringer, *Stability of Relativistic Matter with Magnetic Fields for Nuclear Charges up to the Critical Value*, Commun. Math. Phys. **275**, 479–489 (2007). {pages 139, 157.}

[72] R. L. Frank, E. H. Lieb, and R. Seiringer, *Hardy–Lieb–Thirring inequalities for fractional Schrödinger operators*, J. Amer. Math. Soc. **21**, 925–950 (2008). {pages 142, 143, 232.}

[73] J. Fröhlich, M. Griesemer, and B. Schlein, *Asymptotic completeness for Rayleigh scattering*, Ann. Henri Poincaré **3**, 107–170 (2002). {page 207.}

[74] J. Fröhlich, E. H. Lieb, and M. Loss, *Stability of Coulomb systems with magnetic fields I: The one-electron atom*, Commun. Math. Phys. **104**, 251–270 (1986). {pages 167, 168, 170, 171.}

[75] W. H. Furry, *On Bound States and Scattering in Positron Theory*, Phys. Rev. **81**, 115–124 (1951). {page 188.}

[76] S. Gadre, L. Bartolotti, and N. Handy, *Bounds for Coulomb energies*, J. Chem. Phys. **72**, 1034–1038 (1980). {page 111.}

[77] A. Galindo and P. Pascual, *Quantum Mechanics*, vols. I and II, Springer Texts and Monographs in Physics (1990). {page xiv.}

[78] C. Garrod and J. K. Percus, *Reduction of the N-particle variational problem*, J. Math. Phys. **5**, 1756–1776 (1964). {page 46.}

[79] G. M. Graf, *Stability of matter through an electrostatic inequality*, Helv. Phys. Acta **70**, 72–79 (1997). {page 130.}

[80] M. Griesemer and H. Siedentop, *A minimax principle for the eigenvalues in spectral gaps*, J. London Math. Soc. **60**, 490–500 (1999). {page 191.}

[81] M. Griesemer and C. Tix, *Instability of a Pseudo-Relativistic Model of Matter with Self-Generated Magnetic Field*, J. Math. Phys. **40**, 1780–1791 (1999). {pages 193, 196, 197, 219.}

[82] S. J. Gustafson and I. M. Sigal, *Mathematical Concepts of Quantum Mechanics*, Springer (2003). {page 223.}

[83] C. Hainzl, M. Lewin, and J. P. Solovej, *The thermodynamic limit of quantum Coulomb systems. Part I. General theory*, Adv. Math. **221**, 454–487 (2009). *Part II, Applications, ibid* 488–546 (2009). {pages 127, 221, 258.}

[84] M. Hamermesh, *Group theory and its applications to physical problems*, Addison-Wesley, Reading, London (1962). {page 38.}

[85] G. H. Hardy, J. E. Littlewood, and G. Pólya, *Some simple inequalities satisfied by convex functions*, Messenger of Math. **58**, 145–152 (1929). {page 86.}

[86] G. H. Hardy, J. E. Littlewood, and G. Pólya, *Inequalities*, second edition, Cambridge University Press (1952). {page 251.}

[87] W. Heitler, *The Quantum Theory of Radiation*, third edition, Oxford University Press (1954). {page 206.}

[88] B. Helffer and D. Robert, *Riesz means of bounded states and semi-classical limit connected with a Lieb–Thirring conjecture I*, J. Asymp. Anal. **3**, 91–103 (1990); *II*, Ann. Inst. H. Poincare **53**, 139–147 (1990). {page 68.}

[89] I. W. Herbst, *Spectral theory of the operator* $(p^2 + m^2)^{1/2} - Ze^2/r$, Comm. Math. Phys. **53**, 285–294 (1977). {pages 143.}

[90] F. Hiai, *Log-majorization and norm inequalities for exponential operators,* in: *Linear Operators*, Banach Center Publications, vol. **38**, p. 119, Institute of Mathematics, Polish Acad. Sci. (1997). {page 86.}

[91] G. Hoever and H. Siedentop, *Stability of the Brown–Ravenhall operator*, Math. Phys. Electr. J. **5**, paper 6, 1–11 (1999). {page 194.}

[92] M. Hoffmann-Ostenhof and T. Hoffmann-Ostenhof, *Schrödinger inequalities and asymptotic behavior of the electron density of atoms and molecules*, Phys. Rev. A **16**, 1782–1785 (1977). {page 146.}

[93] D. Hundertmark, *Some bound state problems in quantum mechanics*, in *Spectral theory and mathematical physics*, Proc. Sympos. Pure Math. **76**, Part 1, 463–496, Amer. Math. Soc. (2007). {page 68.}

[94] D. Hundertmark, E. H. Lieb, and L. E. Thomas, *A sharp bound for an eigenvalue moment of the one-dimensional Schrödinger operator*, Adv. Theor. Math. Phys. **2**, 719–731 (1998). {page 67.}

[95] W. Hunziker, *On the spectra of Schrödinger multiparticle Hamiltonians*, Helv. Phys. Acta **39**, 451–462 (1966). {page 223.}

[96] T. Ichinose, *Note on the Kinetic Energy Inequality Leading to Lieb's Negative Ionization Upper Bound*, Lett. Math. Phys. **28**, 219–230 (1993). {page 227.}

[97] J. H. Jeans, *The mathematical theory of electricity and magnetism*, third edition, Cambridge University Press (1915). {page 1.}

[98] J. Karamata, *Sur une inégalité relative aux fonctions convexes*, Publ. Math. Univ. Belgrade **1**, 145–148 (1932). {page 86.}

[99] T. Kato, *Perturbation theory for linear operators*, Springer (1966). See Remark 5.12, p. 307. {page 143.}

[100] M. K. Kwong, *Uniqueness of positive solutions of* $\Delta u - u + u^p = 0$ *in* $\mathbb{R}^n$, Arch. Rat. Mech. Anal. **105**, 243–266 (1989). {page 137.}

[101] J. H. Lane, *On the Theoretical Temperature of the Sun under the Hypothesis of a Gaseous Mass Maintaining its Volume by its Internal Heat and Depending on the Laws of Gases Known to Terrestrial Experiment*, Amer. J. Science and Arts, second series **50**, 57–74 (1870). {page 111.}

[102] A. Laptev and T. Weidl, *Sharp Lieb–Thirring inequalities in high dimensions*, Acta Math. **194**, 87–111 (2000); See also: *Recent results on Lieb–Thirring inequalities*, Journées "Équations aux Dérivées Partielles" (La Chapelle sur Erdre, 2000), Exp. No. XX, Univ. Nantes, Nantes (2000). {page 68.}

[103] J. L. Lebowitz and E. H. Lieb, *Existence of Thermodynamics for Real Matter with Coulomb Forces*, Phys. Rev. Lett. **22**, 631–634 (1969). {pages 3, 248, 255.}

[104] E. Lenzmann, *Uniqueness of Ground States for Pseudo-Relativistic Hartree Equations*, Analysis & PDE **2**, 1–27 (2009). {page 243.}

[105] J. M. Levy-Leblond, *Nonsaturation of Gravitational Forces*, J. Math. Phys. **10**, 806–812 (1969). {page 234.}

[106] R. T. Lewis, H. Siedentop, and S. Vugalter, *The essential spectrum of relativistic multi-particle operators*, Ann. Inst. H. Poincaré Phys. Théor. **67**, 1–28 (1997). {page 223.}

[107] P. Li and S. T. Yau, *On the Schrödinger equation and the eigenvalue problem*, Comm. Math. Phys. **88**, 309–318 (1983). {page 67.}

[108] E. H. Lieb, *The Stability of Matter*, Rev. Mod. Phys. **48**, 553–569 (1976). {pages 5, 127, 131, 132.}

[109] E. H. Lieb, *Existence and Uniqueness of the Minimizing Solution of Choquard's Non-Linear Equation*, Studies in Appl. Math. **57**, 93–105 (1977). {page 243.}

[110] E. H. Lieb, *The $N^{5/3}$ law for bosons*, Phys. Lett. A **70**, 71–73 (1979). {pages 133, 135.}

[111] E. H. Lieb, *A lower bound for Coulomb energies*, Phys. Lett. A **70**, 444–446 (1979). {page 110.}

[112] E. H. Lieb, *The Number of Bound States of One-Body Schrödinger Operators and the Weyl Problem*, Proc. Amer. Math. Soc. Symposia in Pure Math. **36**, 241–252 (1980). {pages 67, 68, 69.}

[113] E. H. Lieb, *Thomas–Fermi and related theories of atoms and molecules*, Rev. Mod. Phys. **53**, 603–641 (1981). {pages 124, 127, 225, 228.}

[114] E. H. Lieb, *Bound on the maximum negative ionization of atoms and molecules*, Phys. Rev. A **29**, 3018–3028 (1984). {pages 223, 224, 225, 227, 228.}

[115] E. H. Lieb, *The stability of matter: From atoms to stars*, Bull. Amer. Math. Soc. (N. S.) **22**, 1–49 (1990). {page 5.}

[116] E. H. Lieb and J. L. Lebowitz, *The Constitution of Matter: Existence of Thermodynamics for Systems Composed of Electrons and Nuclei*, Adv. in Math. **9**, 316–398 (1972). {pages 3, 248, 256, 257.}

[117] E. H. Lieb and M. Loss, *Stability of Coulomb systems with magnetic fields II: The many-electron atom and the one-electron molecule*, Commun. Math. Phys. **104**, 271–282 (1986). {page 171.}

[118] E. H. Lieb and M. Loss, *Analysis*, Second Edition, American Mathematical Society, Providence, Rhode Island (2001). {Multiple citations}

[119] E. H. Lieb and M. Loss, *Stability of a Model of Relativistic Quantum Electrodynamics*, Commun. Math. Phys. **228**, 561–588 (2002). {pages 5, 213, 218, 218, 219, 220.}

[120] E. H. Lieb and M. Loss, *A note on polarization vectors in quantum electrodynamics*, Commun. Math. Phys. **252**, 477–483 (2004). {page 207.}

[121] E. H. Lieb and M. Loss, *The Thermodynamic Limit for Matter Interacting with Coulomb Forces and with the Quantized Electromagnetic Field: I. The Lower Bound*, Commun. Math. Phys. **258**, 675–695 (2005). {page 248.}

[122] E. H. Lieb, M. Loss, and H. Siedentop, *Stability of Relativistic Matter via Thomas–Fermi Theory*, Helv. Phys. Acta **69**, 974–984 (1996). {pages 148, 152, 154, 157.}

[123] E. H. Lieb, M. Loss, and J. P. Solovej, *Stability of matter in magnetic fields*, Phys. Rev. Lett. **75**, 985–989 (1995). {pages 4, 82, 171.}

[124] E. H. Lieb and H. Narnhofer, *The Thermodynamic Limit for Jellium*, J. Stat. Phys. **12**, 291–310 (1975). *Errata* J. Stat. Phys. **14**, 465 (1976). {pages 272, 273, 274, 275.}

[125] E. H. Lieb and S. Oxford, *Improved lower bound on the indirect Coulomb energy*, Int. J. Quantum Chem. **19**, 427–439 (1981). {pages 6, 111.}

[126] E. H. Lieb and R. Seiringer, *Derivation of the Gross–Pitaevskii Equation for Rotating Bose Gases*, Commun. Math. Phys. **264**, 505–537 (2006). {page 60.}

[127] E. H. Lieb, H. Siedentop, and J. P. Solovej, *Stability and instability of relativistic electrons in classical electromagnetic fields*, J. Stat. Phys. **89**, 37–59 (1997). {pages 5, 171, 175, 178, 193, 194, 196.}

[128] E. H. Lieb, I. M. Sigal, B. Simon, and W. Thirring, *Asymptotic Neutrality of Large-Z Ions*, Phys. Rev. Lett. **52**, 994–996 (1984). *Approximate Neutrality of Large-Z Ions*, Commun. Math. Phys. **116**, 635–644 (1988). {page 229.}

[129] E. H. Lieb and B. Simon, *Thomas–Fermi theory revisited*, Phys. Rev. Lett. **31**, 681–683 (1973). {page 3.}

[130] E. H. Lieb and B. Simon, *Thomas–Fermi theory of atoms, molecules and solids*, Adv. in Math. **23**, 22–116 (1977). {pages 3, 127, 138.}

[131] E. H. Lieb and J. P. Solovej, *Ground State Energy of the One-Component Charged Bose Gas*, Commun. Math. Phys. **217**, 127–163 (2001). *Erratum ibid.* **225**, 219–221 (2002). {page 137.}

[132] E. H. Lieb and J. P. Solovej, *Ground State Energy of the Two-Component Charged Bose Gas*, Commun. Math. Phys. **252**, 485–534 (2004). {pages 135, 138.}

[133] E. H. Lieb, J. P. Solovej, and J. Yngvason, *Asymptotics of Heavy Atoms in High Magnetic Fields: II. Semiclassical Regions*, Commun. Math. Phys. **161**, 77–124 (1994). {page 82.}

[134] E. H. Lieb and W. Thirring, *Bound for the Kinetic Energy of Fermions which Proves the Stability of Matter*, Phys. Rev. Lett. **35**, 687–689 (1975). Errata *ibid.*, 1116 (1975). {pages 3, 62, 70, 127, 129.}

[135] E. H. Lieb and W. Thirring, *Inequalities for the Moments of the Eigenvalues of the Schrödinger Hamiltonian and Their Relation to Sobolev Inequalities*, in *Studies in Mathematical Physics*, E. Lieb, B. Simon, A. Wightman, eds., Princeton University Press, 269–303 (1976). {pages 62, 67, 70, 85, 131.}

[136] E. H. Lieb and W. Thirring, *Gravitational Collapse in Quantum Mechanics with Relativistic Kinetic Energy*, Annals of Phys. (N.Y.) **155**, 494–512 (1984). {pages 236, 244.}

[137] E. H. Lieb and H.-T. Yau, *The Chandrasekhar Theory of Stellar Collapse as the Limit of Quantum Mechanics*, Commun. Math. Phys. **112**, 147–174 (1987). A summary is given in *A Rigorous Examination of the Chandrasekhar Theory of Stellar Collapse*, Astrophys. Jour. **323**, 140–144 (1987). {pages 241, 241, 242.}

[138] E. H. Lieb and H. -T. Yau, *The Stability and Instability of Relativistic Matter*, Commun. Math. Phys. **118**, 177–213 (1988). {pages 92, 100, 139, 141, 145, 154, 157, 158, 159, 162.}

[139] M. Loss and H. -T. Yau, *Stability of Coulomb systems with magnetic fields III: Zero energy bound states of the Pauli operator*, Commun. Math. Phys. **104**, 283–290 (1986). {pages 4, 167.}

[140] O. Matte and E. Stockmeyer, *Spectral theory of no-pair Hamiltonians*, preprint arXiv:0803.1652. {page 223.}

[141] K. McLeod and J. Serrin, *Uniqueness of positive radial solutions to* $\Delta u + f(u) = 0$ *in* $\mathbb{R}^n$, Arch. Rat. Mech. Anal. **99**, 115–145 (1987). {page 137.}

[142] M. Mittleman, *Theory of relativistic effects on atoms: configuration-space Hamiltonian*, Phys. Rev. A **24**, 1167–1175 (1981). {page 188.}

[143] S. Morozov, *Essential Spectrum of Multiparticle Brown–Ravenhall Operators in External Field*, Doc. Math. **13**, 51–79 (2008). {page 223.}

[144] I. Newton, *The Principia*, Prometheus Books, Amherst, New York (1995). {pages 5, 91.}

[145] L. Onsager, *Electrostatic interaction of molecules*, J. Phys. Chem **43**, 189–196 (1939). {pages 2, 112, 249, 259.}

[146] L. Onsager, *Crystal statistics. I. A two-dimensional model with an order–disorder transition*, Phys. Rev. **65**, 117–149 (1944). {page 130.}

[147] J. K. Percus, *The role of model systems in the few-body reduction of the N-fermion problem*, Int. J. Quantum Chem. **13**, 89–124 (1978). {page 46.}

[148] A. B. Pippard, *J. J. Thomson and the discovery of the electron*, in: *Electron – A centenary volume*, M. Springfield, ed., Cambridge University Press (1997). {page 2.}

[149] M. Planck, *Zur Theorie des Gesetzes der Energieverteilung im Normalspektrum*, Verhandlung der Deutschen Physikalischen Gesellschaft **2**, 237–245 (1900). {pages 5, 200, 203, 209.}

[150] M. Reed and B. Simon, *Methods of Modern Mathematical Physics*, Vol. 1–4, Academic Press (1978). {pages 8, 18, 83, 146, 183, 209, 212, 223, 252.}

[151] R. T. Rockafellar, *Convex Analysis*, Princeton University Press (1970). {page 48.}

[152] G. V. Rosenbljum, *Distribution of the discrete spectrum of singular differential operators*, Izv. Vyss. Ucebn. Zaved. Matematika **164**, 75–86 (1976); *English transl.* Soviet Math. (Iz. VUZ) **20**, 63–71 (1976). {page 67.}

[153] D. Ruelle, *Statistical Mechanics. Rigorous Results*, Imperial College Press and World Scientific (1999). {pages 253, 255, 257.}

[154] M. B. Ruskai, *Absence of discrete spectrum in highly negative ions, I*, Commun. Math. Phys. **82**, 457–469 (1982). *II*, Commun. Math. Phys. **85**, 325–327 (1982). {page 229.}

[155] E. Rutherford, *The Scattering of α and β Particles by Matter and the Structure of the Atom*, Phil. Mag. **21**, 669 (1911). {page 2.}

[156] E. Schrödinger, *Quantisierung als Eigenwertproblem*, Annalen der Physik **79**, 361–376 (1926). *ibid.* **79**, 489–527, **80**, 437–490 and **81**, 109–139 (1926). {page 2.}

[157] J. Schwinger, *On the bound states of a given potential*, Proc. Nat. Acad. Sci. U. S.A. **47**, 122–129 (1961). {pages 76, 80.}

[158] L. A. Seco, I. M. Sigal, and J. P. Solovej, *Bound on the ionization energy of large atoms*, Comm. Math. Phys. **131**, 307–315 (1990). {page 229.}

[159] I. Segal, *Tensor algebras over Hilbert spaces, I*, Trans. Amer. Math. Soc. **81**, 106–134 (1956). {page 212.}

[160] E. Seiler and B. Simon, *Bounds in the Yukawa$_2$ quantum field theory: upper bound on the pressure, Hamiltonian bound and linear lower bound*, Commun. Math. Phys. **45**, 99–114 (1975). {page 85.}

[161] R. Seiringer, *On the maximal ionization of atoms in strong magnetic fields*, J. Phys. A: Math. Gen. **34**, 1943–1948 (2001). {page 222.}

[162] R. Seiringer, *Ground state asymptotics of a dilute, rotating gas*, J. Phys. A: Math. Gen. **36**, 9755–9778 (2003). {pages 38, 60.}

[163] I. M. Sigal, *How many electrons can a nucleus bind?*, Ann. Physics **157**, 307–320 (1984). {page 229.}

[164] B. Simon, *Quantum Mechanics for Hamiltonians defined as Quadratic Forms*, Princeton University Press (1971). {page 16.}

[165] B. Simon, *Kato's inequality and the comparison of semigroups*, J. Funct. Anal. **32**, 97–101 (1979). {page 82.}

[166] B. Simon, *The Statistical Mechanics of Lattice Gases*, Princeton University Press (1993). {page 48.}

[167] B. Simon, *Functional Integration and Quantum Physics*, second edition, Amer. Math. Soc. (2005). {pages 84, 85.}

[168] B. Simon, *Trace ideals and their application*, second edition, Math. Surveys and Monographs **120**, Amer. Math. Soc. (2005). {page 39.}

[169] A. Sobolev, *Lieb–Thirring inequalities for the Pauli operator in three dimensions*, IMA Vol. Math. Appl. **95**, 155–188 (1997). {page 82.}

[170] J. P. Solovej, *Asymptotics for bosonic atoms*, Lett. Math. Phys. **20**, 165–172 (1990). {page 229.}

[171] J. P. Solovej, *The ionization conjecture in Hartree–Fock theory*, Ann. Math. **158**, 509–576 (2003). {page 228.}

[172] J. P. Solovej, *Upper Bounds to the Ground State Energies of the One- and Two-Component Charged Bose Gases*, Commun. Math. Phys. **266**, 797–818 (2006). {pages 136, 137, 138.}

[173] J. P. Solovej, *Stability of Matter*, Encyclopedia of Mathematical Physics, J.-P. Francoise, G. L. Naber and S. T. Tsou, eds., vol. **5**, 8–14, Elsevier (2006). {page 126.}

[174] H. Spohn, *Dynamics of Charged Particles and their Radiation Field*, Cambridge University Press (2004). {page 200.}

[175] E. C. Stoner, *The Equilibrium of Dense Stars*, Phil. Mag. **9**, 944–963 (1930). {page 234.}

[176] J. Sucher, *Relativistic Many-Electron Hamiltonians*, Physica Scripta **36**, 271–281 (1987). {page 188.}

[177] B. Thaller, *The Dirac Equation*, Texts and Monographs in Physics, Springer (1992). {page 199.}

[178] W. Thirring, *Quantum Mathematical Physics*, second edition, Springer (2002). {pages 131, 223.}

[179] L. H. Thomas, *The calculation of atomic fields*, Proc. Camb. Phil. Soc. **23**, 542–548 (1927). {pages 3, 111.}

[180] J. J. Thomson, *Cathode rays*, Phil. Mag. **44**, 293 (1897). {page 2.}

[181] C. Tix, *Lower bound for the ground state energy of the no-pair Hamiltonian*, Phys. Lett. B **405**, 293–296 (1997). *Strict positivity of a relativistic Hamiltonian due to Brown and Ravenhall*, Bull. London Math. Soc. **30**, 283–290 (1998). {page 192.}

[182] J. H. Van Leeuwen, *Problems of the electronic theory of magnetism*, J. Phys. (Paris) **2**, 361–377 (1921). {page 65.}

[183] C. Van Winter, *Theory of finite systems of particles, I*, Math. -Phys. Skr. Danske Vid. Selsk. **1**(8), 1–60 (1964). {page 223.}

[184] R. Weder, *Spectral properties of one-body relativistic spin-zero Hamiltonians*, Ann. Inst. Henri Poincaré, Phys. Théor. **20**, 211–220 (1974). {page 143.}

[185] T. Weidl, *On the Lieb–Thirring constants $L_{\gamma,1}$ for $\gamma \geq 1/2$*, Commun. Math. Phys. **178**, 135–146 (1996). {page 67.}

[186] S. Weinberg, *Gravitation and Cosmology*, Wiley (1972). {page 242.}

[187] E. P. Wigner, *Effects of the electron interaction on the energy levels of electrons in metals*, Trans. Faraday Soc. (London) **34** 678–684 (1938). {page 271.}

[188] D. Yafaev, *Sharp constants in the Hardy–Rellich inequalities*, J. Funct. Anal. **168**, 121–144 (1999). {pages 143, 143.}

[189] L. Q. Zhang, *Uniqueness of ground state solutions*, Acta. Math. Sci. (English Ed.) **8**, 449–467 (1988). {page 137.}

[190] G. Zhislin, *Discussion of the spectrum of the Schrödinger operator for systems of many particles*, Tr. Mosk. Mat. Obs. **9**, 81–128 (1960). {pages 223, 224, 229.}

Index